Lecture Notes in Geoinformation and Cartography

Publications of the International Cartographic Association (ICA)

Series editors

William Cartwright, Melbourne, Australia
Georg Gartner, Wien, Austria
Liqiu Meng, Munich, Germany
Michael P. Peterson, Omaha, USA

More information about this series at http://www.springer.com/series/10036

Georg Gartner · Haosheng Huang
Editors

Progress in Location-Based Services 2016

 Springer

Editors
Georg Gartner
Research Group Cartography
Vienna University of Technology
Vienna
Austria

Haosheng Huang
Department of Geography
University of Zurich
Zürich
Switzerland

ISSN 1863-2246   ISSN 1863-2351 (electronic)
Lecture Notes in Geoinformation and Cartography
ISSN 2195-1705   ISSN 2195-1713 (electronic)
Publications of the International Cartographic Association (ICA)
ISBN 978-3-319-83700-0   ISBN 978-3-319-47289-8 (eBook)
DOI 10.1007/978-3-319-47289-8

© Springer International Publishing AG 2017
Softcover reprint of the hardcover 1st edition 2016
This work is subject to copyright. All rights are reserved by the Publisher, whether the whole or part of the material is concerned, specifically the rights of translation, reprinting, reuse of illustrations, recitation, broadcasting, reproduction on microfilms or in any other physical way, and transmission or information storage and retrieval, electronic adaptation, computer software, or by similar or dissimilar methodology now known or hereafter developed.
The use of general descriptive names, registered names, trademarks, service marks, etc. in this publication does not imply, even in the absence of a specific statement, that such names are exempt from the relevant protective laws and regulations and therefore free for general use.
The publisher, the authors and the editors are safe to assume that the advice and information in this book are believed to be true and accurate at the date of publication. Neither the publisher nor the authors or the editors give a warranty, express or implied, with respect to the material contained herein or for any errors or omissions that may have been made.

Printed on acid-free paper

This Springer imprint is published by Springer Nature
The registered company is Springer International Publishing AG
The registered company address is: Gewerbestrasse 11, 6330 Cham, Switzerland

Preface

Location-Based Services (LBS) have become a research field since the early 2000s. In recent years, lots of progress has been made in this research field, due to the increasingly maturity of the underpinning communication technologies and mobile devices. LBS have become more and more popular not only in citywide outdoor environments, but also in shopping malls, museums, and many other indoor environments. They have been applied for emergency services, tourism services, intelligent transport services, social networking, gaming, assistive services, etc.

Since its initiation by Georg Gartner from TU Wien (Austria) in 2002, the LBS conference series has become one of the most important scientific events dedicated to LBS. The conferences have been held in Vienna (2002, 2004, 2005), Hong Kong (2007), Salzburg (2008), Nottingham (2009), Guangzhou (2010), Vienna (2011), Munich (2012), Shanghai (2013), Vienna (2014), and Augsburg (2015). Starting from 2015, the LBS conferences have become the annual event of the newly established Commission on Location-Based Services of the International Cartographic Association (ICA). In November 2016, the 13th LBS conference (LBS 2016) will be hosted by TU Wien in Vienna, Austria.

This book contains a selection of peer-reviewed full papers submitted to LBS 2016. All the chapters have been accepted after a rigor reviewing process. The book provides a general picture of recent research activities related to the domain of LBS. Such activities emerged in the last years, especially concerning issues of outdoor/indoor positioning, smart environment, spatial modeling, personalization, context-awareness, cartographic communication, novel user interfaces, crowd sourcing, social media, big data analysis, usability and privacy.

We would like to thank all the authors for their excellent work and all referees for their critical and constructive reviews. We hope you enjoy reading these papers, and look forward to your participation in the future LBS conferences.

Vienna, Austria	Georg Gartner
Zürich, Switzerland	Haosheng Huang
September 2016	

Reviewers

The production of this book would have not been possible without the professional help of our program committee members. We would like to thank all the following experts who have helped to review the papers published in this book.

Rein Ahas, University of Tartu, Estonia
Pouria Amirian, Ordnance Survey, UK
Gennady Andrienko, Fraunhofer IAIS, Germany
Masatoshi Arikawa, University of Tokyo, Japan
Thierry Badard, Laval University, Canada
Kate Beard, University of Maine, USA
William Cartwright, RMIT University, Australia
Christophe Claramunt, Naval Academy Research Institute, France
Weihua Dong, Beijing Normal University, China
Matt Duckham, RMIT University, Australia
David Forrest, University of Glasgow, UK
Peter Froehlich, Austrian Institute of Technology, Austria
Georg Gartner, TU Wien, Austria
Amy Griffin, UNSW Canberra, Australia
Haosheng Huang, University of Zurich, Switzerland
Frédéric Hubert, Laval University, Canada
Mike Jackson, University of Nottingham, UK
Bin Jiang, University of Gävle, Sweden
Markus Jobst, TU Wien, Austria
Hassan Karimi, University of Pittsburgh, USA
Farid Karimipour, University of Tehran, Iran
Pyry Kettunen, Finnish Geospatial Research Institute, Finland
Carsten Keßler, Aalborg University Copenhagen, Denmark
Peter Kiefer, ETH Zurich, Switzerland
Christian Kray, University of Münster, Germany
Jukka Krisp, University of Augsburg, Germany
John Krumm, Microsoft Research, USA

Lars Kulik, University of Melbourne, Australia
Ki-Joune Li, Pusan National University, South Korea
Rui Li, University at Albany, SUNY, USA
Chun Liu, Tongji University, China
Rainer Malaka, Bremen University, Germany
Liqiu Meng, TU Munich, Germany
Xiaolin Meng, University of Nottingham, UK
Harvey Miller, The Ohio State University, USA
Peter Mooney, NUI Maynooth, Ireland
Hans Neuner, TU Wien, Austria
Kristien Ooms, Ghent University, Belgium
Ross Purves, University of Zurich, Switzerland
Martin Raubal, ETH Zurich, Switzerland
Karl Rehrl, Salzburg Research, Austria
Kai-Florian Richter, University of Zurich, Switzerland
Guenther Retscher, TU Wien, Austria
Dimitris Sacharidis, TU Wien, Austria
Tapani Sarjakoski, Finnish Geospatial Research Institute, Finland
Christoph Schlieder, University of Bamberg, Germany
Volker Schwieger, University of Stuttgart, Germany
Jie Shen, Nanjing Normal Unviersity, China
Takeshi Shirabe, KTH Royal Institute of Technology, Sweden
Josef Strobl, University of Salzburg, Austria
Kazutoshi Sumiya, University of Hyogo, Japan
Sabine Timpf, University of Augsburg, Germany
Martin Tomko, University of Melbourne, Australia
Nico Van de Weghe, Ghent University, Belgium
Stefan Van Der Spek, TU Delft, The Netherlands
Corné van Elzakker, University of Twente, The Netherlands
Robert Weibel, University of Zurich, Switzerland
Manfred Weisensee, Jade Hochschule, Germany
Dirk Wenig, University of Bremen, Germany
Stephan Winter, University of Melbourne, Australia
Linyuan Xia, Sun Yat-sen University, China
Kefei Zhang, RMIT University, Australia
Sisi Zlatanova, TU Delft, The Netherlands

Contents

Part I Positioning

Wi-Fi Fingerprinting with Reduced Signal Strength Observations from Long-Time Measurements 3
Guenther Retscher and Florian Roth

Situation Goodness Method for Weighted Centroid-Based Wi-Fi APs Localization 27
Germán M. Mendoza-Silva, Joaquín Torres-Sospedra, Joaquín Huerta, Raul Montoliu, Fernando Benítez and Oscar Belmonte

Smartphone Sensor-Based Orientation Determination for Indoor-Navigation 49
Andreas Ettlinger, Hans-Berndt Neuner and Thomas Burgess

SubwayAPPS: Using Smartphone Barometers for Positioning in Underground Transportation Environments 69
Kris van Erum and Johannes Schöning

Part II Outdoor and Indoor Navigation

Effects of Visual Variables on the Perception of Distance in Off-Screen Landmarks: Size, Color Value, and Crispness 89
Rui Li

Investigation of Landmark-Based Pedestrian Navigation Processes with a Mobile Eye Tracking System 105
Conrad Franke and Jürgen Schweikart

Increasing the Density of Local Landmarks in Wayfinding Instructions for the Visually Impaired 131
Rajchandar Padmanaban and Jakub Krukar

**Identifying Divergent Building Structures Using Fuzzy
Clustering of Isovist Features**................................. 151
Sebastian Feld, Hao Lyu and Andreas Keler

**Generation of Meaningful Location References for Referencing Traffic
Information to Road Networks Using Qualitative Spatial Concepts**.... 173
Karl Rehrl, Richard Brunauer, Simon Gröchenig and Eva Lugstein

Efficient Computation of Bypass Areas........................... 193
Jörg Roth

A Heuristic for Multi-modal Route Planning...................... 211
Dominik Bucher, David Jonietz and Martin Raubal

Part III Spatial-Temporal Data Processing and Analysis

**Development of a Road Deficiency GIS Using Data from Automated
Multi-sensor Systems**... 233
Alexander Mraz and Abdenour Nazef

**Identifying Origin/Destination Hotspots in Floating Car Data
for Visual Analysis of Traveling Behavior**........................ 253
Mathias Jahnke, Linfang Ding, Katre Karja and Shirui Wang

Part IV Innovative LBS Applications

**Enhancing Location Recommendation Through Proximity
Indicators, Areal Descriptors, and Similarity Clusters**........... 273
Sebastian Meier

**Connecting the Dots: Informing Location-Based Services
of Space Usage Rules**... 293
Pavel Andreevich Samsonov

**Concept Design of #hylo—Geosocial Network for Sharing
Hyperlocal Information on a Map**.................................. 309
Hanna-Marika Halkosaari, Mikko Rönneberg, Mari Laakso,
Pyry Kettunen, Juha Oksanen and Tapani Sarjakoski

**Multimodal Location Based Services—Semantic 3D City
Data as Virtual and Augmented Reality**............................ 329
José Miguel Santana, Jochen Wendel, Agustín Trujillo, José Pablo Suárez,
Alexander Simons and Andreas Koch

Part V User Studies, Privacy and Motivation

**Ephemerality Is the New Black: A Novel Perspective on Location
Data Management and Location Privacy in LBS**...................... 357
Mehrnaz Ataei and Christian Kray

Classes for Creating Location-Based Audio Tour Content: A Case of User-Generated LBS Education to University Students............ 375
Min Lu, Masatoshi Arikawa and Atsuyuki Okabe

Gamification as Motivation to Engage in Location-Based Public Participation?.. 399
Sarah-Kristin Thiel and Peter Fröhlich

Part I
Positioning

Wi-Fi Fingerprinting with Reduced Signal Strength Observations from Long-Time Measurements

Guenther Retscher and Florian Roth

Abstract Indoor positioning which uses signal strength values of Wi-Fi networks have become popular as these wireless networks often already exist and many mobile devices, such as smartphones or tablets, have built-in Wi-Fi cards. Usually fingerprinting is employed for positioning which achieves relatively low positioning accuracies on the several meter level. In the scope of this work two methods are presented which have the potential to improve the fingerprinting performance using long-time RSS observations at reference stations. Both methods employ the usage of at least three reference stations surrounding the area of interest on which signal strength observations are continuously performed during the whole measurement process. Thereby the first method uses a 2-D linear plane-interpolation for the deduction of real-time corrections. For that purpose, the measured signal strengths are reduced by the long-time measurements which are interpolated at the approximate position of the measuring point. In the second method the daily average of the long-time measurements is applied and the improvements of the measurements are calculated by the deviation from the daily average. For this method it is conceivable that a single reference station may be sufficient if it is located in the middle of the area of interest. Field tests were performed in an office building and are analyzed. The fingerprinting algorithms reached an averaged positioning accuracy of around 5 m in dependence on the used smartphone. The daily average improvements (DAI) method provided a better performance than the interpolation method which is highly influenced by the required approximate position of the user.

Keywords Location fingerprinting · Reduced RSS observations · Interpolation method · Daily average improvement method

---

G. Retscher (✉) · F. Roth
Department of Geodesy and Geoinformation, Research Group Engineering Geodesy,
TU Wien, Gusshausstrasse 27-29 E120/5 1040, Vienna, Austria
e-mail: guenther.retscher@tuwien.ac.at

F. Roth
e-mail: froth.vienna@gmail.com

© Springer International Publishing AG 2017
G. Gartner and H. Huang (eds.), *Progress in Location-Based Services 2016*, Lecture Notes in Geoinformation and Cartography,
DOI 10.1007/978-3-319-47289-8_1

1 Introduction

The development of indoor navigation systems has become a growing field of research interest in recent years as many applications nowadays require ubiquitous positioning in combined out-/indoor environments. There are still many unresolved challenges in such type of applications as satellite-based GNSS and GNSS-aided inertial navigation systems (INS) are capable methods for mainly outdoor navigation only. Thus, alternative techniques are developed using different signals, such as radio waves, acoustic signals, or other sensory information collected by mobile devices. All of them have their own strengths as well as limitations. Following a classification of Li and Rizos (2014), indoor localization technologies fall into three categories, i.e., designated technologies based on pre-deployed signal transmission infrastructure, technologies based on 'signals-of-opportunity' and technologies not based on signals. To the first categories belong systems using infrared or ultrasonic signals, magnetic fields, Ultra Wide Band (UWB) or other RF-based systems. Signals-of-opportunity include RF signals originally not intended for positioning, for instance, Wi-Fi, digital television, mobile telephony, FM radio and others. Dead reckoning (DR) using inertial sensors (accelerometers and gyroscopes) as well as vision/camera systems belong to the third category. This paper focuses on Wi-Fi location fingerprinting where it is investigated whether long-time signal strength measurements could lead to an improvement as the positioning accuracy is highly related to the fluctuation of signal strength. Because long-time measurements show the trend of the received signal strength (RSS) and contain the fluctuation of the emitted signal they provide the potential to minimize these effects. Two methods are used to reduce the influence of the fluctuations and compared to standard fingerprinting.

The remainder of the paper is organized as follows: Sect. 2 reviews briefly the fundamentals of the Wi-Fi positioning principle and its methods. In Sect. 3 the two new methods for the reduction of the RSS observations for the establishment of a radio map of RSS distributions for fingerprinting are described. Section 4 presents the setup and the results of the experiments which were performed to survey the effect of RSS observations from long-time measurements. Finally, Sect. 5 contains the discussion of the major results and the concluding remarks as well as an outlook on future investigations.

2 Wi-Fi Positioning Principle

Wi-Fi technology uses microwave signals to provide an electronic device access to the internet. The ability to measure the RSS of a certain Wi-Fi access point (AP) and the coherence between this signal strength and the distance to the AP provides the possibility to use Wi-Fi for positioning. For positioning in RSS-based solutions usually two different methods are employed, i.e., location fingerprinting and trilateration. In this study only fingerprinting is considered. In the following, the challenges are briefly reviewed and then fingerprinting is discussed in detail including its performance.

2.1 Challenges of Wi-Fi Positioning

The major aim of research concerning Wi-Fi positioning is to significantly improve positioning accuracy to the one meter or even to sub-meter level. The low positioning accuracy of several meters achieved so far, however, is mainly caused by the fluctuation of the RSS which is caused by the following reasons (Bai et al. 2014).

- Humans consist of a large part of water (around 60 to 70 %) and the signal is absorbed significantly by water,
- The signal is reflected by metal in walls causing multipath propagation which falsifies the observation of the RSS,
- The emitted signal strength of some APs is depending on the number of users which are connected to the network,
- Radio interference, which is caused by other devices like for example microwave ovens or shop door openers, and
- Different smartphones have different in-built hardware and so the measured RSS differs from phone to phone or other mobile device.

2.2 Wi-Fi Location Fingerprinting

Wi-Fi fingerprinting needs reference points (RPs) with uniquely defined RSS values from every received AP. The current position of the user is then the location which has the greatest similarity to the RSS. Fingerprinting works with a radio map (R) which is based on a number of RPs containing the RSS distributions. These RPs are located in the area of interest and distributed in a representative grid. Figure 1 illustrates the operational principle of fingerprinting.

Fig. 1 Operational principle of location fingerprinting

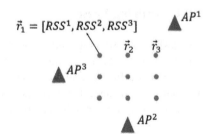

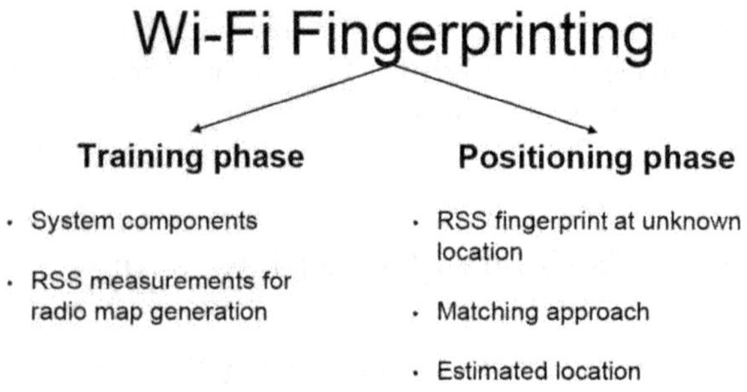

Fig. 2 Characteristics of the two phases of location fingerprinting

Wi-Fi Fingerprinting

Training phase
- System components
- RSS measurements for radio map generation

Positioning phase
- RSS fingerprint at unknown location
- Matching approach
- Estimated location

Every RP has a position and an associated fingerprinting matrix (F) (Huang 2014) as described in:

$$R = \{\vec{p_i}, F_i\} \quad i = 1, 2, \ldots, m \tag{1}$$

$$F_i = [\vec{r}_{RP_i}(T_1), \vec{r}_{RP_i}(T_2), \ldots, \vec{r}_{RP_i}(T_n)] \tag{2}$$

$$\vec{r}_{RP_i}(t) = [RSS_{RP_i}^{AP_1}(t), RSS_{RP_i}^{AP_2}(t), \ldots, RSS_{RP_i}^{AP_k}(t)] \tag{3}$$

where $\vec{p_i}$ is the position of the RP, m the number of RPs, T_n the last time where the sample is collected, t the selected time stamp, k the number of APs, and $RSS_{RP_i}^{AP_k}$ the RSS value.

The vector $\vec{r}_{RP_i}(t)$ in Eq. (2) contains the RSS at the selected RP from all received APs (see Fig. 1). For every time sample there is a $\vec{r}_{RP_i}(t)$ labelled by a unique time stamp. It depends on the type of positioning algorithm if just one value for each AP is needed instead of the whole fingerprinting matrix. In case of a single value per AP the median or mean is calculated and just one vector (location fingerprint) $\vec{r}_{RP_i}(t)$ further exists. The radio map data is saved in a so-called fingerprinting database (DB) which is fed in the training phase (Sect. 2.2.1). Because of the dependency of the location fingerprint and position it is possible to estimate the current user's position with another RSS measurement and the comparison from this measurement with the DB. This is called the positioning phase (see Sect. 2.2.2). Figure 2 shows the main tasks to be carried out in the two phases.

2.2.1 Training Phase

The task of the training, offline or calibration phase is to find the system's design and to collect data for the fingerprinting DB. The RPs form usually a rectangular grid (compare Fig. 1) and the system's designer has to find the suitable number of RPs and the distance between them (i.e., grid spacing). The aim is to provide a suitable positioning resolution, best system performance and the reception of at least three APs at all RPs. It depends on the environment which system parameters are best to improve the system accuracy and precision performance. Because of the long time required to collect the reference RSS data it is not the best way to reduce the grid spacing accordingly (Kaemarungsi 2005). As the human's body absorbs parts of the Wi-Fi signal and falsifies the measured signal strengths multiple measurements in usually four smartphone orientations are performed at every RP. After the grid design is accomplished, the location fingerprints are collected by measuring the RSS from all received APs at every RP. Because of different devices and orientations, multiple DBs are created (for every device and orientation) or models are developed to simulate the influence of different hardware and the user's body.

2.2.2 Positioning Phase

In the positioning or online phase, the mobile device collects the RSS measurements from every receivable AP at the location to be determined. If the fingerprinting DB of the training phase contains for example a statistical distribution the received RSS vector must have the same number of elements. The result is a vector which contains the average RSS from the received AP (location fingerprint) in the form:

$$\tilde{r} = \left[\widetilde{RSS}_{RP_i}^{AP_1}, \widetilde{RSS}_{RP_i}^{AP_2}, \ldots, \widetilde{RSS}_{RP_i}^{AP_k} \right] \qquad (4)$$

Once the measurements are performed and found to be correct, the location fingerprint of the positioning phase $\tilde{r}$ has to be compared with the location fingerprints of the DB. This comparison is performed by a positioning (Kaemarungsi 2005) or calculation (Mok and Retscher 2007) algorithm like the Nearest Neighbour (Sect. 2.2.3) or K-Nearest Neighbour (Sect. 2.2.4).

2.2.3 Nearest Neighbour Algorithm

The simplest calculation algorithm is the Nearest Neighbour (NN) method because it just requires mean or median vectors and no more other RSS data. The fingerprinting matrix becomes then just s single vector which contains one mean (or median) value per AP. The algorithm then calculates most commonly the Euclidean distance between the location fingerprint of the positioning phase and all fingerprints in the DB. The estimated position is the position of the fingerprint with the minimum distance. Equation (5) defines the selection of the fingerprint $\tilde{r}$ and its position ($\hat{p}_{NN}$) with the basic NN algorithm with Euclidean distance (Huang 2014; Kaemarungsi 2005):

$$d(x,y) = \left\| \vec{x} - \vec{y} \right\|_2 = \sqrt{\sum_{j=1}^{l} (x_i - y_i)^2} \tag{5}$$

$$d\left(\vec{\tilde{r}}, \vec{r}_{RP_a}(t)\right) < d\left(\vec{\tilde{r}}, \vec{r}_{RP_b}(t)\right) \quad \forall a \neq b \tag{6}$$

$$\widehat{p}_{NN} = \vec{p}_i \tag{7}$$

where j is the number of elements.

2.2.4 K-Nearest Neighbour Algorithm

An extension of the NN algorithm is the k-Nearest Neighbour (kNN) algorithm. First the fingerprints are sorted by their distance to the location fingerprint of the positioning phase and the first k RPs (k has to be smaller than the number of RPs) with minimal distance are chosen (Eq. (8)). The average over the coordinates of the chosen RPs yields the estimated position ($\widehat{p}_{KNN}$) of the mobile device (Huang 2014).

$$\widehat{p}_{KNN} = \frac{1}{K} \sum_{i=1}^{K} \vec{p}_i \tag{8}$$

with the position $\vec{p}_i$ of the K chosen RPs.

2.3 Performance of Wi-Fi Fingerprinting

It is difficult to state a generally valid positioning accuracy for Wi-Fi fingerprinting because of the dependence on the surrounding environment. Beside the multipath propagation and signal interference, also the design of the grid of RPs influences the accuracy. Usually averaged positioning accuracies in the range of 1-6 m are obtained (Mok and Retscher 2007). Because of errors in such a range the altitude determination is mostly not accurate enough to estimate the correct floor level. In such cases height can be added using barometric measurements. This is possible because air pressure is directly related to height (Retscher 2007).

3 Fingerprinting with Reduced RSS Observations

The spatial coverage of an AP is called radio cell. The size of this cell is defined on the one hand by the surrounding environment and on the other hand by the broadcasting power. An AP reacts to the rising number of connected clients by a regulation of the broadcasting power to supply all clients equally. The more clients are connected in the cell, the RSS is the more decreasing and the cell size shrinking. This effect occurs also as opposite if the number of clients decreases. Long-time measurements could be used to reduce the influence of this fluctuation. Thus, two new methods are developed and investigated which provide the possibility to model these effects. The first one uses interpolation and the second is referred to as Daily Average Improvement (DAI) method.

3.1 Interpolation Method

The concept for the reduction with long-time measurements in case of fingerprinting is illustrated in the flowchart in Fig. 3. It is based on the assumption that a single RSS value measured at a random point $\widetilde{RSS}_{mes}(\vec{p}, t)$ can be separated in two parts as follows:

$$\widetilde{RSS}_{mes}(\vec{p}, t) = RSS_{theo}(\vec{p}, t) + \Delta_{RSS}(\vec{p}, t) \tag{9}$$

Theoretically $RSS_{theo}(\vec{p}, t)$ is the RSS value at a certain point which is caused by the spatial range of the Wi-Fi Signal. It is not affected by any influence of the surrounding environment, but it contains the signal's variation over time. Deviation caused by the surrounding environment $\Delta_{RSS}(\vec{p}, t)$ is the change in RSS (compared to ideal conditions) which is caused by the measurement site assembly and current conditions. So it contains the influences of present people, walls, radio interference, etc. Those influences have a great spatial dependence which means that a point is theoretically uniquely defined by $\Delta_{RSS}(\vec{p}, t)$. Under the assumption (Eq. (10)) that the theoretical signal strength $RSS_{theo}(\vec{p}, t)$ is approximately represented by the

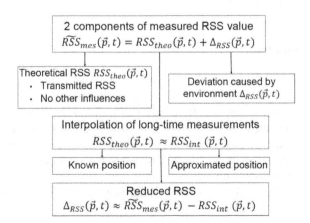

Fig. 3 Calculation steps of the interpolation method

interpolated RSS of long-time measurements $RSS_{int}(\vec{p}, t)$ resulting from Eq. (9) the following subtraction (Eq. (11)) provides the deviation caused by the surrounding environment $\Delta_{RSS}(\vec{p}, t)$.

$$RSS_{theo}(\vec{p}, t) \approx RSS_{int}(\vec{p}, t) \qquad (10)$$

$$\Delta_{RSS}(\vec{p}, t) \approx \widetilde{RSS}_{mes}(\vec{p}, t) - RSS_{int}(\vec{p}, t) \qquad (11)$$

Equation (11) is called reduction and $\Delta_{RSS}(\vec{p}, t)$ is the reduced RSS. The result has a large spatial dependence but it is approximately free from signal's variation over time. Because of that $\Delta_{RSS}(\vec{p}, t)$ should be accurate enough for positioning with Wi-Fi fingerprinting. To be able to compare the elements of the fingerprinting DB from the training phase and the RSS vector with the location fingerprint (from the positioning phase), both have to be reduced.

3.1.1 Reference Measurements

To be able to reduce RSS observations reference measurements have to be carried out which are recorded at the same time as the measurements that have to be reduced. Therefore, long-time measurements are performed which temporally overlap the whole measuring process. Tests performed by Retscher and Tatschl (2016a) have shown that it is important to place the reference measuring units directly below the APs because of following reasons: (1) to get the strongest possible RSS, (2) to have a lower risk of multipath propagation, and (3) the construction of most of the antennas weaken the RSS of more distanced antennas (Tatschl 2016). Another important aspect of the reference location is to choose a useful spatial distribution of these stations. The aim is to be able to interpolate the long-time measurements to every reference point of the fingerprinting DB (see Sect. 3.1.2). So one possibility is to place the reference stations at the borders surrounding the area of interest.

3.1.2 Training Phase

A grid of reference points is measured like in the training phase of the standard Wi-Fi fingerprinting (described in Sect. 2.2.1) but the long-time measurements are performed additionally (see Fig. 4). To get a fingerprinting database with reduced RSS the long-time measurements are interpolated to the reference points as indicated by arrows in Fig. 4 and a reduction is proceeded. For the interpolation of the fingerprinting DB the first step is to find long-time measurements which were recorded simultaneously with the reference points or were recorded only a short time apart. After that a 2-D linear plane-interpolation (Eq. (12)) is calculated to get a theoretical RSS for each reference point.

Fig. 4 Training phase of the interpolation method

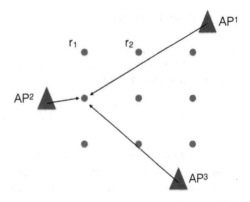

$$RSS_{int}(AP,t) = f(x,y) = a_0 + a_1 x + a_2 y \qquad (12)$$

The coefficients of this polynomial function are dependent on time and a certain AP. They are given by the following equation:

$$\vec{a}(AP,t) = A^{-1} * \vec{z}(AP,t) \qquad (13)$$

or in more detail:

$$\begin{bmatrix} a_1 \\ a_2 \\ a_3 \end{bmatrix} = \begin{bmatrix} 1 & x_1 & y_1 \\ 1 & x_2 & y_2 \\ 1 & x_3 & y_3 \end{bmatrix}^{-1} * \begin{bmatrix} RSS_{lt_1}(AP,t) \\ RSS_{lt_2}(AP,t) \\ RSS_{lt_3}(AP,t) \end{bmatrix} \qquad (14)$$

where $p_j = (xj, yj)$ are the coordinates of the reference station equipment which performed the long-time measurements and $RSS_{lt_1}(AP,t)$ the RSS of the jth equipment of a certain AP at a certain time.

This interpolation described in Eq. (12) produces a RSS value for a certain AP at a certain reference point. After calculating a RSS value for each AP at each reference point the interpolated radio map is completed.

For the reduction of the fingerprinting DB the subtraction (Eq. (11)) of the interpolated RSS of the long-time measurements and the measurement of the reference points results in the reduced RSS. This result is stored in a DB which defines the radio map (Eq. (1)). The following Eq. (15) describes this fingerprinting DB for the i-th reference point.

$$F_i = [\Delta_{RSS,RP_i}^{AP_1}, \Delta_{RSS,RP_i}^{AP_2}, \ldots, \Delta_{RSS,RP_i}^{AP_k}] \qquad (15)$$

where k is the number of APs and $\Delta_{RSS,RP_i}^{AP_k}$ is the single value of the reduced RSS.

3.1.3 Positioning Phase

Figure 5 illustrates the positioning phase. The positioning starts with the measurement of the location fingerprint. So the RSS of each received AP is scanned and stored in a vector (Eq. (4)):

$$\widetilde{\vec{r}}(\tau) = \left[\widetilde{RSS}_{RP_i}^{AP_1}, \widetilde{RSS}_{RP_i}^{AP_2}, \ldots, \widetilde{RSS}_{RP_i}^{AP_k}\right] \quad (16)$$

where τ is the recording time of the location fingerprint and $\widetilde{RSS}_{RP_i}^{AP_k}$ a single RSS value of the location fingerprint of the positioning phase.

The evaluation of the positioning phase consists of two major parts: (1) an approximate position is calculated to know which coordinates (x and y in Eq. (12); see Fig. 5) are used to get the interpolated RSS, and (2) the interpolated RSS and the reduction is calculated. After that the standard fingerprinting positioning algorithm is performed with a DB containing reduced RSS values and a reduced location fingerprint. For the approximate solution of the user's position the not-reduced RSS values of the reference points are needed. The approximate position is calculated using the process described in Sect. 2.2.2. The results are the coordinates of the approximate position $\vec{p}_{app} = [x_{app}, y_{app}]$. Then the interpolation for a location fingerprint is performed. So for the time τ fitting RSS values from long-time measurements exist:

$$\vec{z}(AP, \tau) = [RSS_{lt_1}(AP, \tau), RSS_{lt_2}(AP, \tau), RSS_{lt_3}(AP, \tau)] \quad (17)$$

With Eq. (12) an interpolated fingerprint is calculated:

$$\vec{r}_{vec}(\vec{p}_{app}, \tau) = [RSS_{int}(AP_1, \tau), RSS_{int}(AP_2, \tau), \ldots, RSS_{int}(AP_k, \tau)] \quad (18)$$

and using Eq. (11) the reduction is calculated. In Eq. (19) the reduction of a single RSS value and in Eq. (20) the reduction of the whole fingerprint is described:

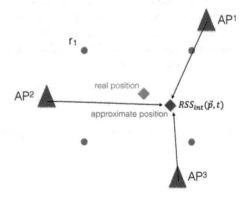

Fig. 5 Positioning phase of the interpolation method

$$\Delta_{RSS}(AP_k, \vec{p}_{app}, \tau) = \widetilde{RSS}_{RP_i}^{AP_k} - RSS_{int}(AP_k, \tau) \tag{19}$$

$$\Delta_{RSS}(\vec{p}_{app}, \tau) = \widetilde{\vec{r}}(\tau) - \vec{r}_{int}(\vec{p}_{app}, \tau) \tag{20}$$

With the reduced fingerprint $\Delta_{RSS}(\vec{p}_{app}, \tau)$ and the reduced fingerprinting DB a position determination can be performed. The resulting positioning vector $\vec{p}_{red}$ gives the estimated coordinates of the location fingerprint.

So far only a simple linear interpolation has been applied. Future work considers the use of other more advanced interpolation methods, such as Kriging or inverse distance weighted (IDW) interpolation. A detailed performance analysis and comparison with the linear interpolation approach will be carried out.

3.2 Daily Average Improvement Method

To get a numerical value which describes the fluctuation of the Wi-Fi signals the difference between the fingerprints and a constant vector are calculated. This constant vector should be representative of the normal or average behaviour of the APs RSS. So the average of all long-time measurements over the whole day is chosen as given in Eq. (21):

$$\widehat{\vec{r}} = \left[\widehat{RSS}^{AP1}, \widehat{RSS}^{AP2}, \ldots, \widehat{RSS}^{APk} \right] \tag{21}$$

To calculate the improvements, the average RSS value of all devices which measure the long-time measurements at the moment when the measurement to be improved was recorded ($\vec{r}_{lt}(t)$), is needed. The following equation yields the improvements of a certain measurement:

$$\vec{r}_{corr}(t) = \vec{r}_{lt}(t) - \widehat{\vec{r}} \tag{22}$$

The improved RSS is then calculated with Eq. (23):

$$\vec{r}_{imp}(t) = \vec{r}(t) + \vec{r}_{corr}(t) \tag{23}$$

with the measured fingerprint $\vec{r}(t)$ to be improved.

The implementation is as follows: Firstly, the reference points and reference long-time measurements are recorded equally like the difference method described in Sect. 3.1 in the training phase. After computing all improvements for the time when the reference points were recorded and improving the measurements, there is an improved fingerprinting database. Then in the positioning phase the process to get an improved location fingerprint is similar to the training phase. In the beginning, temporal suitable long-time measurements are searched and the average is calculated ($\vec{r}_{lt}(t)$). After the subtraction (Eq. (22)) with the daily average of all

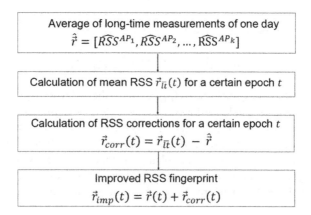

Fig. 6 Calculation steps of the daily average improvement (DAI) method

received APs the improved location fingerprint is calculated using Eq. (23). To get the position with the improved RSS values, a standard fingerprinting algorithm using the NN or kNN is performed (described in Sect. 2.2.3 or 2.2.4). Figure 6 shows the principle of operation illustrated as flowchart.

4 Indoor Experiments

To test the Wi-Fi-based indoor positioning system the ground floor of a multi-storey office building of the TU Wien—Vienna University of Technology has been chosen. Figure 7 shows the area including the entrance to the building, a foyer and class rooms. In this area walls with different wall thickness, different ceiling height, several floor levels, pillars, windows, computers and tables are present. It is public

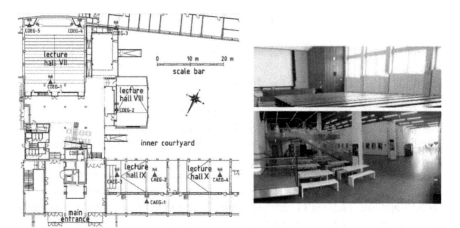

Fig. 7 Floorplan and impressions of the test area

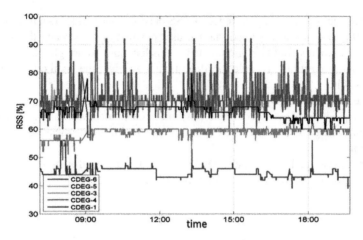

Fig. 8 Long-time reference station RSS observations

space which is highly frequented by people. On 93 reference points in a regular grid with a spacing of around 2.5 m RSS measurements were performed whereby six well distributed APs are receivable at most test points. Three Raspberry Pi's served as reference stations which execute a Python script which records RSS values for all received APs in an infinite loop and an interval of two seconds.

4.1 Long-Time Measurements

Figure 8 shows an example of a long-time variations of the RSS to six APs. As can be seen some RSS scans to certain APs show a good stability, but on the other hand large short-time variations can occur, e.g. for AP 4. Retscher and Tatschl (2016a, b) developed a differential Wi-Fi (termed DWi-Fi) positioning concept which is similar to the well-known DGPS principle. In this case, the RSS measured with the Raspberry Pi's at the reference station locations are used to derive correction parameters and are applied at the mobile client. Thus, a significant improvement of the positioning accuracies is achieved. In the following it is investigated if location fingerprinting can be improved at a similar level.

4.2 Evaluation

To compare the different methods, the measured data was evaluated in four different ways:

(1) Standard Wi-Fi Fingerprinting with the kNN where k was set to 5.

(2) Interpolation method with reduced RSS observations and known user position. In this case it was assumed that the user position is known and the long-time measurements could be interpolated at this location. Although this assumption is nonconvertible it could be determined whether the method is basically possible.
(3) Interpolation method with reduced RSS observations and known approximate user position. The approximate position is calculated with a standard fingerprinting algorithm and is used to interpolate the long-time measurement to this point.
(4) Daily Average Improvement (DAI) method with reduced RSS observations. In this case no approximate solution is needed.

4.3 Resulting Radio Maps of RSS Distribution

The radio maps in Fig. 9 show the spatial distribution of the measured RSS of one AP located inside of the class room VII (compare Fig. 7). As can be seen there are obvious differences in the measured RSS of three smartphones and the recorded measurement runs. For the resulting radio map in Fig. 9a the three reference stations were located at three APs and for the other three Figures the reference stations were placed at the borders surrounding the test site. Figure 9a, b are from the same smartphone. It can be seen that regions of higher RSS are found in an approximate rectangular form which spatially match with the class room (i.e., lecture room VII). This means that the class room's walls cover up a high percentage of the signal.

4.4 Positioning Results

The following three Figs. 10, 11 and 12 show the positioning accuracies of the three used phones in numbers for the four different methods (i.e., standard fingerprinting, interpolation with known position or approximate position and daily average improvement DAI method). Furthermore, the results are shown for four different training and positioning runs. On overview of the characteristics of the four different training and positioning runs can be found in Table 1.

The results of the smartphone 1 in Fig. 10 demonstrate that the interpolation method with known position of the user and the DAI method provide a similar performance as standard fingerprinting. The interpolation method using an approximate position, however, shows lower positioning accuracies. The main reason for this is that the approximate user's position at the ten meter level was not

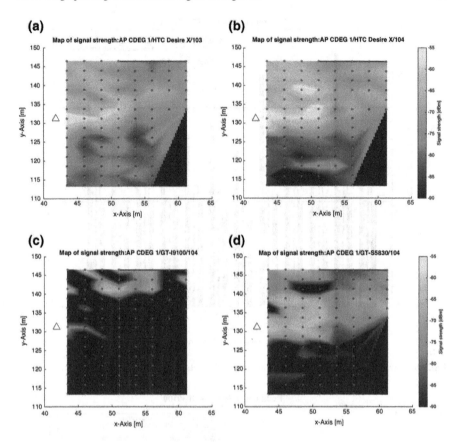

Fig. 9 Radio maps of three smartphones and two different measurement runs

accurate enough. With an improvement of the approximate position higher resulting positioning accuracies are achievable. This can be obtained if the Wi-Fi positioning is augmented by dead reckoning with the smartphone inertial sensors. Then a more precise approximate position is available.

In the results of the other two smartphones (Figs. 11 and 12) some tests are found which show an improvement compared to the standard method. Especially the DAI method's positioning accuracy is higher and provides therefore better performance than the two interpolation methods. Thus, the used phone has a great influence on the positioning result as well. The first smartphone (compare Fig. 10) measured the highest RSS values and shows overall the best positioning results which confirms that a higher signal strength leads to a better positioning accuracy. So the positioning accuracy is also dependent on the built-in hardware of the phone.

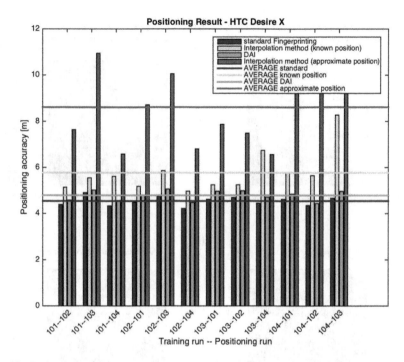

Fig. 10 Comparison of four measurement runs and positioning methods for smartphone 1

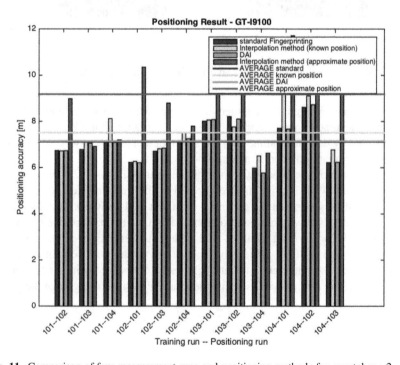

Fig. 11 Comparison of four measurement runs and positioning methods for smartphone 2

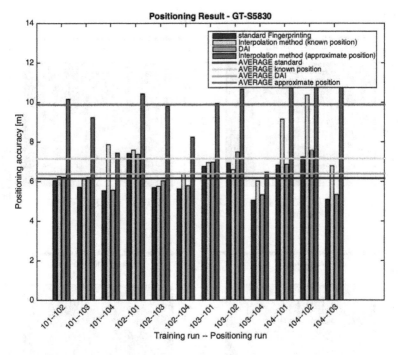

Fig. 12 Comparison of four measurement runs and positioning methods for smartphone 3

Table 1 Overview of the characteristics of the four measurement runs

Measurement run No.	Number of RSS scans	Environmental conditions	Location of reference stations
101	5	Class room doors and blinds closed	Under the AP
102	5	Class room doors and blinds closed	Under the AP
103	9	Class room doors and blinds open	Under the AP
104	5	Class room doors closed and blinds open	At the borders of the test site

4.5 Spatial Distribution of Positioning Results

The maps in Fig. 13 show the spatial distribution of the positioning accuracy for a certain positioning and training run for smartphone 1. It should be noted that the regions of poor accuracy are found at the periphery areas of the test site. The reason for this is the interpolation of the RSS which is used in each method. The triangle which is formed by the three Raspberry Pi's does not cover up the whole measuring

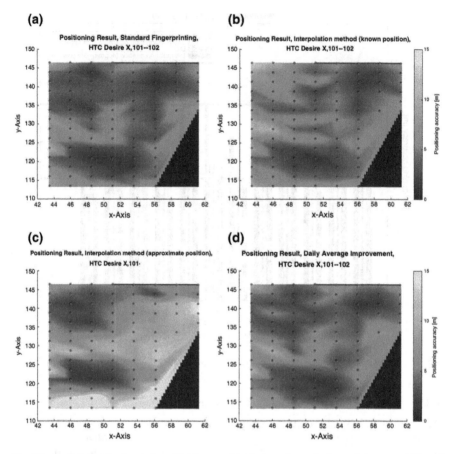

Fig. 13 Spatial distribution of positioning result of smartphone 1 for training run 101 and positioning run 102 for the four methods

area which leads to an extrapolation outside of the triangle. In this area the interpolated RSS values leads to a significant inaccuracy. With a longer distance from the triangle the positioning error increases which is also explained by the extrapolation in this areas.

4.6 Positioning Results Per Measurement Run

In Figs. 14 and 15 the four different measurement runs are compared. Figure 14 shows the accuracy (divided into the results of the four methods) of a measurement run if it is used as training run. Figure 15 on the other hand presents that for the positioning run. It can be seen that no measurement run stands out among the other. Measurement run 104 is the best run if it is used as positioning run but it is the

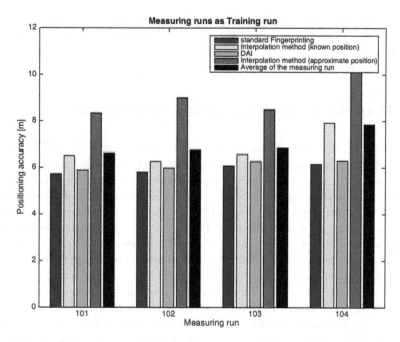

Fig. 14 Positioning result from different measurement runs as training runs for smartphone 1

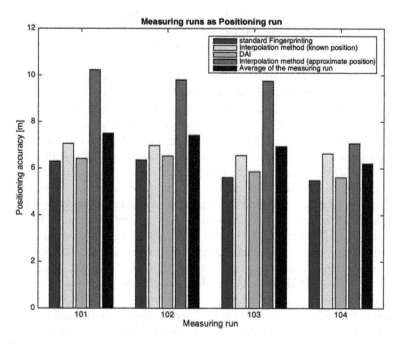

Fig. 15 Positioning result from different measurement runs as positioning runs for smartphone 1

worst run if it is used as training run. There might be great differences in the radio map of the different measurement runs but the positioning result is still quite similar.

5 Discussion of Results and Concluding Remarks

In general, no significant differences between the interpolation and DAI methods in comparison to the standard Wi-Fi fingerprinting was found in this practical scenario of the test site. The DAI method is, independent from the used smartphone, the best new method because it needs no approximate position and models the fluctuation of the Wi-Fi signal very well. The comparison between known and approximate position for the interpolation method shows that the approximate position is an additional source of error which leads to a positioning error which is nearly twice as high as the positioning error of the interpolation method with known positions. One way to use this method and get an improvement of positioning accuracy is to find an alternate way of producing an approximate position. The inertial sensors in the smartphone, such as the accelerometers and gyroscopes as well as the magnetometer, can be used to perform dead reckoning and can yield a better approximate position. Especially, the continuous positions of a user can be determined with low drift rates and suitable positioning accuracies if one enters a building and beforehand the absolute user's location is obtained with GNSS. Thus, we are currently working on an integrated positioning solution combining Wi-Fi fingerprinting with inertial navigation. Then fingerprinting can serve as absolute positioning method and dead reckoning with the inertial sensors for continuous positioning of the mobile client.

Overall the fingerprinting algorithms reached an averaged positioning accuracy of around 5 m dependent on the used smartphone. In the evaluation arithmetic mean and median were compared and the median, which deals better with outliers, shows the better results (i.e., up to 3 m lower positioning errors).

The comparison of the different maps of spatial distribution of the positioning accuracy shows that the regions of poor accuracy are found at the periphery areas of the test site. This is caused by the interpolation and the location of the reference stations. The results can be improved if the reference stations equipped with Raspberry Pi's are placed outside the measuring area and cover up the whole area of interest. Furthermore, different locations for the reference stations in the test site will be investigated. Thereby the location accuracy of the reference stations will be tested more in detail. It is expected that more than three reference stations will yield to better positioning accuracies for the mobile user. On the other hand, it will be investigated if with DAI method a suitable performance is achieved when a single reference station is placed in the middle of the area of interest as theoretically the average of the RSS variations need to be determined only. In addition, the impact of the different Wi-Fi APs in respect to their received signal strength on each reference point is a further task for investigation. Possible solutions might be to exclude certain AP RSS measurements which have a very low signal strength throughout the measurement campaign on a specific reference point or to perform a weighting depending on their RSS magnitude.

Another way to improve the positioning accuracy is to pay attention to the accuracy of the Raspberry Pi's system clock. A temporal drift in time between the smartphones and the Raspberry Pi impedes a potential improvement. Furthermore, another source of error could be eliminated if it would be possible to get the emitted RSS directly from the APs.

The tests have also shown that there are big differences between the used smartphones. This differences become first visible in the radio maps of signal strengths (see Fig. 9 for example) and it is obvious that a higher RSS entails a better positioning result. The smartphone 1 positioning results are the best in every measurement run. Consequentially it is very important that each phone has its own fingerprinting database as the device components of the phone significantly influence the positioning result. It is relevant to check whether a joint database of all phones results in similar positioning accuracy if it is combined with long-time measurements. The DAI method is able to improve the standard fingerprinting in five situations using smartphone 2. This fact proofs that the success of a method is additionally dependent of the used phone.

In current analysis it is investigated if the time interval for averaging over one day in the DAI method is suitable or shall be changed depending on the environmental conditions and occurring fluctuations of the RSS. In the conducted long-time observations, it could be seen that for a number of RSS measurements to certain APs very large fluctuations occurred (compare Fig. 8). Thereof the optimum time interval for averaging for a certain AP has to be found. It is conceivable to reduce the time interval for averaging then for a specific AP showing large fluctuations. The major aim of the investigation in this context is to find the optimum time span for averaging of the reference station observations when applying them to the current user's RSS measurements. Averaging over a whole day might be too long. Due to a successive reduction of this time interval it shall be investigated if an improvement in performance of the algorithm and increase in positioning accuracies for the mobile user is achievable.

In the tests the measurements of the training and positioning phase were made with a maximum time interval of one day. In practice this interval for training is usually many times greater and Ettlinger and Retscher (2016) have shown that this leads to a deterioration of positioning accuracy. In this case it is expected that the new methods are able to achieve an improvement. In future work multiple repeating experiments with a larger time interval between the measurement runs are performed.

Additional future work concerns the applied interpolation strategy and method. Apart from a simple 2-D linear plane-interpolation other more advanced interpolation methods, such as Kriging or inverse distance weighted (IDW) interpolation, will be investigated. For that purpose, a detailed performance analysis and comparison with the linear interpolation approach will be carried out.

To conclude it can be summarized that for this measurement setup and this kind of evaluation the new methods were just able to improve the standard fingerprinting in a few situations. Long-time measurements at the reference stations could lead to a further improvement of standard fingerprinting. Although the positioning accuracy of Wi-Fi fingerprinting is still not competitive with GNSS the rising number of private and public wireless networks makes it a useful and suitable positioning method, especially in the case if GNSS signals are blocked. The further integration with the inertial smartphone sensors is the right direction to obtain a continuous ubiquitous position solution.

Further research is carried out in combination with a new approach for differential positioning using Wi-Fi. A concept has been developed by the first author (see Retscher and Tatschl 2016a, b) based on the well-known DGPS principle where correction parameters are derived from measurements on reference stations. In the case of Wi-Fi positioning differential RSS corrections can be calculated and applied to the current RSS measurements at the mobile client. If the long-time measurements and the thereof derived dynamical radio maps are applied in positioning a significant improvement is achieved. Currently a series of tests is carried out and analyzed. For further information on this approach which integrates Wi-Fi trilateration with fingerprinting, the interested reader is referred to the paper of Retscher and Tatschl (2016b) presenting very promising results. The feasibility of long-time measurements overlapping the whole measurements process is then more realistic in the case if the two positioning methods are combined meaningfully. As already mentioned above a low-cost solution can be realized by using Raspberry Pi's to serve as reference stations. It is then possible to perform continuous localization of mobile users with improved positioning accuracies as in standard Wi-Fi fingerprinting or trilateration as the application of differential corrections yields to a better consideration of large short-time RSS variations in real-time.

References

Bai Y B, Wu S, Retscher G, Kealy A, Holden L, Tomko M, Borriak A, Hu B, Wu H R & Zhang K (2014) A New Method for Improving Wi-Fi Based Indoor Positioning Accuracy. Journal of Location-Based Services LBS, 8:3:135–147.

Ettlinger A F & Retscher G (2016) Positioning Using Ambient Magnetic Fields in Combination with Wi-Fi and RFID. 7th International Conference Indoor Positioning and Indoor Navigation IPIN 2016, October 4–6, Alcalá de Henares, Madrid, Spain.

Li B & Rizos C (2014) Editorial: Special Issue International Conference on Indoor Positioning and Navigation 2012, Part 2. Journal of Location Based Services, 8:1:1–2.

Huang H (2014): Post Hoc Indoor Localization Based on RSS Fingerprinting in WLAN. Master thesis, University of Massachusetts, U.S.A.

Kaemarungsi K (2005): Design of Indoor Positioning Systems Based on Location Fingerprinting Technique. PhD thesis, University of Pittsburgh, U.S.A.

Mok E & Retscher G (2007) Location Determination Using WiFi Fingerprinting Versus WiFi Trilateration. Journal of Location Based Services, 1:2:145–159.

Retscher G. (2007): Augmentation of Indoor Positioning Systems with a Barometric Pressure Sensor for Direct Altitude Determination in a Multi-storey Building. Journal of Cartography and Geographic Information Science (CaGIS), 34: 4:305–310.

Retscher G & Tatschl T (2016a) Differential Wi-Fi – A Novel Approach for Wi-Fi Positioning Using Lateration. FIG Working Week, May 2–6, Christchurch, New Zealand.

Retscher G & Tatschl T (2016b) Indoor Positioning Using Wi-Fi Lateration – Comparison of two Common Range Conversion Models with Two Novel Differential Approaches. IEEE Xplore, 2016 Ubiquitous Positioning Indoor Navigation and Location Based Service (UPINLBS), November 3–4, Shanghai, PR China.

Tatschl T (2016) Indoor Positionierung mit differenziellem WLAN. Master thesis, TU Wien, Vienna, Austria (in German).

Situation Goodness Method for Weighted Centroid-Based Wi-Fi APs Localization

Germán M. Mendoza-Silva, Joaquín Torres-Sospedra,
Joaquín Huerta, Raul Montoliu, Fernando Benítez
and Oscar Belmonte

Abstract Knowing the location of Wi-Fi antennas may be critical for indoor localization. However, in a real environment, their positions may be unknown since they can be managed by external entities. This paper introduces a new method for evaluating the suitability of using the weighted centroid method for the 2D localization of a Wi-Fi AP. The method is based on the idea that the weighted centroid method provides its best results when there are fingerprints taken around the AP. In order to find the probability of being in the presence of such situations, a natural neighbor interpolation method is used to find the regions with the highest signal strengths. A geometrical method is then used to characterize that probability based on the distribution of those regions in relation to the AP position estimation given by the weighted centroid method. The paper describes the testing location and the used Wi-Fi fingerprints database. That database is used to create new databases that recreate different sampling possibilities through a samples deletion strategy. The original database and the newly created ones are then used to evaluate the local-

---

G.M. Mendoza-Silva (✉) · J. Torres-Sospedra · J. Huerta · R. Montoliu · F. Benítez ·
O. Belmonte
Institute of New Imaging Technologies, Universitat Jaume I, Avda,
Vicente Sos Baynat S/N, 12071 Castellón de la Plana, Spain
e-mail: gmendoza@uji.es

J. Torres-Sospedra
e-mail: jtorres@uji.es

J. Huerta
e-mail: huerta@uji.es

R. Montoliu
e-mail: montoliu@uji.es

F. Benítez
e-mail: benitezm@uji.es

O. Belmonte
e-mail: belfern@uji.es

© Springer International Publishing AG 2017
G. Gartner and H. Huang (eds.), *Progress in Location-Based Services 2016*, Lecture Notes in Geoinformation and Cartography,
DOI 10.1007/978-3-319-47289-8_2

ization results of several AP localization methods and the new method proposed in this paper. The evaluation results have shown that the proposed method is able to provide a proper probability for the suitability of using the weighted centroid method for localizing a Wi-Fi AP.

Keywords Indoor localization · Wi-Fi aps localization · Weighted centroid · Interpolation · LBS

1 Introduction

With the widespread presence of mobile devices able to consume online services and provide them the user position in some scenarios, location-based services (LBS) have gained remarkable importance in the recent years. Applications that use LBS can shape the content they provide to the users according to the determined (or estimated) positions of their mobile devices (Werner 2014). While GPS-based techniques provide a reasonably good solution for outdoor localization, they are not suitable for indoor environments. GPS signal strengths inside a building are too low and fluctuate too much to be reliable (Chen and Kotz 2000). Furthermore, the use of the device's GPS sensor is sometimes avoided by the users because of its high power consumption.

Along the years, several methods have been created to provide indoor localization (Al-Ammar et al. 2014; Torres-Solis et al. 2010), based on device sensors other than the GPS one. Though there are device-centric techniques like dead-reckoning (Gutmann et al. 1998), most methods measure a physical quantity from the device surroundings. These physical quantities include sound, light, radio-frequency (RF), magnetic field and others. The methods that focus on RF, especially Wi-Fi, have been very popular. Reasons for this popularity include: (i) the knowledge already existing for the outdoor case and for wireless sensor networks, (ii) the pervasive presence of indoor wireless antennas (IEEE 802.11 and 802.15 standards, i.e., Wi-Fi and others like Bluetooth, respectively) and (iii) the widespread use of smartphones able to connect to those antennas. It is attractive to use already existing building's Wi-Fi infrastructure.

The Wi-Fi based indoor positioning techniques determine the device location based on antennas' signal intensity values (RSSI) that the device receives. Most of these techniques can be grouped according to positioning algorithm and measured property into three main positioning principles: proximity, trilateration, and scene analysis (Farid et al. 2013; Liu et al. 2007; Liu and Yang 2012). For proximity and trilateration techniques, the location of the Wi-Fi emitting antenna or access point (AP) is fundamental. Discovering the AP location is also important for management tasks, such as optimizing AP placement and detecting rogue APs.

For entities whose deployed Wi-Fi networks are unmanaged, or managed by a third party entity, it is usually necessary to estimate AP locations based on the

Wi-Fi signals. The process of measuring those signals and capturing other characteristics of an AP (including its position or coverage) is often termed as war-driving (Berghel 2004). Some studies have addressed the creation of such databases (Ledlie et al. 2011; Moreira and Meneses 2015) and currently several global AP databases exist, such as Wigle.net.[1] With those databases, it is possible to obtain a gross location based only on the MAC address of the strongest AP signal received (nearest neighbor).

Several studies have developed methods for estimating an AP location based on its signal's intensities. The most known of these methods is the weighted centroid. It has been used (i) in user position localization methods (Knauth et al. 2015; Kosović and Jagušt, 2014; Lohan et al. 2015; Wang et al. 2011), (ii) as an AP (and other emitters) localization method or part of it (Blumenthal et al. 2007; Cheng et al. 2005; Cho et al. 2012), or (iii) as a baseline method for new AP localization methods (Han et al. 2009; Ji et al. 2013; Koo and Cha 2011b; Zhao et al. 2014).

Other AP localization methods based on signal strength data have been developed, including the ones presented in Ji et al. (2013); Koo and Cha (2011b); Zhao et al. (2014). According to their published results, their AP localization accuracy ranges from 15 m to less than 2 m. These methods used relatively dense signal measurements, and their robustness to distribution and different numbers of signal measurements was only evaluated in a few cases. Although relatively dense mappings can be practical and they are common in laboratory settings, due to scalability issues, realistic indoor positioning systems avoid dense mappings mainly due to the large time required to generate the reference database or radio map.

This paper's main contribution is a method that, from a set of measured RSSI values of a Wi-Fi antenna, estimates the likelihood of whether the weighted centroid can provide an accurate 2D location estimation for that antenna. We have called this method as the Situation Goodness method. This paper also introduces a new AP localization method based on the natural neighbor interpolation. Furthermore, the paper provides an evaluation of AP localization methods, including the weighted centroid and the new method based on natural neighbor interpolation. The evaluation considers the localization error and considers different signal measurements situations. The Situation Goodness method is evaluated regarding its accuracy to calculate the aforementioned likelihood. This evaluation also considers different signal measurements situations.

The rest of the paper is organized as follows: Sect. 2 addresses a new AP localization method and others found in the literature. Section 3 describes the proposed method for assessing the situation goodness for the weighted centroid method. Section 4 presents experiments done to evaluate the addressed AP location methods and the new Situation Goodness method assessment for the weighted centroid. Finally, Sect. 5 resumes the results obtained in this paper.

---

[1]https://wigle.net/, visited on 10/06/16.

2 Wi-Fi AP Localization Methods

The problem of Wi-Fi AP localization from collected signal strength measurements (RSSI) have been addressed in the literature for more than 10 years, and several methods have been proposed. The following sections describe some of these methods applied to an AP 2D localization. Section 2.1 describes the weighted centroid method, a building block of this paper's main contribution. Section 2.2 introduces a new AP localization method based on natural neighbor interpolation. Section 2.3 addresses another three AP localization methods we have considered relevant. All these methods are the ones used in the evaluation that Sect. 4.2 presents.

2.1 Weighted Centroid Method

The weighted centroid method has been used for Wi-Fi APs in many research works: in user position localization methods, in AP localization methods and as a baseline method for new AP localization methods (Blumenthal et al. 2007; Cheng et al. 2005; Cho et al. 2012; Han et al. 2009; Ji et al. 2013; Knauth et al. 2015; Koo and Cha, 2011b; Kosović and Jagušt 2014; Lohan et al. 2015; Wang et al. 2011; Zhao et al. 2014). The reasons behind its popularity include its simplicity, its low computational complexity, and its relatively low AP localization error in some known situations.

Given a ground truth of n signal strengths (R_i) of an AP, measured at known positions $\boldsymbol{P_i} = (x_i, y_i)$, the weighted centroid method estimates the AP's location using formula (1).

$$P_a = \frac{\sum_{i=1}^{n} P_i W_i}{\sum_{i=1}^{n} W_i}, \qquad (1)$$

where W_i represents the calculated weight of the ith measurement and $\boldsymbol{P_a} = (x_a, y_a)$ is the estimated AP's location. The weights represent the importance given to the position of each measurement and are, ideally, inversely proportional to the distance between the measurement point and the AP. As the real distance is usually unknown, it can be estimated using a power-based propagation model like, e.g., the one presented in Bahl and Padmanabhan (2000). However, the weights are commonly calculated directly from the RSSI values, i.e., without estimating a distance. The calculation is based on the idea that the shorter the distance, the more intense the RSSI. Therefore, the more intense the RSSI, the greater must be its associated weight. The weight calculation used in this paper is presented in Lohan et al. (2015) and it is shown in (2).

$$W_i = 10^{\frac{R_i}{10}} \qquad (2)$$

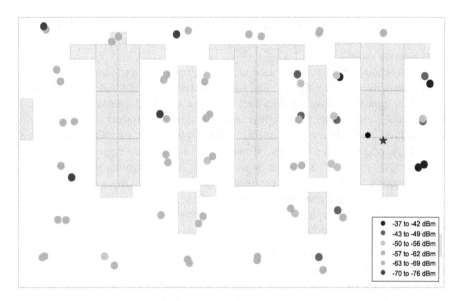

Fig. 1 Weighted centroid method's estimation when there are measurements taken around the AP. The *colored circles* represent the input measurement positions and strengths. The *black* point and the blue star represent the AP location estimation and its actual location, respectively. The *pale blue* and *pink* rectangles are the furniture found in the sampled area

The way this method calculates the AP position determines the situations in which the method provides good location estimations: If (i) there are measurements taken around the real AP position or (ii) the measurements are close to the AP. Figure 1 shows an example of case (i). This figure presents measured signal intensities and the location estimation of the AP. The data are shown over a furniture layout of the sampled area composed by shelves and cabinets (pink) and tables (blue). In this example, the estimated location is closest to the locations of the measurements with the strongest RSSI values, i.e., the measurements with RSSI values stronger than 56 dBm. In the case (ii) the measurements closest to the AP are more likely to be the strongest and to have the highest weights. However, if the measurements were taken only towards one side of the AP, and none of them is close to the AP, the location estimation is poor. Despite those facts, many studies have used the weighted centroid as a baseline comparison method while considering situations where this method does not produce good localization results.

2.2 Interpolation Contours Centroid Method

Interpolation and extrapolation methods have already been used in RSSI-based indoor localization, mainly in the Wi-Fi radio map's database enrichment (Arai and Tolle 2013; Ezpeleta et al. 2015; Lee and Han 2012). Although some research works have also used extrapolation methods in indoor localization-related contexts, like in Talvitie et al. (2015), the interpolation methods are far more accurate that the extrapolation ones.

Interpolation methods can be used to spot regions of the sampled area where the signal intensities are the highest, and use these regions to estimate the AP's location. This approach shares a drawback with the weighted centroid: It is able to locate the AP only inside the sampled area. The drawback results from the fact that interpolation methods provide function values estimations for within the convex hull of the original set of points. However, as the selected regions are the ones closest to AP's location, they can be used to get AP localization results similar to those obtained by the weighted centroid in the cases where the weighted centroid performs the best; and even better results in the other cases.

In particular, the implementation we have used employs the natural neighbor interpolation (Sibson and others 1981). The steps we propose are the following:

1. Calculate an interpolated grid of points using natural neighbor interpolation.
2. Determine the goal intensity level. To find the goal level, three intensity levels ("high", "medium" and "low") are determined so that they split the input RSSI values into four equally spaced intervals. The "high" level is the goal intensity level.
3. Compute the contours corresponding to the goal (highest) intensity level. Contours are calculated using the Marching Squares algorithm (Maple 2003).[2]
4. Obtain the points that represent the contours computed in step 3. If needed, assure two requirements: (i) the point densities along the edges are similar and (ii) the edges' curvature are well described by their points.
5. Using the points from step 4, calculate a centroid by averaging the points' coordinates. The position of this centroid is the AP location estimation.

Figure 2 shows the result of estimating an AP's position using this method on a set of RSSI values. The blue, green and red lines depict the contours corresponding to the "low", "medium" and "high" intensity levels, respectively. The number over a contour shows the corresponding intensity level value. Notice that only the red contours, which are the ones that bound the highest signal strengths regions, are used to estimate the AP location. This method is more computationally complex than the weighted centroid, and its complexity depends on the number of interpolated grid's points. Also, the grid interpolation density influences in the actions needed in step 4. Step 3 already provided the contours as a discretized set of points. If the grid is interpolated with high density, e.g., 0.05 m of separation between points (which is the one used in the experiments presented in Sect. 4), then the interpolation is smooth, and the contours' points satisfy the requirements of step 4. However, if the grid's interpolation density is low, a further interpolation procedure on the contours' points is needed to meet the above requirements.

---

[2]As implemented in Matlab.

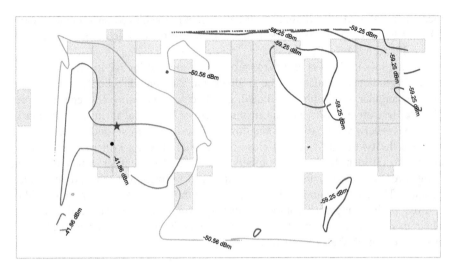

Fig. 2 AP location estimation using the interpolation contours centroid method. The AP location estimation (*red* contours' centroid) is shown by the *black* point and the real AP location by a *blue star*

2.3 Other Methods from Literature

In the literature, many research works have addressed the AP localization (Blumenthal et al. 2007; Cheng et al. 2005; Cho et al. 2012; Cho et al. 2012; Han et al. 2009; Ji et al. 2013; Koo and Cha 2011a, b, 2012; Nam 2014; Varzandian et al. 2013; Zhang et al. 2011; Zhao et al. 2014). In this section, we describe three methods we have considered to be relevant.

The method presented in Koo and Cha (2011b) linearly approximates the exponential relation between distance and signal strength that establishes the well-known distance-power law. They introduce this linear approximation in an equation system in which each equation corresponds to a measurement point. Then, they perform the proper transformations—based on Savvides et al. (2001)—and obtain a system of linear equations that is only dependent on the measurements' positions and intensities, and on one linearization coefficient that does not affect the final AP position estimation. The authors tested their method through simulations and the errors estimation they presented were higher than 5 m.

In Ji et al. (2013), a simulation-like (Monte Carlo test) approach is used to test a set of possible locations of an AP to find the one that better fits the measured RSSI values for the AP. For each possible AP position, and based on the measured RSSI values, the path-loss model parameters are estimated by solving a system of linear equations. Those parameters are in turn used to obtain an estimated RSSI value at each measurement point. The difference between the expected RSSI and the measured RSSI values is then used as a metric to evaluate the goodness of the tested location for the AP. With simulations, their method achieved an accuracy below 10 m in 95 % of cases. With the real samples, the method achieved an accuracy below 12 m in 90 % of cases.

The authors in Zhao et al. (2014) proposed a method for AP localization based on RSSI gradients calculation. The gradient (intensity and direction) for each point on a rectangular, uniformly distributed, grid is calculated using 1D centered, point discrete derivative masks, i.e., taken into account the signal intensity variation in the y-axis—between the two (up and down) closest neighbors—and in the x-axis—between the two (right and left) closest neighbors. Then, a k-means clustering is applied to (i) identify direction outliers and (ii) find the cluster with the highest number of sampling points for using its head's position as the AP estimated position. The authors compared their method with the weighted centroid algorithm and the gradient approach without clustering, using experimentally collected data. Their method outperformed the other two methods and had a mean localization error of 1.5 m on the experimentation setup.

3 Situation Goodness Method

This section presents a method that, given a set of Wi-Fi intensity measurements for an AP, provides a probability that describes whether the weighted centroid method is likely to provide a solution that is close to the actual AP location. The weighted centroid method provides its best solution when some of the intensity measurements are around the AP, and to a lesser extent, when they are close to the AP.

The idea behind the method combines part of the interpolation method presented in Sect. 2.2 and the weighted centroid method. Specifically, the new method explores how the highest intensity level contours explained in Sect. 2.2 are distributed regarding the AP location estimation provided by the weighted centroid.

Figure 3 presents the idea in two typical situations. The top image presents the case where the measurements do not surround the AP (which is located at the bottom, outside the measurements convex hull). For the top image case, the highest level regions are not balanced around the weighted centroid estimation. In the bottom image, on the other hand, the highest level regions are distributed around the weighted centroid estimation. For this second case, the weighted centroid estimation is very close to the actual AP location.

The new method explores the geometrical relation between the weighted centroid solution and the aforementioned highest level regions. That relation is then used to calculate the probability for the weighted centroid method to provide a solution that is close to the actual AP location. The method performs the following steps:

1. Calculate the weighted centroid (WC) AP location estimation $p_{wc} = (x_{wc}, y_{wc})$.
2. Calculate the highest level regions' contours, as set of points $C = \{C_1, ..., C_n\}$, where $C_i = \{c_{i1} = (x_{i1}, y_{i1}), ..., c_{im} = (x_{im}, y_{im})\}$.

Situation Goodness Method for Weighted ... 35

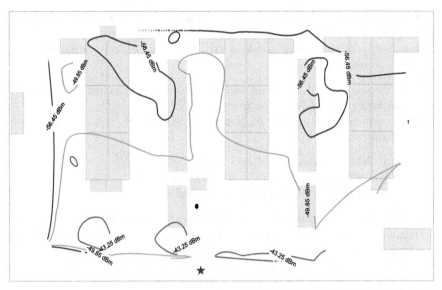

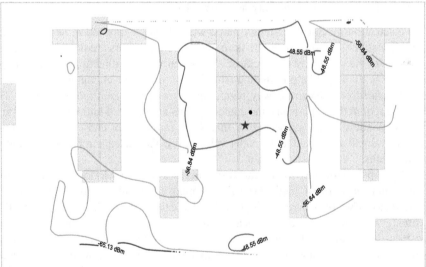

Fig. 3 Two cases of the distribution of highest intensity level contours (drawn in *red*) regarding the weighted centroid AP location estimation (the *black dot*). The real AP location is shown by a *blue star*

3. Calculate the directions (angles) of all contour points in C, taking p_{wc} as origin, as well as the distances to p_{wc}. The direction of the contour point $c_{ij} = (x_{ij}, y_{ij})$, which is the jth point of ith contour, is calculated as $v_{ij} = \tan^{-1}((y_{wc} - y_{ij})/(x_{wc} - x_{ij}))$. The distance associated to c_{ij} is calculated as the Euclidean distance between c_{ij} and p_{wc}.

4. Group the contours' points according to their directions in windows with a small size ($\pi/60$).
5. For each pair of opposing windows (W_f, W_b), e.g., $W_f = [0, \pi/60]$ and $W_b = [\pi, \pi + \pi/60]$, do:
    a. Calculate s_{wf} and s_{wb} as the sum of all distances (calculated in step 3) of points in W_f and W_b, respectively.
    b. Determine the dominance d_{wf} and d_{wb} of windows W_f and W_b, respectively, as follows:
        i. When $(s_{wf}/s_{wb}) > 2$, $d_{wf} = 1$ and $d_{wb} = 0$.
        ii. When $(s_{wf}/s_{wb}) < 0.5$, $d_{wf} = 0$ and $d_{wb} = 1$.
        iii. When $0.5 \leq (s_{wf}/s_{wb}) \geq 2$, $d_{wf} = 1$ and $d_{wb} = 1$.
6. Fill the dominance gaps. When two windows whose dominance is 1 are separated by no more than 3 (grace gap) windows whose dominance is 0 (gap windows), set the dominance of the separating windows to 1.
7. Set u as the number of opposing windows whose dominance is different and t as the total number of windows whose dominance is 1.
8. Provide the output probability as: $prob = 1 - u/t$.

The calculation methods for steps 1 and 2 were already explained in Sects. 2.1 and 2.2, respectively. Step 2 takes into account how the contours are described as a set of points in the same way that the Sect. 2.2 presents it. The directions and distances obtained in the step 3 help in finding whether the highest level contours are located in a balanced or unbalanced way regarding the weighted centroid estimation. To ease the required comparisons, the contour points are grouped into windows in step 4. The window's size presented in that step, ($\pi/60$), was chosen because it is relatively small and round, but smaller values were also tested and provided equally good results.

The step 5 associates values called dominances to each window. Each of these values indicates, for a particular pair of opposing windows (opposing sets of directions), towards where the highest level areas are located. To determine the winning window(s), the sum of distances from the window's points to the centroid position is calculated. If the sum of a windows is twice or more the opposing window's sum, the first window is the dominant one. Figure 4 shows the dominances of two sample pairs of opposing windows for a given highest level contour. Notice how the number of points found inside the blue triangles and the distances between these and the weighted centroid's AP location estimation are similar. Thus, the windows corresponding to these triangles are dominant. In the case of the windows represented by the orange triangles, it can be noticed that only one is dominant.

In step 6, as the contours can be very irregular, the gaps (windows) between dominant windows that are close to each other are filled (make them dominant). The value determining the gap maximum size is called grace gap. This value is directly related to the window size. In our tests, we found than grace gap value of 3 worked well for a windows size of ($\pi/60$).

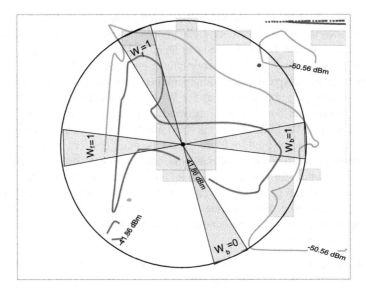

Fig. 4 The concepts of opposing windows and dominance. The *orange triangles* represent two opposing windows, as well as the *blue* ones. The blue windows are equally dominant, but only one of the *orange* ones is

Step 7 uses the dominance values to find unbalanced situations, i.e., where one window is dominant and the opposing one is not. This unbalanced situations count is finally used in step 8 to calculate the intended probability. Section 4.3 presents the evaluation of the above method.

4 Experiments

This section presents an evaluation of the AP localization methods described in Sect. 2, as well as an evaluation of the method presented in Sect. 3. Before the evaluations, the Wi-Fi measurements datasets used in the evaluations are presented.

4.1 Test Locations and Wi-Fi Samples Databases

The GeotecLab Wi-Fi database[3] was used to perform the experiments. That database and its related test location are already described in Torres-Sospedra et al. (2016). In short, the samples in the database were taken in the headquarters' lab of

---

[3]Available at http://indoorloc.uji.es, visited on 10/06/16.

the Geospatial Technologies (GEOTEC) research group, in two separate moments, by two persons using two different mobile phones, respectively. Although plenty of details can be found in Torres-Sospedra et al. (2016), it is worth mentioning that the lab is a typical office working environment—wooden and metallic furniture, working computers, typical office equipment, people working and moving, and was not intentionally prepared in any way for Wi-Fi related experiments. Also, the lab occupies 260 m^2 and thus can be considered as a medium-sized facility.

The GeotecLab Wi-Fi database is divided into two datasets, one containing samples that are meant to be used for training, and the other one for validation purposes. This configuration is typically used in the machine learning techniques applied in the most popular method for Wi-Fi based indoor localization: Wi-Fi fingerprinting. Both datasets contain data regarding the intensities of the APs detected by the used phones.

The two mentioned datasets were used to create the one dataset used in work described in our paper. The new dataset was created by selecting samples using the following three criteria:

1. Combine the training and validation datasets. As this work does not use machine learning approaches, the datasets were combined in order to have a large number of samples to work with.
2. Select the samples corresponding to APs whose positions are known. The selected antennas are the ones located inside the GEOTEC's lab and knowing their locations was mandatory for the evaluation of AP localization methods.
3. Use only the samples taken using the Samsung S3 (Android 4.3) phone. The Samsung S3 phone is able to detect 2.4 GHz and 5.0 GHz networks, while the LG Spirit (Android 5.0.1) is only able to detect 2.4 GHz networks. By using only these samples, it was possible to test the AP localization methods for 2.4 GHz and 5.0 GHz emitting antennas and also to have a similar number of samples for each AP.

The new dataset, from now on called the root dataset and shown in Fig. 5a, was in turn used to create new datasets by applying a sample elimination strategy in order to recreate different sampling alternatives. The used elimination strategy is a uniform random elimination. This strategy led to obtaining 10 datasets, which contain 10, 20, 30, ..., 100 percent of the root dataset samples, respectively. Figure 5b, c present sample datasets with 50 and 10 % of the root dataset samples, respectively.

4.2 AP Localization Methods Evaluation

The methods evaluated in this section were already described in previous sections:

- Weighted centroid, as described in Sect. 2.1,
- Interpolation contours centroid, as explained in Sect. 2.2,
- Linearization, Monte Carlo, and Gradients as shown in Sect. 2.3.

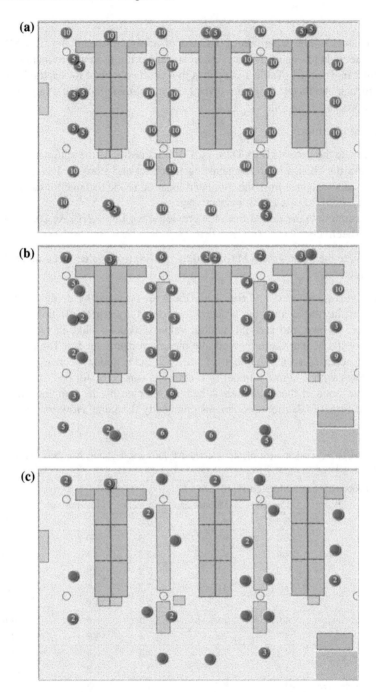

Fig. 5 The original database samples **a**, and the results of eliminating 50 % **b** and 90 % **c** of measurements. The *numbered circles* represent the amount of samples that are located at that position (or very close to it)

Each localization experiment consisted of applying one of the five AP localization methods to one of the datasets created according to the elimination strategy described in Sect. 4.1. Table 1 describes the results obtained for each of the above experiments. As that strategy involves random elimination, the evaluation for a method proceeded as follows:

1. Repeat 50 times:
    a. Create the datasets D10, D20, ..., D100 according to the elimination strategy.
    b. Run the chosen method against each dataset and store the localization error results (distance from the estimated location to the real one for each antenna) in E10, E20, ..., E100, respectively.
    c. Accumulate the results from the previous step into Ac10, Ac 20, ..., Ac100, respectively.
2. Calculate the mean values M10, M20, ..., M100 from values accumulated in the previous step.

The figures presented in Table 1 are the ones obtained by applying the above steps. Despite the evaluation process considered to repeat 50 times the same experiment to take into account random value selections—like in the elimination strategy or the Monte Carlo method, the mean error presented in Table 1 does not steadily increase as the number of measurement points decreases. Although experiments with a higher number of repetitions can be performed, the above fact means that none of these methods is heavily affected by the sampling density, as long as the point's distribution remains practically the same. However, considering

Table 1 Mean estimation error of the selected AP location methods. Numbers are rounded to decimeter level

Method dataset	Mean error (m)				
	Centroid	Monte carlo	Linear	Interpolation	Gradient
D10	3.8	8.2	7.5	3.5	11.9
D20	4.7	8.1	7.3	4.0	13.0
D30	4.4	8.1	7.3	4.3	13.6
D40	4.2	8.0	7.2	2.9	14.0
D50	3.7	8.0	7.2	3.6	14.3
D60	3.8	8.0	7.1	3.9	14.4
D70	4.0	7.9	7.1	3.4	14.6
D80	4.3	7.8	7.1	3.3	14.4
D90	4.1	7.8	7.1	3.4	14.4
D100	4.0	7.7	7.1	2.8	14.6
Overall mean error	4.1	8.0	7.2	3.5	13.9
Overall variance	0.1	0.0	0.0	0.2	0.7

Table 2 Differences in mean error localization when differencing between cases where there are samples taken around the AP, and those that do where no samples are taken around the AP

Method dataset	Mean error (m)	
	Centroid (surrounded)	Centroid (non-surrounded)
D10	2.2	5.2
D20	2.8	6.3
D30	2.0	6.4
D40	2.6	5.6
D50	1.3	5.7
D60	2.1	5.2
D70	2.3	5.3
D80	2.0	6.0
D90	1.9	5.9
D100	1.9	5.6
Overall mean error	2.1	5.7

the overall mean error, the best two are the weighted centroid and interpolation methods. If we consider the overall variance for these two methods, the weighted centroid is the best one.

Table 2 deepens into the fact that the weighted centroid method provide better localization results when fingerprints have been collected around the target AP. The column "Centroid (surrounded)" presents the mean localization error for such cases, i.e., considering only antennas for which there are fingerprints taken around them. The overall localization error, which considers all datasets, for the "surrounded" cases is 3.7 m better than in the "non-surrounded" cases.

4.3 Situation Goodness Method Evaluation

In this section, we describe the results of evaluating the Situation Goodness method. For the evaluation, we used similar approaches to the ones used in Sect. 4.2. In particular, for the evaluation that used the datasets created by applying the elimination strategy, the following steps were used:

1. Repeat 50 times:

    a. Create the datasets D10, D20, ..., D100 according to the elimination strategy.
    b. Run the method against each dataset and store the probability results in P10, P20, ..., P100, respectively.
    c. Accumulate the results from the previous step into Ac10, Ac 20, ..., Ac100, respectively.

2. Calculate the mean values M10, M20, ..., M100, and variance values V10, V20, ..., V100 from values accumulated in the previous step.

The evaluation results are presented in Tables 3 and 4. Hereinafter, we call "surrounded cases" the cases in which the weighted centroid method provides a good AP location estimation (as a result of the existence of measurements taken around the AP). We have called the other cases as "non-surrounded cases". In Tables 3 and 4, the column "WC good solution" indicates the surrounded cases with a "yes" value and non-surrounded cases with a "no" value.

The figures in Table 3 show how the Situation Goodness method provides low probability values for non-surrounded cases. For these cases, the probability values are always low (below or equal to 0.3). As can be expected, the most confident results correspond to the datasets with the highest number of measurements.

For the surrounded cases, the probability values are above 0.3 for datasets from D20 to D100, i.e., if the dataset is not much degraded, the method does identify the surrounded cases taken into account the mean value. The most confident results correspond to the datasets with the highest number of measurements.

The mean probability values of antennas 70, 74, 80 and 84 are similar. This is a likely result of the fact that those are actually one antenna that transmits at 2.4 GHz and 5.0 GHz, providing 2 networks at each frequency. Considering the surrounded cases, the antenna 68 is the one with lower probability values. Antennas 95, 96 and 97 are located on top of desks, while antenna 68 is located in the ceiling. The antenna 67 is located inside a room adjacent to the sampled area.

Table 4 presents the variability of the probability values across the 50 experiment repetitions. As expected, there is no variation for the dataset D100, due to no samples elimination was performed on it. For datasets D50 to D90, the variance value is no higher than 0.07, which, paired with the mean probability values for those datasets (presented in Table 3), indicates that the probability value will likely be above 0.3 for the surrounded cases, and below that value for the non-surrounded cases.

Table 3 Mean values of the probability provided by the situation goodness method

Dataset antenna	Probability mean										WC good solution
	D10	D20	D30	D40	D50	D60	D70	D80	D90	D100	
67	0.19	0.29	0.27	0.28	0.30	0.24	0.22	0.28	0.26	0.13	No
68	0.18	0.36	0.42	0.43	0.42	0.44	0.50	0.53	0.52	0.52	Yes
70	0.14	0.22	0.15	0.19	0.09	0.13	0.13	0.09	0.08	0.00	No
74	0.14	0.17	0.21	0.21	0.15	0.20	0.19	0.11	0.19	0.00	No
80	0.15	0.20	0.17	0.17	0.12	0.13	0.14	0.10	0.10	0.00	No
84	0.16	0.17	0.18	0.15	0.11	0.11	0.14	0.10	0.12	0.00	No
95	0.42	0.54	0.59	0.53	0.62	0.63	0.70	0.71	0.66	0.85	Yes
96	0.17	0.31	0.36	0.40	0.43	0.48	0.42	0.46	0.51	0.66	Yes
97	0.46	0.65	0.71	0.71	0.79	0.82	0.82	0.86	0.83	0.91	Yes

Table 4 Variance values of the probability provided by the situation goodness method

Dataset antenna	Probability variance									WC good solution
	D10	D20	D30	D40	D50	D60	D70	D80	D90	D100	
67	0.08	0.08	0.08	0.06	0.07	0.06	0.04	0.05	0.04	0.00	No
68	0.05	0.08	0.07	0.06	0.07	0.05	0.02	0.02	0.02	0.00	**Yes**
70	0.04	0.07	0.05	0.06	0.01	0.02	0.02	0.01	0.01	0.00	No
74	0.06	0.07	0.08	0.10	0.04	0.07	0.07	0.04	0.06	0.00	No
80	0.05	0.06	0.06	0.06	0.02	0.02	0.03	0.01	0.01	0.00	No
84	0.06	0.06	0.07	0.06	0.03	0.03	0.06	0.03	0.04	0.00	No
95	0.11	0.11	0.08	0.08	0.07	0.05	0.04	0.04	0.06	0.00	**Yes**
96	0.05	0.05	0.04	0.05	0.05	0.04	0.04	0.02	0.03	0.00	**Yes**
97	0.12	0.08	0.06	0.07	0.02	0.02	0.01	0.01	0.02	0.00	**Yes**

Based on the previous analyses, the Situation Goodness method can be used to combine the weighted centroid method with another AP localization method. We tested the following way of combination: Given a set of RSSI measurements for an AP, if the probability given by the Situation Goodness method is lower than 0.31, use the main method (in our case, one of the tested ones); otherwise, use the weighted centroid method. Table 5 presents, as an example, the localization errors when the above methods combination is applied to dataset D100. The weighted centroid method was combined with the linear approximation method. The probability provided by the Situation Goodness method was correctly used to identify the surrounded cases, and for the AP associated to those cases, the localization solution provided by the weighted centroid was used instead of the one provided by the linear approximation method.

Table 6 presents the localization errors when the AP localization evaluation is performed over the methods combination as explained above. When compared with

Table 5 Localization mean errors on dataset D100 when the situation goodness method is applied to combine the linear and weigthed centroid methods

Dataset antenna	Mean error (m) for D100		WC good solution
	Linear	W.C.-Linear	
67	15.0	15.0	No
68	7.8	5.2	**Yes**
70	5.7	5.7	No
74	5.8	5.8	No
80	5.6	5.6	No
84	5.8	5.8	No
95	3.5	0.6	**Yes**
96	10.2	0.8	**Yes**
97	4.1	1.0	**Yes**

Table 6 Mean localization error for combinations of weighted centroid and other methods

Case dataset	Mean error (m)				
	W. C.-W. C.	W. C.-Monte Carlo	W. C.-Linear	W. C.-Interpolation	W. C.-Gradient
D10	4.7	6.5	6.2	4.5	8.3
D20	4.4	5.9	5.8	3.9	9.2
D30	4.2	5.2	5.6	3.8	9.1
D40	4.2	5.3	5.4	3.6	9.9
D50	4.0	5.0	5.5	3.5	9.9
D60	4.0	4.8	5.2	3.5	10.2
D70	4.1	4.8	5.1	3.4	9.8
D80	4.1	4.7	5.4	3.3	10.3
D90	4.1	4.8	5.2	3.2	10.4
D100	4.0	4.3	5.1	2.8	11.1
Overall mean error	4.2	5.1	5.4	3.5	9.8
Overall variance	0.0	0.4	0.1	0.2	0.6

the data presented by Tables 1 and 6 shows how the mean localization error is mostly similar for the interpolation method, and is greatly reduced for the Monte Carlo, Linear and Gradient methods (which have poor performances on our datasets).

According to the results of the experimentation presented in this paper, if the probability calculated with our Situation Goodness method is reasonably high, the weighted centroid's estimation can be considered as a proper AP location solution for other RSSI measurement datasets. Thus, we recommend the usage the Situation Goodness method to combine the weighted centroid method with another AP localization method for determining several APs' locations in other datasets. Despite that the probability threshold value we used in our experimentation is 0.31, we recommend using at least a value of 0.5, because further experimentation in other datasets is required to provide a more accurate threshold value.

5 Conclusion

This paper has presented a new method that, given a set of Wi-Fi intensity measurements for an AP, provides a probability that describes whether the weighted centroid method is likely to provide a solution that is close to the actual AP location. We have called this method as the Situation Goodness method. To spur the interest in this method, we have provided facts from literature and an AP localization methods evaluation to show the goodness of using the weighted centroid method. In the AP localization evaluation, we included a new method that we have developed and which is based on natural neighbor interpolation. Through evaluation, we have shown that the Situation Goodness method we propose can effectively achieve its intended goal in a range of Wi-Fi measurement collection alternatives.

Using the Situation Goodness method, it is possible to choose when to use the weighted centroid method for an actual AP localization or as a baseline comparison method. For those cases where the Situation Goodness method indicates that the weighted centroid method is not the best choice, a different AP localization method should be used. Among the alternatives, the AP localization method we have proposed can be used.

Acknowledgments The authors gratefully acknowledge funding from the European Union through the GEO-C project (H2020-MSCA-ITN- 2014, Grant Agreement Number 642332, http://www.geo-ceu/).

References

Al-Ammar, M. A., Alhadhrami, S., Al-Salman, A., Alarifi, A., Al-Khalifa, H. S., Alnafessah, A., & Alsaleh, M. (2014). Comparative Survey of Indoor Positioning Technologies, Techniques, and Algorithms. In Cyberworlds (CW), 2014 International Conference on (pp. 245–252). http://doi.org/10.1109/CW.2014.41.

Arai, K., & Tolle, H. (2013). Color radiomap interpolation for efficient fingerprint wifi-based indoor location estimation. International Journal of Advanced Research in Artificial Intelligence, 2(3), 10–15.

Bahl, P., & Padmanabhan, V. N. (2000). RADAR: An in-building RF-based user location and tracking system. In INFOCOM 2000. Nineteenth Annual Joint Conference of the IEEE Computer and Communications Societies. Proceedings. IEEE (Vol. 2, pp. 775–784).

Berghel, H. (2004). Wireless infidelity I: War driving. Communications of the ACM, 47(9), 21–26.

Blumenthal, J., Grossmann, R., Golatowski, F., & Timmermann, D. (2007). Weighted Centroid Localization in Zigbee-based Sensor Networks. In Intelligent Signal Processing, 2007. WISP 2007. IEEE International Symposium on (pp. 1–6). http://doi.org/10.1109/WISP.2007.4447528.

Chen, G., & Kotz, D. (2000). A Survey of Context-Aware Mobile Computing Research. Hanover, NH, USA: Dartmouth College.

Cheng, Y.-C., Chawathe, Y., LaMarca, A., & Krumm, J. (2005). Accuracy Characterization for Metropolitan-scale Wi-Fi Localization. In Proceedings of the 3rd International Conference on Mobile Systems, Applications, and Services (pp. 233–245). New York, NY, USA: ACM. http://doi.org/10.1145/1067170.1067195.

Cho, Y., Ji, M., Lee, Y., Kim, J., & Park, S. (2012). Improved Wi-Fi AP position estimation using regression based approach. In Proc. of the International Conference on Indoor Positioning and Indoor Navigation.

Cho, Y., Ji, M., Lee, Y., & Park, S. (2012). WiFi AP position estimation using contribution from heterogeneous mobile devices. In Position Location and Navigation Symposium (PLANS), 2012 IEEE/ION (pp. 562–567). http://doi.org/10.1109/PLANS.2012.6236928.

Ezpeleta, S., Claver, J., Pérez-Solano, J., & Martí, J. (2015). RF-Based Location Using Interpolation Functions to Reduce Fingerprint Mapping. Sensors, 15(10), 27322–27340. http://doi.org/10.3390/s151027322.

Farid, Z., Nordin, R., & Ismail, M. (2013). Recent advances in wireless indoor localization techniques and system. Journal of Computer Networks and Communications, 2013.

Gutmann, J.-S., Burgard, W., Fox, D., & Konolige, K. (1998). An experimental comparison of localization methods. In Intelligent Robots and Systems, 1998. Proceedings., 1998 IEEE/RSJ International Conference on (Vol. 2, pp. 736–743 vol.2). http://doi.org/10.1109/IROS.1998.727280.

Han, D., Andersen, D. G., Kaminsky, M., Papagiannaki, K., & Seshan, S. (2009). Access Point Localization Using Local Signal Strength Gradient. In S. B. Moon, R. Teixeira, & S. Uhlig (Eds.), Passive and Active Network Measurement: 10th International Conference, PAM 2009, Seoul, Korea, April 1–3, 2009. Proceedings (pp. 99–108). Berlin, Heidelberg: Springer Berlin Heidelberg. http://doi.org/10.1007/978-3-642-00975-4_10.

Ji, M., Kim, J., Cho, Y., Lee, Y., & Park, S. (2013). A novel Wi-Fi AP localization method using Monte Carlo path-loss model fitting simulation. In 2013 IEEE 24th Annual International Symposium on Personal, Indoor, and Mobile Radio Communications (PIMRC) (pp. 3487–3491). http://doi.org/10.1109/PIMRC.2013.6666752.

Knauth, S., Storz, M., Dastageeri, H., Koukofikis, A., & Mähser-Hipp, N. A. (2015). Fingerprint calibrated centroid and scalar product correlation RSSI positioning in large environments. In Indoor Positioning and Indoor Navigation (IPIN), 2015 International Conference on (pp. 1–6). http://doi.org/10.1109/IPIN.2015.7346968.

Koo, J., & Cha, H. (2011a). Autonomous construction of a WiFi access point map using multidimensional scaling. In Pervasive Computing (pp. 115–132). Springer.

Koo, J., & Cha, H. (2011b). Localizing WiFi access points using signal strength. IEEE Communications Letters, 15(2), 187–189. http://doi.org/10.1109/LCOMM.2011.121410.101379.

Koo, J., & Cha, H. (2012). Unsupervised locating of WiFi access points using smartphones. Systems, Man, and Cybernetics, Part C: Applications and Reviews, IEEE Transactions on, 42(6), 1341–1353.

Kosović, I. N., & Jagušt, T. (2014). Enhanced Weighted Centroid Localization Algorithm for Indoor Environments. World Academy of Science, Engineering and Technology, International Journal of Computer, Electrical, Automation, Control and Information Engineering, 8(7), 1184–1188.

Ledlie, J., g. Park, J., Curtis, D., Cavalcante, A., Camara, L., Costa, A., & Vieira, R. (2011). Mole: A scalable, user-generated WiFi positioning engine. In Indoor Positioning and Indoor Navigation (IPIN), 2011 International Conference on (pp. 1–10). http://doi.org/10.1109/IPIN.2011.6071942.

Lee, M., & Han, D. (2012). Voronoi Tessellation Based Interpolation Method for Wi-Fi Radio Map Construction. IEEE Communications Letters, 16(3), 404–407. http://doi.org/10.1109/LCOMM.2012.020212.111992.

Liu, H., Darabi, H., Banerjee, P., & Liu, J. (2007). Survey of Wireless Indoor Positioning Techniques and Systems. Systems, Man, and Cybernetics, Part C: Applications and Reviews, IEEE Transactions on, 37(6), 1067–1080. http://doi.org/10.1109/TSMCC.2007.905750.

Liu, H.-H., & Yang, Y.-N. (2012). Study on the use of a weighted screening method for indoor positioning systems. In The 15th International Symposium on Wireless Personal Multimedia Communications.

Lohan, E. S., Talvitie, J., e Silva, P., Nurminen, H., Ali-Loytty, S., & Piche, R. (2015). Received signal strength models for WLAN and BLE-based indoor positioning in multi-floor buildings. In Localization and GNSS (ICL-GNSS), 2015 International Conference on (pp. 1–6).

Maple, C. (2003). Geometric design and space planning using the marching squares and marching cube algorithms. In Geometric Modeling and Graphics, 2003. Proceedings. 2003 International Conference on (pp. 90–95). http://doi.org/10.1109/GMAG.2003.1219671.

Moreira, A., & Meneses, F. (2015). Where@UM - Dependable organic radio maps. In Indoor Positioning and Indoor Navigation (IPIN), 2015 International Conference on (pp. 1–9). http://doi.org/10.1109/IPIN.2015.7346751.

Nam, S. Y. (2014). Localization of Access Points Based on Signal Strength Measured by a Mobile User Node. Communications Letters, IEEE, 18(8), 1407–1410.

Savvides, A., Han, C.-C., & Strivastava, M. (2001). Dynamic fine-grained localization in Ad-Hoc networks of sensors. Proceeding MobiCom '01 Proceedings of the 7th Annual International Conference on Mobile Computing and Networking, 166–179. http://doi.org/10.1145/381677.381693.

Sibson, R., & others. (1981). A brief description of natural neighbour interpolation. Interpreting Multivariate Data, 21, 21–36.

Talvitie, J., Renfors, M., & Lohan, E. S. (2015). Distance-Based Interpolation and Extrapolation Methods for RSS-Based Localization With Indoor Wireless Signals. IEEE Transactions on Vehicular Technology, 64(4), 1340–1353. http://doi.org/10.1109/TVT.2015.2397598.

Torres-Solis, J., Falk, T. H., & Chau, T. (2010). A review of indoor localization technologies: towards navigational assistance for topographical disorientation. INTECH Open Access Publisher.

Torres-Sospedra, J., Montoliu, R., Mendoza, G., Belmonte, O., Rambla, D., & Huerta, J. (2016). Providing Databases for Different Indoor Positioning Technologies: Pros and Cons of Magnetic field and Wi-Fi based Positioning. Mobile Information Systems.

Varzandian, S., Zakeri, H., & Ozgoli, S. (2013). Locating WiFi access points in indoor environments using non-monotonic signal propagation model. In Control Conference (ASCC), 2013 9th Asian (pp. 1–5).

Wang, J., Urriza, P., Han, Y., & Cabric, D. (2011). Weighted Centroid Localization Algorithm: Theoretical Analysis and Distributed Implementation. IEEE Transactions on Wireless Communications, 10(10), 3403–3413. http://doi.org/10.1109/TWC.2011.081611.102209.

Werner, M. (2014). Indoor Location-Based Services: Prerequisites and Foundations. Springer.

Zhang, Z., Zhou, X., Zhang, W., Zhang, Y., Wang, G., Zhao, B. Y., & Zheng, H. (2011). I am the antenna: accurate outdoor ap location using smartphones. In Proceedings of the 17th annual international conference on Mobile computing and networking (pp. 109–120).

Zhao, F., Luo, H., Geng, H., & Sun, Q. (2014). An RSSI gradient-based AP localization algorithm. China Communications, 11(2), 100–108. http://doi.org/10.1109/CC.2014.6821742.

Smartphone Sensor-Based Orientation Determination for Indoor-Navigation

Andreas Ettlinger, Hans-Berndt Neuner and Thomas Burgess

Abstract Many methods of indoor navigation for smartphones are augmented with Pedestrian Dead Reckoning (PDR) to improve accuracy and to reduce latency. PDR requires an accurate estimate of the device orientation. From the pitch and roll angles the sensor readings can be rotated to the horizontal plane, and with the yaw angle the direction of movement can be determined. While a simple implementation using only accelerometer and magnetometer is possible, more accurate results may be obtained by also including the gyroscope measurements. The approach in this paper uses a Kalman filter to fuse gyroscope with accelerometer and magnetometer readings. The system equation uses random walk on straight trajectories and additional gyroscope readings on turns. Turns are detected using a statistical test on the innovation of the Kalman filter as well as a condition on the estimated yaw-rate from the gyroscope. A second Kalman filter separates gravity from specific force by processing acceleration measurements. The estimated gravity is used in the orientation filter to determine pitch and roll. The filter has been tested using trajectories with known ground truth taken with off the shelf mobile devices in corridor and office environments. The outer heading accuracy approaches 10°, dominated by systematic effects, largely due to magnetic disturbances. The achieved inner accuracy for the heading is 4°.

Keywords Indoor-navigation · Orientation determination · Smartphone sensors · Kalman filter · Innovation test

---

A. Ettlinger (✉) · H.-B. Neuner
Research Group Engineering Geodesy, TU Wien, Vienna, Austria
e-mail: andreas.ettlinger@tuwien.ac.at

H.-B. Neuner
e-mail: hans.neuner@geo.tuwien.ac.at

T. Burgess
indoo.rs GmbH, Vienna, Austria
e-mail: thomas@indoo.rs

1 Introduction

Outdoor positioning and navigation on modern smartphones typically reaches an accuracy of a few meters. This level of accuracy of the users' 3D-Position is enabled through GNSS (Zhu et al. 2013). Additionally, most smartphones have a built in magnetometer and accelerometer to supply the positioning and navigation process by PDR (Pedestrian Dead Reckoning). In PDR the travelled distance and the heading are used to calculate the horizontal coordinates of the users' position. The height has to be determined in another way, for example by using a barometer. A gyroscope is nowadays also a common sensor in smartphones. With this sensor it is possible to determine the orientation rate of a smartphone independently from accelerometer and magnetometer.

As GNSS requires a clean signal path to the satellite, it quickly becomes unreliable or even unavailable in urban and indoor environments. Hence, a PDR system which computes the orientation and position accurately would be advantageous to support the navigation of pedestrians through buildings. As in other fields of application also in indoor navigation with smartphones the determination of the orientation (mainly the heading respectively the yaw angle) is a challenging task. In (Kang et al. 2012) the heading is calculated with magnetometer and gyroscope. The different results for the heading are fused with linear combinations. (Abadi et al. 2014) proposes a collaborative algorithm which uses machine learning to detect magnetic perturbations which affect the heading derived from the magnetometer. Only perturbation-free headings from multiple users are used to calculate the current heading. Another approach is shown in (Renaudin and Combettes 2014), where a Kalman filter is used to fuse accelerometer, magnetometer and gyroscope. The accelerometer and the magnetometer measurements as well as the orientation are propagated by using the gyroscope measurements. By detecting static gravity and magnetic field, sensor biases can be computed. Here quaternions are used to represent the orientation of the smartphone.

The approach proposed in this paper also uses a Kalman filter to fuse the inertial sensors of a smartphone. The results from accelerometer and magnetometer are used as observations. The heading is assumed to be constant, except when a turn of the user is detected. Then the gyroscope is used to stochastically propagate the heading. Hence, the main component is the detection of turns. The detection of turns respectively straight walking phases was already topic of research in pedestrian navigation (Zhu et al. 2013; Borenstein et al. 2009; Jiménez et al. 2010). All these approaches are using the gyroscope measurements to detect turns of the user. (Zhu et al. 2013) smooths the accelerometer measurements while the user is walking straight to derive gravity. (Borenstein et al. 2009) is correcting the gyroscope measurements for the bias in straight walking phases and (Jiménez et al. 2010) feds an extended Kalman filter with corrections to make the resulting trajectory straight. The approach proposed in this paper also uses the gyroscope measurements to detect turns. The novel part is the additional use of a statistical test on the innovations of the Kalman filter and if a turn is detected the gyroscope measurements are included in the calculation of the yaw in the system equation.

In contrast to (Renaudin and Combettes 2014) the Euler angles are used to represent the orientation of the smartphone. This has advantages in comparison to quaternion representation (quaternion-based state-space formulation is not observable and can cause numerical problems) but also disadvantages ("gimbal lock") (Särkkä et al. 2015).

2 Smartphone Sensors

The orientation of a mobile device can in general be described by three angles, the Euler angles pitch θ, roll ϕ and yaw ψ. The positive rotation of all angles is defined to be a clockwise rotation. Figure 1 shows the smartphone coordinate system with the corresponding Euler angles. The aim of PDR is to calculate positions in a specified coordinate system. In navigation tasks often a "NED"-system (North, East, Down) is used, which is a global system. Here the X-axis points to the north direction, the Y-axis points to east and the Z-axis points to the direction of gravity. The Euler angles are used to transform the sensor measurements from the sensor coordinate system into the specified global system. Figure 2 shows the relationship between the sensor- and the global system. In this paper lower case letters (y, x, z)

Fig. 1 Smartphone coordinate system and corresponding orientation angles. Pitch (theta) is the rotation around the width axis (y) relative to the horizontal plane, roll (phi) is the rotation around the height axis (x) relative to the horizontal plane, and yaw is the rotation around its depth axis (z) relative to geographic north

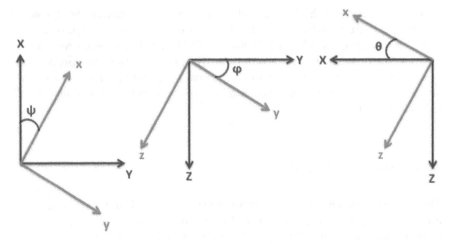

Fig. 2 Relationship between sensor (*green*)—and global (*blue*) coordinate system

always indicate the sensor coordinate system and upper case letters (Y, X, Z) always indicate the global or the navigation coordinate system.

Modern smartphones come with many built in sensors. For orientation determination at least an accelerometer and a magnetometer are necessary. The following procedure to calculate the Euler angles from magnetometer and accelerometer measurements is depicted from (Ozyagcilar 2012). In the following it is assumed that the sensor axis of accelerometer, magnetometer and gyroscope are aligned in the same manner as the smartphones coordinate system (Fig. 1). To calculate pitch and roll the accelerometer has to sense the gravity (2.1) and (2.2). This is the case when the smartphone is static. If the smartphone is moved, additional acceleration will be sensed by the accelerometer. These additional accelerations have to be removed. A detailed explanation how this is done here can be found in Sect. 3.2. Pitch and roll are then used to transform the magnetometer measurements to the horizontal plane (2.3). By using the horizontal components of the magnetic field the yaw can be calculated (2.4). Here it is important that only the geomagnetic field is sensed. If there are objects present which produce also magnetic fields, the geomagnetic field is perturbed and therefore the yaw computation is biased.

$$\tan(\varphi) = \frac{a_y}{a_z} \qquad (2.1)$$

$$\tan(\theta) = \frac{-a_x}{a_y \sin(\varphi) + a_z \cos(\varphi)} \qquad (2.2)$$

$$\begin{pmatrix} m_X \\ m_Y \\ m_Z \end{pmatrix} = \begin{pmatrix} \cos\theta & \sin\theta \sin\varphi & \sin\theta \cos\varphi \\ 0 & \cos\varphi & -\sin\varphi \\ -\sin\theta & \cos\theta \sin\varphi & \cos\theta \cos\varphi \end{pmatrix} \begin{pmatrix} m_x \\ m_y \\ m_z \end{pmatrix} \qquad (2.3)$$

$$\tan(\psi) = \frac{-m_Y}{m_X} \tag{2.4}$$

$$\dot{\varphi}_k = \left(\omega_y \sin(\varphi_{k-1}) + \omega_z \cos(\varphi_{k-1})\right) \tan(\theta_{k-1}) + \omega_x$$
$$\dot{\theta}_k = \omega_y \cos(\varphi_{k-1}) - \omega_z \sin(\varphi_{k-1}) \tag{2.5}$$
$$\dot{\psi}_k = \left(\omega_y \sin(\varphi_{k-1}) + \omega_z \cos(\varphi_{k-1})\right) \cos(\theta_{k-1})^{-1}$$

$$\varphi_k = \varphi_{k-1} + \Delta t \cdot \dot{\varphi}_k$$
$$\theta_k = \theta_{k-1} + \Delta t \cdot \dot{\theta}_k \tag{2.6}$$
$$\psi_k = \psi_{k-1} + \Delta t \cdot \dot{\psi}_k$$

If the device also has a built in gyroscope, the orientation can be calculated in a second way, independently from accelerometer and magnetometer. The gyroscope senses the angular rate ω of its sensitive axis. If initial values for pitch, roll and yaw are available, the Euler rates can be calculated (formula (2.5)) according to (Titterton and Weston 2004). By multiplying the Euler rates with the time interval Δt and adding them to the previously calculated Euler angles, the actual orientation can be determined (2.6). Figure 3 shows the results for yaw from magnetometer and gyroscope. To calculate yaw, pitch and roll are needed (2.1) and (2.2). For yaw out of magnetometer measurements as well as for the yaw from gyroscope, the same pitch and roll angles are used (calculated according to Sect. 3.2).

As seen in Fig. 3 the result derived from the gyroscope drifts away with time whereas the result derived from magnetometer stays rather constant. But in the beginning of the recording a huge peak occurs in the yaw from the magnetometer. Here the user passed a door with a metal frame and made two small turns. Thus, also the yaw derived from the gyroscope is changing (but not as much as the yaw from magnetometer). The pattern of both results is quite the same (except the first peak). In the result from gyroscope the variations are smaller. As mentioned above for both calculations the same pitch and roll angle is used and they are calculated out of estimated accelerations. Thus, the similar pattern in both yaw results could yield from steps of the user because also in the estimated accelerations (and respectively in pitch and roll) are still periodic patterns (Sect. 3.2). To check this assumption, the power spectra of both results for yaw are calculated (Fig. 4). In the power spectrum of the yaw from gyroscope the maximum amplitude is at a frequency of about 1.9 Hz (frequencies lower than 1 Hz are not considered as step frequencies). In the power spectrum of the yaw from magnetometer, the maximum amplitude is also at 1.9 Hz. But there are several amplitudes which have nearly the same magnitude. If only the part between the two peaks is used to calculate the power spectrum (green line in Fig. 4), the maximum is now distinctly at 1.9 Hz. 1.9 Hz is a plausible value for the step frequency. Hence, the similar patterns in Fig. 3 are due to the users' steps.

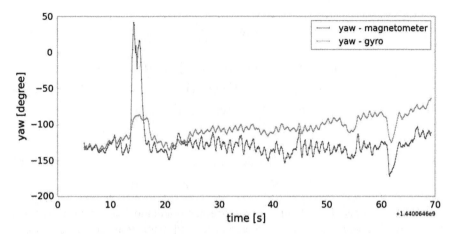

Fig. 3 Yaw calculated with magnetometer and gyroscope. As user walked straight in this recording, yaw is expected to be constant. Instead, we see significant drift in the gyroscope and large magnetic disturbance

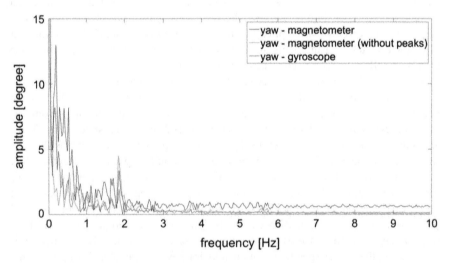

Fig. 4 Power spectrum calculated out of the yaw from magnetometer and gyroscope. Without peaks means that only the part between the two peaks (Fig. 3) is used

In this recording the main disadvantages of both sensors can be seen. The gyroscope has a drift due to the sensor bias and the magnetometer is of course sensitive to magnetic perturbations. So the idea of this paper is to create an algorithm which fuses the smartphone sensors in such a way that these drawbacks are minimized.

3 Orientation Filter

To fuse accelerometer, magnetometer and gyroscope data a Kalman filter is used. A Kalman filter mainly consists of two parts, the system equation and the measurement equation (Gelb 1974). The system equation is used to predict the state vector by using a physical model. Here the state vector consists of the three Euler angles pitch, roll and yaw. The measurement equation uses observations of the state vector to "refine" respectively to update the predicted state. Thus, the equations to estimate the state vector are also called update equations (fourth equation in (3.1) according to Gelb 1974). First the innovation d has to be calculated. This is the difference between measurement and system equation on observation level. With the observation matrix A the predicted state $\bar{x}$ is transformed to predicted observations. Additionally the covariance matrix D of the innovation and the gain matrix K have to be calculated. Here the covariance matrix of the predicted state $\Sigma_{\bar{x}\bar{x}}$ and the covariance matrix of the observations Σ_{ll} are needed. With the innovation and these two matrices the estimated state vector $\widehat{x}$ and the corresponding covariance matrix $\Sigma_{\widehat{x}\widehat{x}}$ can be calculated.

$$\begin{aligned}
d_k &= l_k - A_k \bar{x}_k \\
D_k &= \Sigma_{ll,k} + A_k \Sigma_{\bar{x}\bar{x},k} A_k^T \\
K_k &= \Sigma_{\bar{x}\bar{x},k} A_k^T D_k^{-1} \\
\widehat{x}_k &= \bar{x}_k + K_k d_k \\
\Sigma_{\widehat{x}\widehat{x},k} &= \Sigma_{\bar{x}\bar{x},k} - K_k D_k K_k^T
\end{aligned} \quad (3.1)$$

In the here proposed Kalman filter two system equations are used for the yaw angle. In Sect. 3.1 the system equation is explained in detail. The decision, on which system equation will be used for yaw, is made by observing two indicators (Sects. 3.1 and 3.3). The measurement equation will be explained in Sect. 3.2.

3.1 System Equation

The system equations for pitch and roll are very simple because they are assumed to be constant (3.2) and (3.3). The $w_{k-1,i}$ are the disturbances of the corresponding angle. This is a reasonable assumption because pedestrians usually hold the smartphone rather static in their hands while navigating. So there should be only slow changes of pitch and roll.

$$f_1: \varphi_k = \varphi_{k-1} + w_{k-1,\varphi} \tag{3.2}$$

$$f_2: \theta_k = \theta_{k-1} + w_{k-1,\theta} \tag{3.3}$$

To avoid a drift in the results of the filter, the gyroscope is only used when the user is turning. The advantage of the gyroscope is its high measurement rate. Thus, the gyroscope is very useful when the user is turning. Turning usually lasts not longer than some seconds, so gyro drift should not influence the result of the filter. If the user is walking straight again, the gyroscope is not taking part in the system equation. This behavior can be achieved by using two system equations for the yaw. In the first system equation the yaw is assumed to be constant (3.4). This is the case when the user is walking straight. If the user is turning the gyro measurements are used as control inputs (3.5). This is the same approach as calculating the yaw only with the gyroscope (Sect. 2). The orientation filter processes with a frequency of 20 Hz, so Δt is 0.05 s.

$$f_3: \psi_k = \psi_{k-1} + w_{k-1,\psi} \tag{3.4}$$

$$f_3: \psi_k = \psi_{k-1} + \frac{\Delta t}{\cos \theta_{k-1}} \left[\sin \varphi_{k-1} \omega_{y,k} + \cos \varphi_{k-1} \omega_{z,k} \right] + w_{k-1,\psi} \tag{3.5}$$

A decision has to be made on what system equation will be used for yaw-propagation. Therefore, two indicators are chosen. If one of the indicators responds, it will be assumed that the orientation of the smartphone is changing. The first indicator is the yaw rate from Sect. 2 (2.5). If the current yaw rate is bigger than 30°/s, (3.5) will be used as system equation. Figure 5 shows the calculated yaw rate from the recording where the user was walking straight. It contains a "base noise" of approximately 30°/s and there are peaks when the user is turning. So the

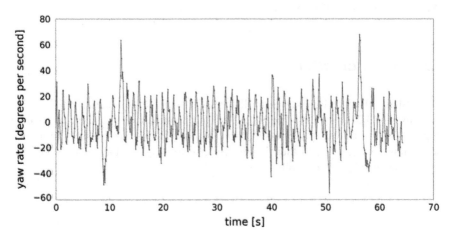

Fig. 5 Yaw rate calculated with gyroscope

threshold of 30°/s is a reasonable choice. The second indicator is the result of a statistical test on the innovation d from the Kalman filter. A detailed explanation of this statistical test follows in Sect. 3.3. The reason for using two indicators is to detect fast turns (yaw-rate) but also slow turns (statistical innovation test).

For the update equations also the covariance of the predicted state is needed (3.6). T, B and C contain derivatives of the system Eqs. (3.7) and are Jacobi matrices. The covariance matrix of the control inputs is equal to the covariance matrix of thy gyro measurements. The covariance matrix of the system noise is equal to the covariance matrix of the estimated state at time k = 0 divided by Δt^2. To derive an initial state the formulas (2.1), (2.2) and (2.4) are used. By error propagation (Niemeier 2008) also the corresponding covariance matrix can be computed. Additionally the third diagonal element of C will be multiplied with 0.01 if (3.4) is used as system equation for the yaw. This value is empirically motivated. By doing this, the system equation gets more weight in the update equations. Thus, the noise of the estimated yaw will be decreased in phases when the user is walking straight.

$$\Sigma_{\bar{x}\bar{x},k} = T_k \Sigma_{\hat{x}\hat{x},k-1} T_k^T + B_k \Sigma_{uu} B_k^T + C_k \Sigma_{ww} C_k^T \tag{3.6}$$

$$T_k = \begin{pmatrix} 1 & 0 & 0 \\ 0 & 1 & 0 \\ \frac{\delta f_3}{\delta \varphi} & \frac{\delta f_3}{\delta \theta} & 1 \end{pmatrix}$$

$$B_k = \begin{pmatrix} 0 & 0 & 0 \\ 0 & 0 & 0 \\ 0 & \frac{\delta f_3}{\delta \omega_y} & \frac{\delta f_3}{\delta \omega_z} \end{pmatrix} \tag{3.7}$$

$$C_k = \Delta t \cdot T_k$$

3.2 Measurement Equation

The observations are introduced as direct observations of the state. Hence, the design matrix A is equal to the identity matrix. However, only the covariance matrices of the measured accelerations and magnetic flux densities are known. By doing error propagation according to (Niemeier 2008) the covariance matrix of the observed angles can be calculated (3.8). The covariance matrix of the accelerations is derived from a second Kalman filter which tracks the direction of gravity by fusing accelerometer and gyroscope. In the system equation the direction of gravity (respectively the accelerometer readings from a static smartphone) are predicted by using the rotation rates from the gyroscope. The observations are the accelerometer measurements. The more the total amount of the measured accelerations differ from the known value of gravity, the more the observation noise will be increased.

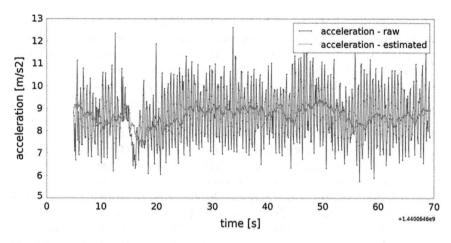

Fig. 6 Raw and estimated acceleration in z-direction of smartphone coordinate system

By doing this additional accelerations should be removed and only gravity remains in the estimated acceleration measurements. A detailed explanation of this Kalman filter can be found in (Särkkä et al. 2015). Figure 6 shows the result for the measured acceleration in z-direction of the smartphone coordinate system (Fig. 1). It can be seen that there are still systematic effects due to the steps of the user (Sect. 2) but they are reduced severely. Out of the estimated accelerations respectively gravity, pitch and roll are calculated now (2.1) and (2.2). Afterwards yaw is calculated with the horizontal magnetic field values (2.3) and (2.4).

Also the covariances of the magnetometer measurements are adapted similar to (Sabatini 2006) or (Tian et al. 2014). Equation (3.9) shows how this is done. The measurement noise of the magnetometer will be increased if the total amount of the magnetic field is changing. The total amount of the geomagnetic field should be—related to the dimensions of a building—constant. So if the magnetic field is changing, this is used as trigger that magnetic perturbations are present. By increasing the measurement noise of the magnetometer, the weight of the observed yaw angle will be decreased in the update equations. Thus, the influence of magnetic perturbations on the filter result should be minimized.

$$\Sigma_{ll,k} = F \begin{pmatrix} \Sigma_{\widetilde{aa},k} & 0_{3\times 3} \\ 0_{3\times 3} & \bar{\sigma}_m^2 * I_{3\times 3} \end{pmatrix} F^T \qquad (3.8)$$

$$\bar{\sigma}_m^2 = (\sigma_m + \Delta m_{k-1,k})^2$$
$$\Delta m_{k-1,k} = \left|\|\vec{m}_k\| - \|\vec{m}_{k-1}\|\right| \qquad (3.9)$$

3.3 Statistical Test

The result of this statistical test is used as indicator if the user is turning or not. In the classical sense this test is used to verify if the system equation and the measurement equation are compatible. If they are not compatible another system equation has to be used or the system noise has to be increased (Heunecke et al. 2013). Another reason for a significant test result could be an outlier in the measurement data. Thinking on the problem of identifying turns of the user, it is the first case of intervention we want to apply. If (3.4) is used as system equation but the user is turning, system- and measurement equation will give statistically significant, different results. Then (3.5) should be used as system equation for yaw. Because of outliers or short term magnetic perturbations, the change between the system equation is not done when the statistical test responds the first time. In the later on evaluations the change to (3.5) is always done after 6 responds of the statistical test. This value corresponds to 0.3 s. Thus, if the predictions and the observations significantly differ longer than 0.3 s, the system equation will be changed.

The considered value is the innovation of the yaw. The null hypothesis is that the innovation of the yaw is zero or in other words that the difference of system- and measurement equation is not significantly different from zero (3.10). The test value is given in (3.11) and follows a chi-square distribution with one degree of freedom. The level of significance is chosen to 95 % and the corresponding quantile of the chi square distribution is 3.81.

$$H_0: E(d_k) = 0$$
$$H_A: E(d_k) \neq 0 \quad (3.10)$$

$$P\left\{ \frac{d_\psi^2}{\sigma_{d_\psi}^2} \leq \chi^2_{n_{l,k};1-\alpha} | H_0 \right\} = 1 - \alpha \quad (3.11)$$

4 Results

To evaluate the achievable accuracies of the proposed orientation filter three scenarios where chosen. First the user is walking straight while the recording (scenario 1). In the second recording the user makes five turns (scenario 2) and in the third recording the user is walking a polygonal trajectory repeatedly (scenario 3). The three trajectories can be seen in Fig. 7. For these experiments path recordings were made using indoo.rs Mobile Toolkit[1] which runs on a Samsung Galaxy S4. With this app the user can mark his current position on a map. Within the experiments this was done every time when the user changed his walking direction. By using the

---

[1]Indoo.rs: https://play.google.com/store/apps/details?id=com.customlbs.android.mmt.

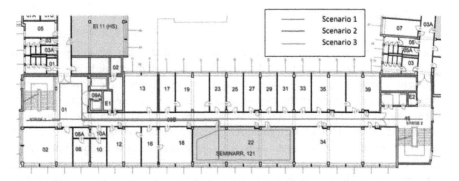

Fig. 7 Rooms of the department of geodesy and geoinformation with the 3 trajectories respectively scenarios

coordinate differences between such two user positions a reference can be obtained for the calculated yaw. It has to be said, that this reference for yaw is only accurate to a few degrees due to the imperfect marking of the current position on the map. The difference between the reference and the calculated yaw gives the orientation error. The orientation error is not calculated all the time because there is no accurate reference available while the user is turning. Thus, half a second before and half a second after the turning the orientation error is not calculated. The test measurements were done in the rooms of the department of geodesy and geoinformation, which are part of TU Wien.

The recording where the user is walking straight was already used to calculate the yaw from magnetometer and gyroscope. Figure 8 additionally shows the result derived by the orientation filter. It can be seen that the filter result is not affected by magnetic perturbations in the beginning of the recording where the result from the magnetometer has a peak. There is also no drift in the filter result in comparison to

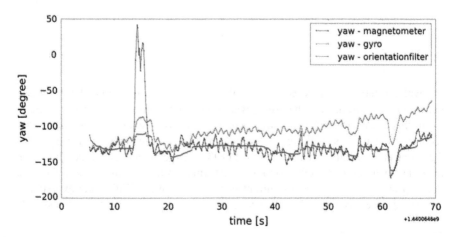

Fig. 8 Yaw calculated with gyroscope, magnetometer and orientation filter

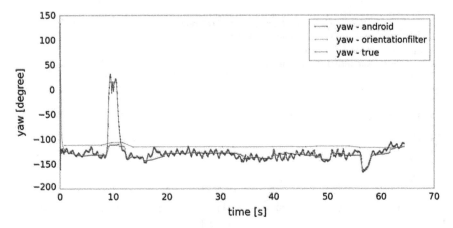

Fig. 9 Filter result, android result and reference for the yaw (scenario 1)

the gyroscope result. So the drawbacks of each sensor are eliminated. Additionally the filter result is much smoother than the results from magnetometer and gyroscope. Figure 9 shows the filter result in comparison to the android intern solution for the yaw and the reference yaw calculated out of the positions marked from the user on a map while the recording. The android solution is derived from the orientation sensor, which uses accelerometer and magnetometer to calculate the orientation of the smartphone.[2] There is still a constant offset to the reference yaw. The reason could be due to magnetic perturbations. The mean difference between the reference yaw and the filter result is 16.2° (android solution: 18.5°). In the following the mean difference between reference and filter result will be designated as outer accuracy. If the initial heading of the user is known quite accurate, one could use the changes of the yaw to calculate PDR positions. In this case the inner accuracy is a more suitable accuracy measure. In this paper the inner accuracy is defined as the mean standard deviation of the residuals. The residuals are the differences between filter result and reference.

Figure 10 shows the residuals with the corresponding mean values and standard deviations for scenario 1. In the following the mean of the residuals will be designated as offset. The offset and the standard deviation are calculated for each part of the trajectory where the yaw is constant. As mentioned above half a second before and after the user is turning, the residuals are not calculated. The reason is that no reference is available while the turning phases of the user. Thus, in the following figures which show the residuals, no values are in this parts of the recording were the user was turning. The standard deviations are quite constant, except in the parts where the user was passing the doors. The mean standard deviation respectively the inner accuracy in this scenario is 5.4°. The offsets are

---

[2]Android: https://developer.android.com/guide/topics/sensors/sensors_position.html, retrieved on 13.6.2016.

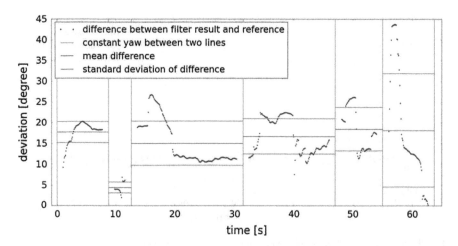

Fig. 10 Differences between the filter result for yaw and the reference yaw with mean and standard deviation of this difference (scenario 1)

varying in the same dimension as the standard deviations of the residuals, except in the part where the user passed the first door.

Figures 11 and 12 show the results for roll and pitch. Here no reference values are available. But the smartphone was held in texting mode while the recordings were done. So roll should be approximately zero and pitch should be somewhere between 20° and 60°. Both angles should not vary much because the smartphone was held rather static. This is the case for the filter results. Additionally the filter result is rather smooth in comparison to the results using the estimated accelerations (Sect. 3.2) because they are modeled as random walk in the system equation (formulas (3.2) and (3.3)). Hence, remaining systematic effects due to the steps of

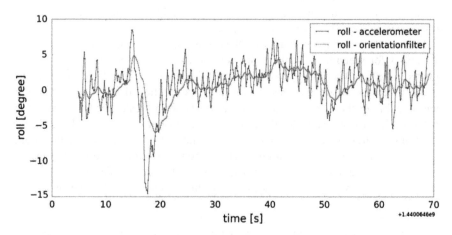

Fig. 11 Roll calculated with the accelerometer and the orientation filter (scenario 1)

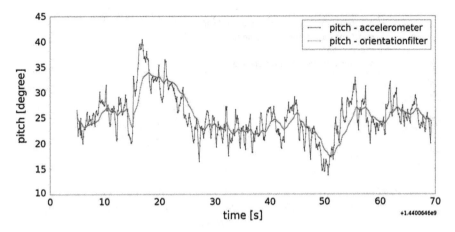

Fig. 12 Pitch calculated with the accelerometer and the orientation filter (scenario 1)

the user are removed. In normal walking behavior, the user holds the smartphone quite static in its hands and the changes in pitch and roll are long-periodic. Thus, the random walk model should be sufficient. But if pitch and roll change fast, the filter result will need some time to follow. In scenario 1 this time shift is approximately two seconds. The results for pitch and roll in the other test scenarios are quite the same.

In the scenario where the user made five turns the filter result is also better than the result from android (Figs. 13 and 14). The mean orientation error is 8.9° by using the orientation filter in comparison to 12.4° by using the result from android. Also here systematic deviations from the true yaw are present. Scenario 2 is also used to evaluate the reliability of detecting turns made by the user (Fig. 15).

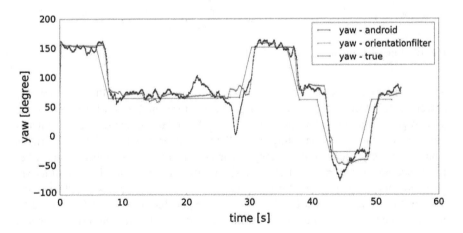

Fig. 13 Filter result, android result and reference for the yaw (scenario 2)

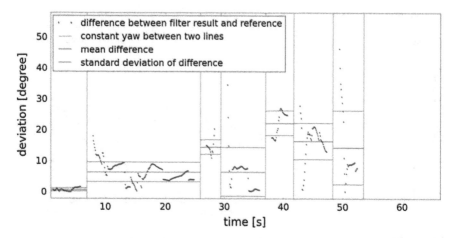

Fig. 14 Differences between the filter result for yaw and the reference yaw with mean and standard deviation of this difference (scenario 2)

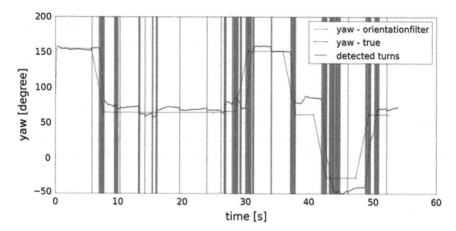

Fig. 15 Turns detected by the orientation filter

The vertical blue lines mark the turns detected by the orientation filter. All turns are detected. But there are also some wrong detected turns and most of them are in the second part of the recording (between 10 and 25 s). Hence, the orientation filter is too sensitive in case of detecting turns. If the number of wrong detected turns is small (relatively spoken) and if they occur detachedly, the influence on the filter result should be negligible. The high sensitivity will be a problem, if turns are detected continuously wrong for more than 20 s. Then gyro drift will influence the filter result because system Eq. (3.5) is used longer than it is needed. It would be worse if true turns are not detected. In this case the filter result would show latency.

The offset is varying more than 20° in scenario 2 (Fig. 14). In the beginning it is nearly zero and in the end of the trajectory it varies between 15 and 20°. Also the variations of the standard deviations are higher than ~7°. The inner accuracy in this scenario is 5.1°. The trajectories of scenario 1 and scenario 2 are overlapping in the corridor (Fig. 7). In the third part of Fig. 10 and the second part of Fig. 14 are the corresponding offsets and standard deviations. Though the user walked the same path, the offsets differ about 8° and the standard deviations differ about 3°.

The filter result of scenario 3, where the user was walking a polygonal trajectory, shows a periodic pattern which matches well with the true yaw (Fig. 16). Again, systematic offsets are present, mainly in the parts of the trajectory where the true yaw is approximately −120°. The deviations are constant and bigger than 20°. But in the other parts the estimation for yaw is satisfying. The mean error is 12.2°. For clarity reasons the android result is not plotted here. Figure 17 shows the residuals in this scenario. The variations of the offsets and the standard deviations are in the same order as in scenario 2. The inner accuracy is 3.6°. Because the user walked this polygonal trajectory repeatedly one could assume that the offsets and standard deviations are the same when the user passes the same part of the trajectory again. However, this is not the case. Solely the offset in the part of the trajectory where the true yaw is about −120° stays quite constant.

Figure 18 is a detailed view of Fig. 16 with the android result. Here the main advantages of the orientation filter can be seen and summarized. The filter result is much smoother than the result from android respectively magnetometer. In phases where the user is walking straight, the heading stays also constant. Additionally it is (in this case) independent from magnetic perturbations. This can be seen in the right part of Fig. 18 where the filter result matches with the true yaw. But the result from magnetometer has an offset.

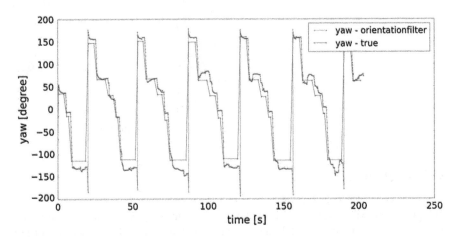

Fig. 16 Filter result and reference for the yaw (scenario 3)

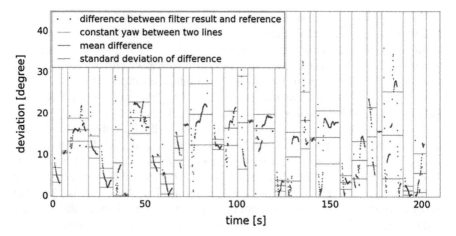

Fig. 17 Differences between the filter result for yaw and the reference yaw with mean and standard deviation of this difference (scenario 3)

Fig. 18 Detail of the recording from scenario 3

5 Conclusions

The proposed orientation filter fuses the inertial sensors (accelerometer, gyroscope and magnetometer) of a smartphone in a way that the drawbacks of each sensor are minimized. To avoid gyro drift, the measurements from the gyroscope are used only when turns are detected by this filter. The detection is done with two indicators, the yaw rate from the gyroscope and with the result of the statistical innovation test.

The turns of the user were detected correctly in the carried out measurements. Though, the turn detection of the filter is too sensitive. Even when the user is walking straight, detached turns are detected. Detailed analyses have to be done when such cases occur. The drawback of the magnetometer is its sensitivity to magnetic perturbations. By adapting the measurement noise of the magnetometer, this influence is reduced. To calculate pitch and roll the accelerometer has to sense only gravity. Due to the steps of the user the raw accelerometer measurements are not feasible to calculate pitch and roll. In the evaluated test measurements the influence of the users' motion is greatly reduced. It has to be admitted that there were no accurate reference values available but in the results of pitch and roll is no periodic pattern visible which could be induced by the steps of the user.

In the test measurements the motion direction coincided with the pointing direction of the smartphone. Because of the changing system equation for the yaw and the increase of the measurement noise of the magnetometer if the magnitude of the magnetic field is changing, the results of the orientation filter are more accurate than the results only from gyroscope or magnetometer. The results are also better than the results from the android intern solution for the orientation. Also the differences between calculated yaw and reference yaw (respectively the residuals) were analyzed. The variations of the means and standard deviations of the residuals while straight walking phases are quite high. Neither have they been rarely reproducible when parts of the trajectories where measured repeatedly. The mean of the standard deviations of the residuals was used as a measure for the achieved inner accuracy. The inner accuracy is about 60 % better than the outer accuracy. Table 1 summarizes the achieved accuracies for yaw of the three chosen scenarios.

The results of the orientation filter are related to the attitude of the smartphone. So if the smartphone is not pointing into the direction of the users' motion the result from subsequently PDR will be false. Due to this fact, additional algorithms which overcome this problem have to be considered. The test measurements were only done with one smartphone, the Samsung Galaxy S4. In following investigations the orientation filter has to be also tested on other devices from different manufacturers. Additionally test measurements should be done in other indoor environments and better reference values for the users' position should be included. The residuals calculated in Sect. 4 are not normally distributed. The Kalman filter and the statistical innovation test require normally distributed data though. Hence, in future work other evaluation approaches have to be considered or the statistical test has to be adapted.

This paper has focused on the presentation of the algorithm of the orientation filter and first evaluations on its performance. The orientation filter will be part of a

Table 1 achieved outer and inner accuracy

Scenario	Outer accuracy		Inner accuracy
	Orientation filter	Android	Orientation filter
1 (straight)	16.2°	18.5°	5.4°
2 (straight with turns)	8.9°	12.4°	5.1°
3 (polygonal, repeatedly)	12.2°	14.5°	3.6°

mobile terminal which will be running on the smartphone, so battery life is an important issue. Because of only one matrix inversion and saving only the last state, the orientation filter should not influence the battery life in a negative way. In the mobile terminal, additional sensors, signal sources and positioning approaches will be used to derive the users' position. This is necessary because the orientation filter is still influenced by magnetic perturbations. If magnetic perturbations are present all time, the result for the yaw will be falsified respectively biased. In this case orientation filter and the subsequently done PDR would fail. To avoid such cases additional orientation or position information is required. So following investigations have to be focused on the combination of PDR with other indoor positioning approaches to eliminate the influence of magnetic perturbations.

References

Abadi MJ, Luceri L, Hassan M, et al (2014) A Collaborative Approach to Heading Estimation for Smartphone-based PDR Indoor Localisation. In: International Conference on Indoor Positioning and Indoor Navigation (IPIN), pp 554–563.
Borenstein J, Ojeda L, Kwanmuang S (2009) Heuristic reduction of gyro drift in IMU-based personnel tracking systems. In: SPIE Defense, Security, and Sensing. International Society for Optics and Photonics, pp 73061H–73061H.
Gelb A (1974) Applied optimal estimation. MIT press, Cambridge
Heunecke O, Kuhlmann H, Welsch W, et al (2013) Auswertung geodätischer Überwachungsmessungen, 2nd edn. Wichmann, Berlin.
Jiménez AR, Seco F, Prieto JC, Guevara J (2010) Indoor pedestrian navigation using an INS/EKF framework for yaw drift reduction and a foot-mounted IMU. In: Positioning Navigation and Communication (WPNC), 2010 7th Workshop on. IEEE, pp 135–143.
Kang W, Nam S, Han Y, Lee S (2012) Improved heading estimation for smartphone-based indoor positioning systems. In: Personal Indoor and Mobile Radio Communications (PIMRC), 2012 IEEE 23rd International Symposium on. IEEE, pp 2449–2453.
Niemeier W (2008) Ausgleichungsrechnung: Statistische Auswertemethoden. Walter de Gruyter, Berlin.
Ozyagcilar T (2012) Implementing a tilt-compensated eCompass using accelerometer and magnetometer sensors. Freescale semiconductor Application Note, Volume 3.
Renaudin V, Combettes C (2014) Magnetic, Acceleration Fields and Gyroscope Quaternion (MAGYQ)-Based Attitude Estimation with Smartphone Sensors for Indoor Pedestrian Navigation. Sensors 14:22864–22890.
Sabatini AM (2006) Quaternion-based extended Kalman filter for determining orientation by inertial and magnetic sensing. Biomed Eng IEEE Trans On 53:1346–1356.
Särkkä S, Tolvanen V, Kannala J, Rahtu E (2015) Adaptive Kalman Filtering and Smoothing for Gravitation Tracking in Mobile Systems. In: International Conference on Indoor Positioning and Indoor Navigation (IPIN), pp 1–7.
Tian Z, Zhang Y, Zhou M, Liu Y (2014) Pedestrian dead reckoning for MARG navigation using a smartphone. EURASIP J Adv Signal Process 2014:1–9.
Titterton D, Weston J (2004) Strapdown inertial navigation technology, 2nd edition. Institution of Engineering and Technology, United Kingdom.
Zhu X, Li Q, Chen G (2013) APT: Accurate outdoor pedestrian tracking with smartphones. In: INFOCOM, 2013 Proceedings IEEE. IEEE, pp 2508–2516.

SubwayAPPS: Using Smartphone Barometers for Positioning in Underground Transportation Environments

Kris van Erum and Johannes Schöning

Location information that is crucial for all location-based services is almost always available due to a number of different positioning techniques and technologies such as GPS and WiFi positioning. However, positioning technologies cannot provide sufficient position information when a user is underground, e.g. travelling with a car through a tunnel or on subways on an underground public transportation network. While there have been a number of attempts to utilize expensive infrastructure and smartphone sensors to address this situation, all of these techniques are either limited in scope, very expensive, or somewhat limited in accuracy. In this paper, we present a novel smartphone-based approach called SubwayAPPS (Subway Air Pressure Positioning System) that for the first time utilizes relative air pressure changes as detected by smartphone barometers to position a user. We first demonstrate the feasibility of this approach by comparing the depth characteristics of five major underground transportation networks across the globe and show that our novel approach is feasible for positioning users while they are underground in these networks. Second, we show with two user tests in Brussels and London that our lightweight approach works well as other more complex techniques, e.g. techniques that rely on pattern matching using the build-in accelerometers or gyroscopes, under realistic conditions.

K. van Erum · J. Schöning (✉)
Hasselt University, Hasselt, Belgium
e-mail: johannes.schoening@uhasselt.be

K. van Erum
e-mail: kris.vanerum@student.uhasselt.be

1 Introduction and Motivation

The ubiquitous availability of positioning techniques has accelerated the spread of location-based services (LBS). To navigate, most can rely on their smartphone as most modern smartphones provide positioning information using GPS, WiFi (Kawaguchi et al. 2009; Fernandes et al. 2014) and cell information with high accuracy (Zandbergen 2009). Together with popular map applications (e.g. Google Maps, Apple Maps or Bing Maps) and other LBS applications (Krammer et al. 2014; Schall et al. 2011), ranging from map-based services (Schöning et al. 2009) to LBS that help one to be alone on a hike (Posti et al. 2014), they have become the de facto standard for everyday navigation outdoors as well as partially indoors. Beyond navigation, LBS provide a rich set of other services from personalized weather forecasts (e.g. OpenWeatherMap,[1] Yahoo! Weather[2] to location-based social networks (e.g. Foursquare[3]). While these LBS now work quite well in most environment above ground, they basically fail to provide positioning information underground as standard positioning technologies fail in those environments. The signals emitted by GPS satellites and cellular base stations have difficulties traveling through concrete and other particulate matter and often no positing infrastructure, such as a continuous set of WiFi access points are available undergrounds. As common response to this problem is that current navigation systems or LBS fall back to a dead reckoning mode (Ojeda and Borenstein 2007), e.g. when a car is traveling through a tunnel. Nevertheless the lack of location information is crucial for all LBS; e.g. a LBS that assist users in an underground public transportation network to answer the following example questions: Which station do I have to leave? Where is the restaurant I was looking for on Yelp? In this paper we address this problem, by developing and evaluating a lightweight positioning technique that relies on detecting relative air-pressure changes by using the build in barometers of smartphones. In particular, we have developed the technique to provide positioning to the millions of people who utilize underground public transportation networks each day, which is the primary use case considered in this paper. In London, 3.5 million subway rides are made each weekday (Transport for London 2016a), in New York it is 5.5 million rides, and in Paris it is 4.2 million. While we focus on underground public transportation networks in this paper, our proposed technique could also applied for other transportation modes, such as traveling with a car through a tunnel. This paper is not the first to address the challenge of underground positioning (Stockx et al. 2014; Sankaran et al. 2014; Thiagarajan et al. 2010; Watanabe et al. 2012). This paper adds to the body of related work by providing a novel smartphone-based approach named SubwayAPPS that uses the build-in barometer to precisely determine the position of a user in an underground public transportation system by detecting relative air pressure changes without the need of additional infrastructure. In addition, the paper makes the following three contributions: First we study the structure of underground

---

[1] http://openweathermap.org/.
[2] https://www.yahoo.com/news/weather/.
[3] https://foursquare.com/.

transportation networks by analyzing the height differences between stations for five cities. Next, we present a novel algorithm for detecting underground stations using only a smartphone's air pressure sensor. This algorithm relies on relative height differences between adjacent stations and the *piston effect*. Further, we implement this algorithm for the Android platform. Finally, we analyze the accuracy of smartphone air pressure sensors for height difference detection under realistic conditions.

2 Related Work

Most current smartphones are equipped with various sensors, such as accelerometers, magnetometers, RFID and barometers, to support user input and output. In addition to their main purpose the sensors can also be used to support positing, using inertial navigation techniques (e.g. Ascher et al. 2001; Fischer et al. 2008; Robertson et al. 2009; Ruiz et al. 2012; Ortakci et al. 2014), when GPS or WiFi positioning are not available.

For example, SubwayPS (Stockx et al. 2014), presented at Sigspatial 2014, uses the accelerometer of a smartphone to detect if a subway train is stationary or not. By integrating the schedules information the algorithm can provide location information to its users. The technique relies on the fact that while a subway train is moving, the total acceleration (in all directions while removing the gravity component with the gyroscope) is higher than when the train is stationary. The algorithm uses a hard threshold to determine if the subway train is riding or stationary. If a number of samples is above this threshold, the algorithm decides that the train is moving. Similarly, if a number of samples is below this threshold, the algorithm decides that the train is stationary. These parameters (threshold and number of samples) can be tweaked for specific subway networks. If the parameters were tweaked, around 85 % of all stops were classified correctly. With the standard, world-wide parameters, 75 % of the stops were classified correctly.

Similar to Stockx et al. (2014), the approach of Thiagarajan et al. (2010) uses the accelerometer to make a prediction whether the train is riding or stationary. The accelerometer samples of stationary and riding periods are modelled as Laplace distributions. Bayes' theorem is then used to calculate the probability that the train is riding. A 30 s time window is used to calculate the mean accelerometer value. During the interval when a subway train is riding and arrives at a station, a probability peak (train is moving) and valley (train is stationary) will occur, which is detected by a peak detector. The average difference between the time that the train arrives at a station and the prediction of the algorithm is 41 s.

Another technique that uses the accelerometer is the StationSense passenger tracking system (Higuchi et al. 2015). In addition to the accelerometer, it utilises the magnetometer, to complement the station detection algorithm. Typically, motors and electrical inverters emit magnetic noise when the train accelerates and decelerates. During stationary periods, the values of the magnetometer have a lower variance. A stop probability is calculated using Bayes' theorem, similar to Thiagarajan et al.

(2010). However, the stop probability can sometimes increase due to inertia driving of the trains. To filter out these false positives, readings of the accelerometer are used. With this method, 72 % of all detections were indeed a stop and 82 % of all stops were classified correctly.

Similar to the work presented in this paper, Watanabe et al. (2012) used an air pressure sensor to detect subway stations. They assume that the height profile of every subway line is unique and these height differences could be measured by an air pressure sensor (see Sect. 4.1). They also take into account the structural factors (e.g. ventilation shafts) of each line. They assume that the change of air pressure is unique for each ride between adjacent subway stations. First, a template is generated for each ride between adjacent stations by recording the air pressure while riding the train. The algorithm then tries to match the incoming air pressure values to one of the templates. Because the train speed varies (e.g. because of the crowdedness on the train), the variation of air pressure is elastic in time. To overcome this problem, Dynamic Time Warping (DTW) is used. Furthermore, because the absolute air pressure values change over time (e.g. due to weather differences), this technique only takes into account the relative change of air pressure. This method was tested in the Tokyo Metro (9 lines with 192 stations) and 85 % of all stops were classified correctly.

In contrast to related work, we are the first to use an air pressure sensor in a mobile phone to provide positioning in underground public transport systems, by relying on relative height differences between adjacent stations and the *piston effect*. That goes beyond basic station detection or simple approaches that detect if a train is moving or not. By looking at the relative height differences between subway stations and the *piston effect*, we overcome the problems of template matching approaches (Watanabe et al. 2012), which look at the height difference pattern over multiple stations. This method needs trips of multiple stations to achieve decent accuracies, as we show in Sect. 4.1.

3 Sensing Air Pressure on Mobile Devices

Before we describe the technical implementation of the SubwayAPPS application in the next section, we briefly describe the current built-in air pressure sensors and their functionality as well as the needed physical principles our approach relies on. Nowadays, most mid-priced smartphones are equipped with barometers mainly to improve the accuracy of the z-coordinate of their built-in localization systems, which is a typical problem of satellite based navigation systems such as GPS or WiFi based solutions.

The sensors used in the smartphones are small microelectromechanical systems (MEMs) ranging from 0.02 to 1.0 mm size and use piezoresistor to detect changes in the atmospheric pressure. The electrical resistivity of the piezoresistor changes when mechanical strain is applied, e.g. due to changes of the atmospheric pressure. The most common sensors are the Bosch BMP180 (Samsung Galaxy, S4, Google Nexus 4; quadratic temperature compensation in driver), Bosch BMP280 (Google Nexus 5,

Apple iPhone 6 and higher, quadratic temperature compensation in driver and noise filter) and the STM LPS331 AP (Samsung Galaxy S3, linear temperature compensation on chip). All have an accuracy of around ±0.12 hPa, which corresponds to ±1 m.

The relation between pressure and altitude is described with the hypsometric formula (Portland State Aerospace Society 2015).

$$h = \frac{((\frac{P_0}{P})^{\frac{L \cdot R}{g}} - 1) \cdot (T + 273.15)}{L}$$

where P_0 and P are pressures at different altitudes (in hPa), h is the height different between these points (in meter), L is the lapse rate of the temperature (-0.0065 K/m), R is the universal gas constant (287.053 J/kg.K), g is earth's gravity constant (9.81 m/s^2) and T is the temperature in (°C).

If we assume a constant temperature of 15 °C, we can use the following formula to calculate the relative height difference between two points:

$$h = \frac{((\frac{P_0}{P})^{\frac{-0.0065 \cdot 287.053}{9.81}} - 1) \cdot (15 + 273.15)}{-0.0065} = ((\frac{P_0}{P})^{-0.19} - 1) \cdot -44330)$$

3.1 Air Pressure in Underground Public Transportation Systems

Our method relies on detecting relative pressure differences and the *piston effect* while travelling on public transportation systems. Therefore it is important to rely on the fact that if the air pressure changes (for example, due to changing weather conditions) are transmitted to all depths, as described in Pascal's law (Young and Freedman 2000).

Almost all human made tunnels have a certain slope. When building a tunnel, engineers need to take into account the geological factors of the environment (e.g. ground type and stability). Having a slope in a tunnel also has some advantages. When a tunnel is flat, water may get stuck in the tunnel and can not drain automatically (Gargini et al. 2008). Furthermore, a slope in a tunnel can influence the movement of smoke in the tunnel in case of a fire (Wu e al. 1997).

While car tunnels have slopes around 2 % from start to end, tunnels of public underground transportation systems have typically slopes below 1 % from station to station, as we outline below in detail. Therefore we have decided to tailor our work to the more complex scenario of providing location information in underground transportation networks.

In subway networks we could also reply on a second important effect; the so called *piston effect* (Pan et al. 2013). In contrast to trains travelling through open space, a train travelling in a tunnel compresses the air in front of the vehicle. Air pressure gets higher in front of the train, whereas a vacuum is created behind the moving vehicle. Therefore, when a train enters a tunnel, the air pressure in and behind the train will drop while the air pressure in front of the train increases. When the subway train arrives at a station, the air pressure normalises again. The *piston effect* also depends on the environment. A change in width of the tunnel affects the air flow around the train and thus influences the measured air pressure. Additionally, when the train passes a ventilation shaft in the tunnel, the air in front of the train can escape and air can be sucked in from the ventilation shaft at the back of the train. This reduces the *piston effect*. Finally, when another train passes on another track, the *piston effect* of the other train will also affect the air pressure measurements.

As said, we rely on detecting relative air pressure changes due to the depth structure of the underground transportation network, therefore we have to correct for the *piston effect*. Nevertheless, we can also make use of the *piston effect* while not on the train, e.g. to detect when a subway train arrives or departs at a station.

4 Depth Structure of Underground Transportation Networks

Our method relies on detecting relative height difference between adjacent stations and using the *piston effect* to detect if the train is stationary or not. Therefore, we analyse in a first step the relative height differences of adjacent subway stations across five different subway networks (London, UK; Moscow, Russia; Tokyo, Japan; Vienna, Austria; Brussels, Belgium) to find out if our approach would work across major underground networks. All cities are among the largest subway networks in the world (Wikipedia 2016) and the depth information of the stations where available for us. In Sect. 4.1 we also analyse the hypothesis of Watanabe et al. (2012) if template matching of air pressure data, could be reliably used across different lines, cities or global networks.

The height information and the geolocations of the stations were extracted from *Transport for London (TfL)*, the company that operates the London underground transportation network. With 270 subway stations, 10 lines and spanning 402 km, London has the biggest subway network in Europe (Transport for London 2016b).

The depth information for the Moscow subway system was extracted from the 3D visualization of Varfolomeev (2015). For this visualization, the depth information of Goncharov (2015) was used.

The depth information of the Tokyo subway system was available on the website of Tokyo Metro (http://www.tokyometro.jp/en/). The depths of all stations were given in meters relative to sea level.

The data of the Vienna UBahn was obtained from Wiener Linien, the company that operates the subway network in Vienna. The locations of the stations were available via the Open Data Portal from the city of Vienna.[4]

---

[4]https://open.wien.gv.at/site/open-data/.

Depth information of the Brussels subway network is not publicly unavailable. Therefore, we have recorded air pressure data of multiple subway trips in Brussels with an air pressure logger application. This data was recorded on two different days, using a Google nexus 4 (with a Bosch BMP180 barometer) and an Apple iPhone 6 (with an Bosch BMP280 barometer). From this data, we calculated the depth information of the Brussels subway stations using the hypsometric formula. If height data between two stations from multiple recorded trips was available, the average value of these recordings was used. We make all of the data public for fellow researchers after publication.

To analyse the different characteristic of the subway networks, we used the following measures:

- **HeightDiff$_{station}$**: The height difference between two adjacent stations
- **HeightDiffTrip$_n$**: The height difference for a trip of length n, where n ranges from 1 to the maximum number of stations in a line.
- **HeightDistDiff$_{station}$**: The height difference between two adjacent stations, divided by the distance between the stations.
- **HeightDistDiffTrip$_n$**: The height difference divided by the distance for a trip of length n, where n ranges from 1 to the maximum number of stations in a line.

Notice that *HeightDiff$_{station}$* is equal to *HeightDiffTrip$_n$* for $n = 1$. The same is true for *HeightDistDiff$_{station}$* and *HeightDistDiffTrip$_n$*. 0

As can be seen from Fig. 1, Moscow, the oldest subway system in the world, has the biggest average height differences (**HeightDiff$_{station}$**) between stations, while Tokyo has the smallest height differences (we also noticed very different characteristics between lines within a city).

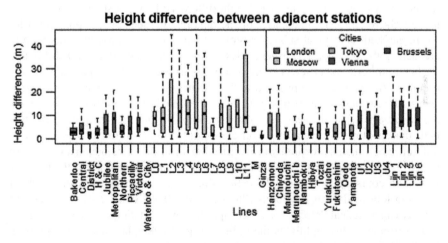

Fig. 1 Comparison of height differences between adjacent subway stations in London, Tokyo, Moscow, Vienna and Brussels per line showing max, min, quartiles and average

Fig. 2 Average height differences and slopes for trips of different lengths across all networks

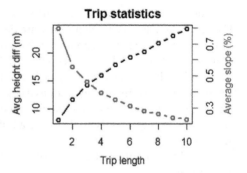

This also is reflected in the slope difference (***HeightDistDiff***$_{station}$). With an average slope of 0.80 %, Moscow's average slope is nearly twice as big as London's (0.44 %) or Tokyo's (0.46 %). As expected, the average height difference between stations increases when the trip length increases (Fig. 2). The average slope decreases when the trip length increases. A negative slope is often followed by a positive slope (e.g. when the track needs to run temporarily deeper underground). Over longer trips, this evens out. Our analysis shows that an approach using the smartphone barometer needs to take the variety of the depth characteristics of the networks into account to work on a global scale.

4.1 Technical Evaluation

In general, we see that the average height difference between stations for all lines is well above 2 m (the granularity of the air pressure sensors is around 1 m) for the researched five global subway systems, therefore our approach works in general across all networks. To further analyse this, we calculated at the probability that the height difference between two adjacent stations exceeds the air pressure sensors accuracy. As expected, the probability that the height difference between two adjacent stations is bigger than 0 m is 100 % and the probability naturally decreases when the height difference between two adjacent stations increases. The probability that two adjacent stations have a height difference of 0.5 m is still 91.38 %. For a height difference of 1m, this probability is decreased to 82.11 %. Therefore, in a worse case scenario, our approach can detect 82.11 % of all 1-stations rides across the five networks. Of course, this is a very theoretic value, as the typical subway ride is about 8.75 min long and takes 6 stations (Stockx et al. 2014). For a 2-stations ride, we already can technical detect 88 % of all trips and for a 4-station ride nearly 92 %.

Nevertheless, this highlights the difference to the template matching approach by Watanabe et al. (2012). We have also calculated how unique the height difference patterns are across different lines, cities and across the five networks we have investigated. On average, 54 % of 1-station trips are unique in an individual line.

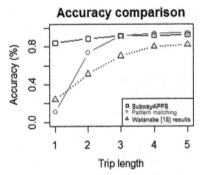

Fig. 3 Accuracy of pattern matching algorithm (*red*) (Watanabe et al. 2012) compared to our approach SubwayAPPS (*black*). Please note, that in this case we assume that the pattern matching approach (Watanabe et al. 2012) works with a 100 % accuracy. The *blue dotted line* shows the accuracy achieved by Watanabe et al. (2012)

This increases to 96 % for 2-station trips. If we look at the uniqueness of a height difference pattern at the level of a city, we see that only 10 % of 1-station patterns are unique. This increases quickly, with 94 % uniqueness for a 4-station trip. This is the same as the uniqueness of 4-station trips at worldwide level. This explains why the template matching approach performs bad on short trips compared to our approach presented in this paper. Nevertheless, they perform similar when the trip length increases. The depth data of the networks is available upon request (Fig. 3).

5 SubwayAPPS

With the SubwayAPPS technique, we show that it is possible to achieve reasonable positioning accuracy on underground transportation networks. SubwayAPPS makes use of the smartphone's barometer by relying on the *piston effect* and height differences between subway stations. In our current implementation, we also expect the user to provide a start station. This could also be done automatically by selecting the closest subway station near the last known GPS coordinate.

As described earlier, some smartphones already filter the air pressure data on the chip. If not, our software applies an alpha filter to do that.

We use the *piston effect* to detect whether the train is stationary or not. When the train is moving, the *piston effect*, together with variations in the environment (width of tunnel, ventilation shafts etc.) cause the air pressure to be unstable. When the train has arrived at a station, the air pressure stabilizes again. To detect if the air pressure is currently stable we look at the variance of the air pressure in a sliding time window. To decide whether the air pressure is currently stable or not, two parameters are needed: the size of the sliding time window and a threshold that determines the boundary between stable and unstable air pressure. We experimentally found the optimal values for the parameters by using different values and count the number of undetected station arrivals and false positives in our dataset that work across all networks. The optimal values for the sliding time window and threshold are 12 s and 0.0015 hPa respectively.

To provide the positioning information, we make use of the height differences between the stations. As mentioned in Sect. 3, we can calculate the height difference between two points by their air pressure at these two points by using the *hypsometric formula*. Nevertheless, when we detect that the air pressure is stable as described above, we can not assume directly that the train is stopping in a station. It could also be the case, that the train stops within a tunnel or similar.

Instead, we first look if the relative height difference between the current position and the previous station, measured by the smartphone's air pressure sensor, matches the real height difference between those stations. Only when the air pressure is stable and the height difference check succeeds, we can conclude that the subway train has arrived at a station. We experimentally found the optimal height tolerance (2 m) between the actual height difference and the measured height difference to detect most stations and eliminate false positives.

Additionally, the height difference check prevents that an unscheduled stop between two stations is classified as an arrival at a station. When the subway train has stopped between two stations, the air pressure can become stable. In this case, the height difference check will fail and the period of stable air pressure will rightfully not be seen as an arrival at a station.

A visualisation of the algorithm with height difference check is shown in Fig. 4. The variance of the air pressure, height difference and result of the algorithm are shown below each other. The blue dotted lines on the height difference chart shows the threshold for the height difference check. The height difference check succeeds if the measured height difference between the previous stop and the current position is

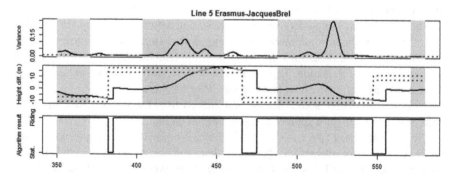

Fig. 4 Example of the SubwayAPPS algorithm. The *top* chart shows the variance of the air pressure in the time window. When the variance is below the threshold (*blue dotted line*), the air pressure is stable. The chart in the *middle* shows the relative height difference measured between stations. The *blue dotted line* shows the expected height difference ± tolerance. When the air pressure is stable and the height difference is as expected, the algorithm concludes that the train has arrived at a station (*bottom* chart)

between the two dotted lines. We can see how the potential false positive between 450 and 500 s is filtered out. After the first stable air pressure period ends, the algorithm only expects a new stop at a relative height difference 9 m below the current stop. Because of this, the second period of stable air pressure is not seen as a stop, because the relative height difference is ±0 m.

6 Implementation

The SubwayAPPS algorithm is implemented on the Android platform, for versions 4.0.4 and higher. The application is mainly based on the existing MetroNavigator application, developed by Stockx et al. (2014) in their Sigspatial 2015 paper. To access the air pressure sensor of the smartphone, we use Android's built-in API to access environment sensors (Android Developer Guide 2015). Once registered, this API returns the air pressure in hPa at around 5 Hz.

As described in the Sect. 5 we check if the air pressure data is already filtered on the chip, or if additional filtering is needed. A boolean variable keeps track whether smoothing is recommended. By default, the value is set to *true*. If an air pressure sensor has built-in smoothing (e.g. Bosch BMP280), this is turned off. We implemented the filter as described in Sankaran et al. (2014), with $\alpha = 0.1$. This gives smoother air pressure readings, comparable to air pressure sensors with built-in filtering.

Throughout the duration of the algorithm, we keep a time window to check whether the air pressure is considered *stable* or not.

We keep a boolean, *stopDetected*, to save the current status of the train (riding or stationary). When the algorithm has just started and the time window is not full yet, the variance of the values in the time window will be low. The algorithm decides that the air pressure is stable. This is acceptable because we assume that the user starts the application when he boards the subway train and it is still waiting to depart. The value of *stopDetected* is thus initialised as true.

6.1 MetroNavigator+

The algorithm described above runs as a background service on Android. When an event, e.g. train arrived at a station, happens, the service sends an *intent* (message) to the Android OS. Every application on the OS can register to receive these intents. The receiving application can then use this information for navigational purposes etc.

One such application that listens for these intents is MetroNavigator+, an application for Android and Android Wear based on the MetroNavigator developed by Stockx et al. (2014). The application similar to the MetroNavigator application the MetroNavigator+ application lets users select a start and destination station. It then tracks the subway trip and shows number of stations left, time left to next station, the

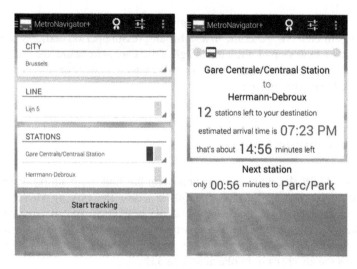

Fig. 5 Screenshots of the MetroNavigator+ application

name of the next station, the estimated arrival time and a visualization of the progress between the previous and next station. While riding, the user can share tweets (if cellular network is available at the stations) about his ride and see Wikipedia information about landmarks that he passes. Finally, a smartwatch extension allows the user to follow the progress of his trip on his watch.

The MetroNavigator+ also stores the relative height differences between the adjacent stations of the trip. The data is stored in the JSON format. To facilitate the querying of the data, a datastore class was implemented to retrieve the information from the web as needed (Fig. 5).

7 Evaluation

To evaluate the SubwayAPPS algorithm, a real-world study was done in the subway networks of Brussels and London. A logger application was developed to record the air pressure, along with information about the current station and the status of the train, while riding the train. The logs were recorded with a Google Nexus 4 (Bosch BMP180 pressure sensor) and an iPhone 6 (Bosch BMP280 pressure sensor). The logs were recorded at a random position in the train (front, back, middle), to prevent that the position of the traveller on the train is influencing the evaluation results. The Nexus 4 and the iPhone used to record was in a *natural position* (e.g. kept on the lap or in the pocket of the user) while recording.

In Brussels, the algorithm was tested on two separate days in May 2016. In total, 20 recordings were made with an average time of 22 min. The average number of stations per trip was 14. The BMP180 sensor was used for 12 recordings, the other were recorded by the BMP280 sensor. All lines were recorded completely at least once.

A second test was done in London on a single day in May 2016. Nine logs were recorded with the BMP180 sensor and 10 with the BMP280 sensor. The trips were randomly chosen on the London underground network in zones 1 and 2. The goal of this test was to board the train, ride it for 7 min and then exit at the next stop. The average trip time of this test is 9 min. The average length of a trip was 4 stations.

7.1 Results

With the BMP180 pressure sensor in Brussels, 162 out of 170 stations (95.29 %) were detected correctly with our approach. 8 out of 12 (66.66 %) complete trips were detected correctly. In London, it classified 20 out of 30 stations (66.66 %) correctly. 5 out of 9 complete trips (55.55 %) were successfully classified.

Using the BMP280 pressure sensor in Brussels, 92 out of 114 stations (80.70 %) were detected correctly. 5 out of 8 (62.5 %) complete trips were detected correctly. In London, it classified 21 out of 36 stations (58.33 %) correctly. It could only detect 1 complete trip correctly.

The results for the Brussels underground network are slightly better than Stockx' (2014) method, which uses the accelerometer and achieved an accuracy of 85 % with tweaked parameters and 75 % with world-wide parameters. However, this method performs better for the London underground network. Further, it performs better than Watanabe et al. (2012), which achieved an accuracy of up to 85 % for a trip of length 5. For a 1 station trip, this method had only 24 % accuracy.

We expect that the performance of our method improves when the depth structure of the network is mapped using the same air pressure sensors used for the SubwayAPPS algorithm. Our method does not depend on the type of train or the smoothness of the tracks, something that negatively influenced Stockx' algorithm. Another advantage of our method is that we can rely on two properties of air pressure: the *piston effect* and the changing air pressure at different altitudes. This allows us to filter out most false positives. A downside of SubwayAPPS is that the depth structure of the underground network must be known prior to tracking. This information is not publicly available for all underground networks. Preferably, the depth structure is mapped with an air pressure sensor, for accurate results.

From this results, we can conclude that there is a noticeable difference between the Bosch BMP180 and BMP280 sensors. From the analysis, we see that some station stops are not detected because the air pressure readings are not stable enough to be under the stable threshold. A possible cause for this, is the built-in filter of the BMP280 sensor. This filter is not as strong as the alpha filter we use to filter the readings of the BMP180 sensor. Because the filter of the BMP280 is not as strong, the variance of the values in the time window does not decrease enough when the train is stationary.

Further, we notice a difference in performance between the tests in Brussels and London. From the analysis, we see two causes for this. First, we notice that the air pressure while stationary does not become stable enough for the algorithm to detect a stop. The subway stations in London are bigger and in general more busy than those in Brussels. They also often connect multiple lines, which is not the case in Brussels. When a train arrives or departs from the same station but at a different platform, this destabilizes the air pressure in the whole station. Another difference between the test in Brussels and London, is the source of the data of the height differences between the stations. The data for London comes from TfL and dates from 2011. We have no control over the accuracy of the data. Changes in the depth structure of some lines may have happened. The data for Brussels is derived from measurements done by the same pressure sensors that we use to do the test. This is more suitable for the execution of the algorithm.

8 Conclusion and Future Work

In this paper, we presented SubwayAPPS, a novel method that uses the air pressure sensor in a smartphone to provide underground location information. SubwayAPPS uses the characteristics of air pressure in closed environments (like subway tunnels) and the height differences between adjacent stations to execute the detection. It does not need any additional instrumentation of the environment, sparing the operators of subway networks who are often on a low budget. The only information the algorithm needs, is a start and end station and a depth map of the subway line (gathered manually using an air pressure sensor or otherwise available as public data).

In the first study, we showed that the height differences between adjacent subway stations are big enough for air pressure sensors in modern smartphones to detect. For this, the subway networks of five major cities (London, Moscow, Tokyo, Vienna and Brussels) were analyzed. Therefore, our approach could also be applied to other underground settings as described earlier. Furthermore, we tested the accuracy of air pressure sensors to detect height differences. From this, we saw that we could theoretically classify 82.11 % of subway stations by looking at their height differences alone. Our test shows that our method, when the depth structure of the underground network is mapped with an air pressure sensor, can get a 10 % better accuracy than existing solutions. Despite this, the algorithm is simple and can be implemented in less than 100 lines of code and therefore inform any LBS as it runs as a background service on Android any existing LBS could easily subscribe to. As a proof of concept, the algorithm is implemented on the Android platform with the MetroNavigator+ application.

However, there might be various reasons why the algorithm might fail at some times. As mentioned before in our evaluation of the algorithm in London, arriving or departing trains on other platforms in the same station can influence the stability of the air pressure.

When a train brakes before it arrives at a station and then drives slowly into the station, the algorithm may be detecting the stop too early. This can be solved by using additional sensors like the accelerometer or a brightness sensor that detects whether the train is still in the tunnel or not.

Furthermore, drastic changes of the weather change the air pressure universally. During an intense storm, an air pressure change which corresponds to a height difference of 3–4 m can happen in only 10 min. The probability that this happens is still relatively small, as the average riding time between adjacent stations is 1–2 min. We can resolve this issue by involving weather information (e.g. base pressure at sea level) for the calculations of the algorithm. This way, we can compensate for the drift in air pressure due to weather conditions.

During rush hours, trains can be very crowded. The user might only be able to sit down during a part of the trip. Alternating between a sitting and standing position changes the height of the smartphone. Even when the user remains in the same position during the whole trip, putting the phone from his hands into his pockets or vice versa changes the height of the phone. This can be resolved by using an external air pressure sensor. This sensor can be attached to a body part that remains at the same height, e.g. the user's shoe.

Finally, the algorithm fails when the services of the subway network change. An example of this, is the temporary closure of a station due to adjustments of the station. The algorithm will expect the train to stop at the closed station, but can not detect this when the train rides past the station. Consequently, all next stations can not be detected by the algorithm, because the expected height difference is not adjusted. To resolve this, the application should frequently receive service updates from the underground network operator.

The algorithm can not recover when it misses a station stop. Consequently, all subsequent stations are not detected as well. A recalibration function could put the algorithm back on track.

For future work, a neural network can be used to detect the patterns in air pressure that occur when the subway train is riding. Further, an algorithm that detects the arrival of a train at a station can be developed to notify the user and aid the SubwayAPPS algorithm to start its calculations. To get a better mapping of the depth structure of subway networks, crowdsourcing can be used to gather information while the SubwayAPPS algorithm is running. Finally, a new algorithm can be developed using the air pressure sensor, accelerometer and magnetometer of the smartphone together. This could greatly improve accuracy as the different sensors overcome each others shortcomings.

Acknowledgments We would like to thank Brent Hecht for his comments on this draft. Furthermore we would like to thank Thomas Stockx for sharing his implementation of the MetroNavigator application (Stockx et al. 2014).

References

Android Developer Guide: Android: Environment sensors. http://developer.android.com/guide/topics/sensors/sensors_environment.html, consulted on October 20, 2015

Ascher, C., Kessler, C., Wankerl, M., Trommer, G.: Using orthoslam and aiding techniques for precise pedestrian indoor navigation. In: Proc. of ION GNSS '09. pp. 743–749 (2001)

Fernandes, L., Barata, F., Chaves, P., Inovaç ao, I.I.: Indoor position method using wi-fi. pp. 187–202 (2014)

Fischer, C., Muthukrishnan, K., Hazas, M., Gellersen, H.: Ultrasound-aided pedestrian dead reckoning for indoor navigation. In: Proc. of MELT '08. pp. 31–36. ACM (2008)

Gargini, A., Vincenzi, V., Piccinini, L., Zuppi, G.M., Canuti, P.: Groundwater flow systems in turbidites of the northern apennines (italy): natural discharge and high speed railway tunnel drainage. Hydrogeology journal 16(8), 1577–1599 (2008)

Goncharov, A.: Transport schemes. http://www.alexeygoncharov.com/index1-eng.html, consulted on November 25, 2015

Higuchi, T., Yamaguchi, H. and Higashino, T.: Tracking motion context of railway passengers by fusion of low-power sensors in mobile devices. In: Proceedings of the 2015 ACM International Symposium on Wearable Computers. pp. 163–170. ACM (2015)

Kawaguchi, N., Yano, M., Ishida, S., Sasaki, T., Iwasaki, Y., Sugiki, K., Matsubara, S.: Underground positioning: Subway information system using wifi location technology. In: 2009 Tenth International Conference on Mobile Data Management: Systems, Services and Middleware. pp. 371–372. IEEE (2009)

Krammer, M., Bernoulli, T., Walder, U.: All you need is contentcreate sophisticated mobile location-based service applications without programming. In: Proceedings of the 11th International Symposium on Location-Based Services. pp. 80–85 (2014)

Ojeda, L. and Borenstein, J.: Personal dead-reckoning system for gps-denied environments. In: 2007 IEEE International Workshop on Safety, Security and Rescue Robotics. pp. 1–6. IEEE (2007)

Ortakci, Y., Demiral, E., Karas, I.: Rfid based 3d indoor navigation system integrated with smart phones. In: Proceedings of the 11th International Symposium on Location-Based Services. pp. 80–85 (2014)

Pan, S., Fan, L., Liu, J., Xie, J., Sun, Y., Cui, N., Zhang, L., Zheng, B.: A review of the piston effect in subway stations. Advances in Mechanical Engineering 5, 950205 (2013)

Portland State Aerospace Society: A quick derivation relating altitude to air pressure. http://psas.pdx.edu/RocketScience/PressureAltitude_Derived.pdf, consulted on October 21, 2015

Posti, M., Schöning, J., Häkkilä, J.: Unexpected journeys with the hobbit: the design and evaluation of an asocial hiking app. In: Proceedings of the 2014 conference on Designing interactive systems. pp. 637–646. ACM (2014)

Robertson, P., Angermann, M., Krach, B.: Simultaneous localization and mapping for pedestrians using only foot-mounted inertial sensors. In: Proc. of Ubicomp '09. pp. 93–96. ACM (2009)

Ruiz, A., Granja, F., Prieto Honorato, J., Rosas, J.: Accurate pedestrian indoor navigation by tightly coupling foot-mounted imu and rfid measurements. IEEE Transactions on Instrumentation and Measurement 61(1), 178–189 (2012)

Sankaran, K., Zhu, M., Guo, X.F., Ananda, A.L., Chan, M.C., Peh, L.S.: Using mobile phone barometer for low-power transportation context detection. In: Proceedings of the 12th ACM Conference on Embedded Network Sensor Systems. pp. 191–205. ACM (2014)

Schall, G., Schöning, J., Paelke, V., Gartner, G.: A survey on augmented maps and environments: approaches, interactions and applications. Advances in Web-based GIS, Mapping Services and Applications 9, 207 (2011)

Schöning, J., Krüger, A., Cheverst, K., Rohs, M., Löchtefeld, M., Taher, F.: Photomap: using spontaneously taken images of public maps for pedestrian navigation tasks on mobile devices. In: Proceedings of the 11th international Conference on Human-Computer interaction with Mobile Devices and Services. p. 14. ACM (2009)

Stockx, T., Hecht, B., Schöning, J.: Subwayps: towards smartphone positioning in underground public transportation systems. In: Proceedings of the 22nd ACM SIGSPATIAL International Conference on Advances in Geographic Information Systems. pp. 93–102. ACM (2014)

Thiagarajan, A., Biagioni, J., Gerlich, T., Eriksson, J.: Cooperative transit tracking using smartphones. In: Proceedings of the 8th ACM Conference on Embedded Networked Sensor Systems. pp. 85–98. ACM (2010)

Transport for London: Annual report and statement of accounts. http://content.tfl.gov.uk/annual-report-2013-14.pdf, consulted on June 11, 2016a

Transport for London: Tfl: Facts & figures. https://tfl.gov.uk/corporate/about-tfl/what-we-do/london-nderground/facts-and-figures, consulted on April 11, 2016b

Varfolomeev, E.: 3d-model of moscow metro. http://varf.ru/metro3d/?en=1Źp=-90Źt=45Źd=41.05255888325765, consulted on November 25, 2015

Watanabe, T., Kamisaka, D., Muramatsu, S., Yokoyama, H.: At which station am i?: Identifying subway stations using only a pressure sensor. In: Wearable Computers (ISWC), 2012 16th International Symposium on. pp. 110–111. IEEE (2012)

Wikipedia: List of metro systems — wikipedia, the free encyclopedia (2016), https://en.wikipedia.org/w/index.php?title=List_of_metro_systemsŹoldid=724270425, consulted on June 08, 2016

Wu, Y., Atkinson, G., Stoddard, J., James, P.: Effect of slope on control of smoke flow in tunnel fires. Fire Safety Science 5, 1225–1236 (1997)

Young, H., Freedman, R.: Sears and Zemansky's University Physics. Addison-Wesley series in physics, Addison-Wesley (2000), https://books.google.be/books?id=GEEbAQAAIAAJ

Zandbergen, P.A.: Accuracy of iphone locations: A comparison of assisted gps, wifi and cellular positioning. Transactions in GIS 13(s1), 5–25 (2009)

Part II
Outdoor and Indoor Navigation

Effects of Visual Variables on the Perception of Distance in Off-Screen Landmarks: Size, Color Value, and Crispness

Rui Li

Abstract Maps on mobile devices provide the convenience of accessing spatial information of the surroundings. They also raise the concern that the small screen size impacts user's acquisition of spatial knowledge as the small size causes the fragmentation of spatial knowledge. This fragmentation leads to the expense of more cognitive efforts and degradation of spatial knowledge. Following some effective approaches of representing distance to off-screen locations as contextual cues, this paper reports a design that incorporates not only the direction but also the distance using symbols. In addition, this study uses three different visual variables including size, color value, and crispness at the ordinal level of measurement and then compares their effectiveness on the perception of distance. Results show that color value contributes the least to the perception of distance. Size leads to slightly higher accuracy than crispness in comparing distances of off-screen landmarks. This study provides valuable information to further explore the impacts of different visual variables that could facilitate the acquisition of spatial knowledge on mobile devices. This study also points out the necessity of follow-up studies to clarify some issues in the current design as well as its impact on actual wayfinding performance.

Keywords Off-screen landmarks · Visual variables · Distance · Spatial knowledge

1 Introduction

The use of mobile devices in navigation raises the concerns including the inefficiency of cognitive mapping, the degradation of spatial knowledge, and the poor awareness of space (Parush et al. 2007; Speake and Axon 2012). The main reason is that the small size of mobile screen impacts the acquisition and the integration of

R. Li (✉)
Department of Geography and Planning, State University of New York at Albany, Albany, NY, USA
e-mail: rli4@albany.edu

spatial knowledge. In order to learn a relatively large environment, a user has to make more cognitive efforts and interactions such as zooming and panning to compensate the fragmentation caused by small screen size (Schmid 2008). To overcome this, researchers suggest possible approaches to visualize unseen locations beyond the mapped area (off-screen locations) at the edge of a screen to facilitate the learning of a large area. These approaches utilize properties of geometric shapes to let a user mentally infer the location beyond the screen (Burigat and Chittaro 2011). For example, Baudisch and Rosenholtz (2003) used a portion of a circle and Gustafson et al. (2008) used a portion of triangle to indicate the locations that are off screen. When a user looks at the visualized partial geometry at the edge of screen, he or she can mentally complete the geometry and infer the approximate location based on the variation of arc or triangle. These approaches provide effective ways to visualize distance of off-screen locations for users, but overlook the importance of the identities of these off-screen locations by only using generic geometry for visualization.

The identity of a location is useful for persons as landmarks facilitate spatial orientation and wayfinding efficiency (Michon and Denis 2001; Raper et al. 2007; Waters and Winter 2011). Not just as an anchor point that a person can use to efficiently integrate their acquired spatial knowledge, landmarks also provide easy indication of a location where the change of direction is needed. These landmarks can inform wayfinders in advance that if a change of direction or continuation is needed (Lovelace et al. 1999). To examine the effectiveness of visualizing landmarks on mobile devices, Li and colleagues (2014) used symbols to represent off-screen locations with their identities (off-screen landmarks) at the edge of mobile screen. Since only considering the direction to those off-screen landmarks, their results showed reduced frequency of zooming and better acquired spatial knowledge at the survey level. It is clear that this approach does not present the distance to those off-screen landmarks, which could be more informative and helpful to users. Later Li and Zhao (2016) investigated the effects of visualizing distance in symbols at both ratio and ordinal level and suggested that ordinal level of measure is more effective for perceiving off-screen distance.

The current study further investigates the possible ways of visualizing distance in symbols of off-screen landmarks. By integrating the information of direction and distance in the symbols for off-screen landmarks, this study examines their effects on perception of distance. Adapting from cartographic theories of visual variables (MacEachren 1995; Robinson et al. 1995), this study selects size, color value, and crispness as the three visual variables used for design and assessment.

Following this introduction, the next section introduces related approaches that address representing distances to off-screen locations. Section three describes the method that this study uses to design three scenarios involving the selected visual variables. Section four presents the design and evaluation of designed scenarios on perceiving distance, and is followed by the sections of results, discussion, and conclusion respectively.

2 Related Work

2.1 Limit of Small Screen

While bringing the convenience, the small screen of mobile devices unfortunately limits the acquisition of spatial knowledge comparing to desktop screen and paper maps (Burigat and Chittaro 2011; Münzer et al. 2006). Regarding user interaction, the small size results in an increased amount of zooming and panning on mobile devices. For example, when information is displayed at a small cartographic scale, users acquire an overview with less details. Likewise, when users look at displayed information with a greater level of details at a larger cartographic scale, they lose the general overview of the entire surrounding. If important points of interest fall beyond the shown area on the screen, the user will have to pan or zoom to find them. Regarding the acquisition of spatial knowledge, the displayed map on small mobile screens further impedes users from learning the surrounding configurationally. As only a small portion of area is displayed on screen, it takes a user more cognitive efforts and time to integrate those separately learned areas into one mental representation. This mental representation, furthermore, is prone to errors in spatial knowledge.

Studies have been carried out to investigate the influence of small screen on performance related to wayfinding and acquisition of spatial knowledge. For instance, Dillemuth (2005) investigated the effects of maps at different scales on acquisition of spatial knowledge. In particular, participants estimated distance and direction based on different viewable map extents on mobile devices. Results showed that smaller map size was prone to lower accuracy and longer response time. Participants' acquired spatial knowledge was also different across conditions. Later, Ishikawa and Kiyomoto (2008) compared users' wayfinding performance using different forms of navigation support including maps on mobile devices and paper maps. Results also showed that small map size on mobile devices led to longer travel time and more errors in directional estimation and configurational knowledge. The authors suggested that the size of map was a major obstacle to the acquisition of spatial knowledge. Therefore it is necessary to identify an optimal size and extent to convey spatial information of surrounding areas. The findings of Willis and colleagues (2009) also coincide with those above. They pointed out that users who learned the environment through mobile devices had poorer spatial knowledge than those who learned from traditional paper maps. In particular, participants who used mobile maps made more errors in their estimated distances.

Techniques have been suggested such as visualizing contextual cues indicating locations that are beyond the mapped area as a way to facilitate the acquisition of spatial knowledge (Burigat and Chittaro 2011; Baudisch and Rosenholtz 2003). For example, the Overview + Details approach (Plaisant et al., 1995) displays multiple views at different scales on one display at the same time. This method raises the concern that those views at different scales would still require users to mentally integrate all of them into a single mental representation. Consequently, it still requires more users' cognitive efforts switching between views in order to reorient themselves (Willis et al. 2009). Considering the important roles of landmarks in wayfinding, the Off-screen landmark approach (Li et al. 2014) directly uses cartographic symbols of off-screen landmarks to indicate their directions on screen. This approach significantly reduces the zooming frequency and contributes to a higher accuracy of spatial orientation. It, however, does not consider the distance to those off-screen landmarks. Some well-known approaches adapt properties of geometry to visualize off-screen locations. For instance, the Halo approach (Baudisch and Rosenholtz 2003) uses arc as a part of circle, while the Wedge approach (Gustafson et al. 2008) uses a part of triangle to infer distance and directions to off-screen locations. These types of approach consider distance and direction to a distant location, but neglect the identity of the location which could serve as landmark. These methods of representing off-screen objects are reviewed in more details in the following section.

2.2 Related Approaches

While acknowledging the positive roles of landmarks in human navigation (Raubal and Winter 2002; Winter et al. 2008; Anacta et al. 2016), it is a necessary step to embed both direction and distance information in the design of symbols for off-screen landmarks. As mentioned earlier, some approaches have already visualized information to infer the distance to off-screen locations, while no identities of these locations are visualized. The Halo approach (Baudisch and Rosenholtz 2003) indicates the distance to off-screen locations based on the size of an arc as part of a circle. Users can mentally visualize the circle based on the displayed arc at the edge of the screen. If an arc indicates a larger circle, its distance to a user is further. The problem of this approach is that smaller arcs become more difficult for users to infer their distances.

Similarly, the Wedge method (Gustafson et al. 2008) uses partial triangles to visualize distance and direction. Users would mentally complete the triangle and infer the location where the tip of the triangle would form as the location of an off-screen location. Burigat and Chittaro (2011) carried out evaluation and suggested that the Wedge is more effective than Halo to convey distances to off-screen locations. In addition, even though there are clustered partial triangles at the edge of a screen, users can still distinguish different wedges that point to off-screen locations. Similar to Halo, the Scaled Arrow approach (Burigat et al. 2006) uses the orientation and size of an arrow to indicate direction and distance to an off-screen location. By using the generic geometric forms, all Scaled Arrow, Halo, and Wedge approaches do not consider the identities of those off-screen locations as well as on-screen locations that may also be important.

Some other approaches add additional information on top of geometric shapes. For example, the Hop approach introduced by Irani et al. (2006) is built based on Halo with added interaction. While indicating locations that are beyond the mapped area, displayed windows at the edge provide interaction that users can click to view details of the off-screen area. Because it has the same mechanism as the Halo approach, Hop still makes the understanding of distance difficult when crowding happens at the edge of the screen or when the arcs are too small.

Edge Radar (Gustafson and Irani 2007) is an approach similar to bifocal display with symbols at edges of the screen to indicate the existence of off-screen locations. This approach scales both dimension and representation to project symbols at the edge to resemble their original topological relationship in environment that do not overcrowd. This approach, however, requires a user to acquire overall spatial distribution of those off-screen locations before using it, otherwise the projected symbols do not portray their distances between locations to the user. This method is more suitable for small areas.

More recently, Gollenstede and Weisensee (2014) introduced an approach of visualizing straight lines from user's location to off-screen locations. The visualization at the edge represents network and then connect to users' location. When users interact with this representation through zooming and panning, the size and color of the visualized lines would change associated with their distances to and types of distant locations within network. The lines connecting users and off-screen network locations provide users orientation to pan along a specific line for their specific need within this network.

In short, almost all approaches effectively consider the visualization of distance and direction to an off-screen location. Many approaches, however, neglect to identify the object (e.g. gas station) that could serve as important cues for cognitive mapping and spatial orientation. This is the main purpose of this study to investigate the design of symbols for off-screen landmarks to convey distance information.

3 Representing Distance in Landmark Symbols

Many of these reviewed approaches utilize the size of a geometric form as the indicator of distance. In particular, the size of those geometric shapes changes proportionally in terms of actual distances to a user in environment. This representation of distance is hence based on the ratio level of measurement. A recent evaluation on the effects of level of measurement and size on mobile device (Li and Zhao 2016) suggested that both ordinal and ratio levels of measurement result in similarly low accuracy of perceiving furthest distance among all off-screen landmarks. This study, however, found out that visualization of distance based on the ordinal level contributes to higher accuracy of comparing distance between off-screen landmarks than that at ratio level. This is the main reason that this study choose to use ordinal level of measurement in its designs. While acknowledging the limit of the level of measurement and use of size in this design, this study intends to extend the investigation on the impact of more specific visual variables on the perception of distance.

Based on the primary visual variables suggested in cartographic literature (Robinson et al. 1995), the variable *color value* is chosen in this study to compare with the variable size. The color value refers to the lightness of a color, so that a lighter color indicates a longer distance while a darker color of the same hue indicates a shorter distance. Additionally, based on the introduction of three new visual variables related to dynamic and digital mapping (MacEachren 1995), the variable *crispness* is chosen as the third visual variable in the design of symbols for off-screen landmarks. The crispness represents the vagueness of boundary in map symbol. That is to say, the crisper a map symbol is, the shorter distance it indicates.

The size of symbols is based on their distance to a user's location. For distance shorter than 1 km, the distance for off-screen landmarks is considered nearby. For distance between 1 and 3 km, the distance for off-screen landmarks is considered medium. For distance longer than 3 km, the distance for off-screen landmarks is considered far. Particular to the variable size, 100×100, 60×60, and 30×30 pixels sizes are assigned to nearby, middle, and far off-screen landmarks respectively. Regarding the variable color value, the online tool Colorbrewer (Harrower and Brewer 2003) was used to determine values that could generate distinguishable categories. As a result, three values (RGB: 0, 0, 0), (RGB: 89, 89, 89), and (RGB: 189, 189, 189) were chosen to be assigned to symbols representing nearby, middle, and far off-screen landmarks. Regarding the variable crispness, open-source tool Inkscape (inkscape.org) was used to create blurred version of symbols. In particular, symbols that represent nearby off-screen landmarks have no blurred effect. Symbols representing middle off-screen landmarks are 3 % blurred horizontally and vertically. Symbols representing far off-screen landmarks are 6 % blurred horizontally and vertically, in order to create distinctive yet legible symbol categories. All off-screen landmarks have bounding square to indicate its status of being off-screen. Figure 1 shows examples of off-screen landmarks used in each scenario.

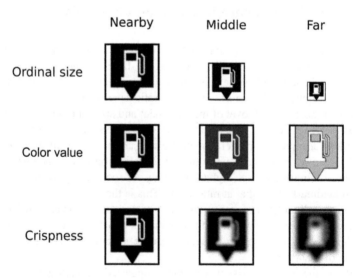

Fig. 1 Examples of gas station symbols used in each scenario representing nearby, middle, and far distance to off-screen landmark locations (Symbols source: www.flaticon.com)

4 Evaluation

To assess the effects of each visual variable on perception of distance to off-screen landmarks, the online platform Amazon Mechanical Turk (MTurk) was used to carry out experimental tasks. MTurk allows any requester to post human intelligence tasks (HITs) online and let workers around the world to participate. This population of volunteers represent a very diverse group that could shed light on the effectiveness of designs across different platforms and devices. In total 171 requesters participated in these experiments, while 6 out of 56 responses in the ordinal size scenario, 7 out of 58 responses in the color value group, and 6 out of 57 responses in the crispness group were unqualified and hence excluded. Consequently data of 50 participants from ordinal size scenario, 51 participants from color value group, and 51 participants from crispness group were entered for analyses.

This study used the same landmarks and rational selected in the earlier experiment of assessing the effect of level of measurement (Li and Zhao 2016). These selected landmarks include two on-screen landmarks and six off-screen landmarks which do not overlap on screen. The actual Android-based application addresses the issue of overlapping, which is not reported here as it is not the emphasis of this report. The mapped area represents part of the author's main campus. There are no text labels represented in the App so a participant can only see the landmarks and the base map. To further exclude similarity among all three designs, all landmarks' identities are pseudo in the color value and crispness scenarios. For example, the actual location of gas station is represented by the symbol of hospital. Figure 2 shows the screen shots of all three scenarios.

Once a participant selected one published task, he or she would have access to the screenshot of this mobile app as shown in Fig. 2. Each participant was informed that human symbol at the center indicated his or her location in this new environment. Two local and six landmarks were displayed to help them familiarize this

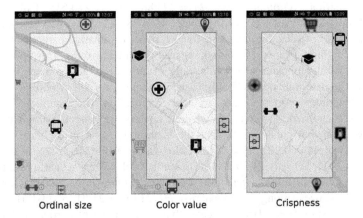

Fig. 2 Screenshots of three scenarios used in evaluation

surroundings. Each participant had to answer ten questions according to four categories of tasks: (1) naming the closest and furthest landmark; (2) comparing distance between two on-screen landmarks and naming the closer one; (3) comparing distance between one on-screen and one off-screen landmark and naming the closer one; and (4) comparing distance between two off-screen landmarks and naming the closer one. These tasks were designed to investigate if participants could distinguish the qualitative differences between on-screen and off-screen landmarks and further evaluate if they could understand the distance information in the specific design of off-screen symbols. For participants in the ordinal size scenario, they had to compare the size of symbols to determine the relative distances to off-screen landmarks. For participants in the color value scenario, they had to compare the darkness or lightness of symbols to determine relative distances to off-screen landmarks. For participants in the crispness scenario, they had to compare the vagueness of symbols to determine the relative distance to off-screen landmarks. Except for the question in task 1 that participants had to compare all eight symbols of landmarks, all other tasks only required participants to compare a pair of two symbols. The time that each participant took and the accuracy in each category of tasks were recorded.

5 Results

The time and accuracy in each category of tasks are used as dependent variables while the scenario serves as the independent variable in a one-way ANOVA. Results are presented in the following order: time, accuracy of selecting the closest and furthest landmark, the accuracy of comparing two on-screen landmarks, the accuracy of comparing one on-screen and one off-screen landmark, and the accuracy of comparing two off-screen landmarks. Participants in all three scenarios took similar time (sec) to complete each experiment resulting no differences among all three scenarios (Ordinal size: $M = 293.86$, $SD = 240.51$; Color value: $M = 290.96$, $SD = 321.97$; and Crispness: $M = 284.22$, $SD = 453.75$), $p = 0.99$.

In the tasks of selecting the closest and furthest landmarks, participants had no difficulty finding the closest landmark regardless of their groups. This finding is not surprising as it replicates a person's daily experience of using maps on mobile devices. The determination is simply based on looking for the symbol of on-screen landmark with the shortest distance to the user's location. The accuracy of selecting the furthest landmark, however, shows great contrast. In general, participants seem to have somewhat challenge finding the furthest location in all three scenarios (Ordinal size: $M = 0.22$, $SD = 0.42$; Color value: $M = 0.10$, $SD = 0.30$; Crispness: $M = 0.61$, $SD = 0.49$). It is clear to see that color value scenario results in the lowest accuracy in finding the furthest landmark. One-way ANOVA shows significant differences among these scenarios, $F(2, 149) = 21.31, p < 0.001$. Post hoc comparison using Tukey-HSD indicates that the accuracy of participants in crispness scenario is significantly higher than that in color value scenario, as well as than

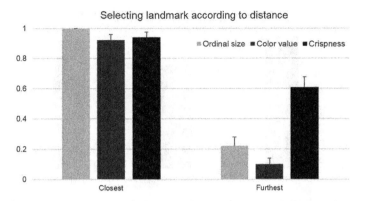

Fig. 3 Participants' accuracy of selecting the closest and furthest landmarks

Table 1 Furthest landmarks selected by participants (Bold number indicates the correct selection in each scenario. Cell with no value indicates no landmark selected)

Scenario	Selected furthest landmark							
	Gas station	Gym	Hospital	Soccer field	Supermarket	Train station	Bus stop	School
Ordinal size	13	4	16	6	**1**	10		
Color value	11	3			2	22	11	2
Crispness			3		11	**28**	1	8

that in ordinal size scenario. Color value scenario, however, does not lead to different accuracy from ordinal size scenario. Figure 3 shows the results of participants' performance in this category.

To find out the errors made by participants in this task, the specific selection of landmark is further examined. Since ordinal level of measurement is used in each scenario, there can be more than one correct response for the furthest landmark. Participants who selects either would receive score in this task. Table 1 shows the specific landmarks that participants selected for the furthest in each scenario. It is worthy to note that in the ordinal size scenario, many participants considered the hospital ($n = 16$) and then gas station ($n = 13$) as the furthest landmarks. In the color value scenario, the most selected landmarks by participants were train station ($n = 22$) and then gas station ($n = 11$) and bus stop ($n = 11$). Only 5 participants selected the correct furthest landmark. In the crispness scenario, more than half participants chose the correct landmarks. Eleven participants, however, considered supermarket as the furthest location. Possible reasons leading to these results are present in the discussion.

In the task of comparing two on-screen landmarks, participants have similar performance as in the task of selecting the closest landmark in task 1. In particular, all participants in the ordinal size scenario had the comparison correct ($M = 1.00$,

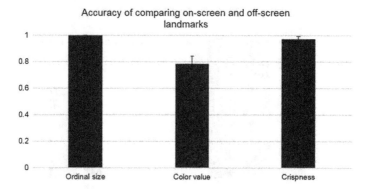

Fig. 4 Participants' accuracy in the task of comparing one on-screen and one off-screen landmark

$SD = 0.00$) while participants in the other two groups had similar accuracy (Color value: $M = 0.98$, $SD = 0.14$; Crispness: $M = 0.94$, $SD = 0.24$), p = 0.17.

Regarding the performance of comparing one on-screen and one off-screen landmark, significant difference exists among these scenarios, $F(2, 149) = 10.52$, $p < 0.001$. Post hoc comparison using Tukey-HSD indicates that the accuracy in color value scenario ($M = 0.78$, $SD = 0.42$) is significantly lower than that in ordinal size scenario ($M = 1.00$, $SD = 0.00$) as well as than that in crispness scenario ($M = 0.97$, $SD = 0.16$). There is no difference in accuracy between the ordinal size and the crispness scenario. Figure 4 also shows these results.

In the task of comparing two off-screen landmarks, significant difference also exists among three scenarios, $F(2, 149) = 8.10$, $p < 0.001$. Post hoc comparison using Tukey-HSD shows that participants in the color value group have the lowest accuracy ($M = 0.45$, $SD = 0.30$) which is significantly lower than that in ordinal size scenario ($M = 0.64$, $SD = 0.21$) as well as than that in crispness scenario ($M = 0.55$, $SD = 0.22$). Accuracy from both ordinal size and crispness scenarios does not show difference. These results are also shown in Fig. 5.

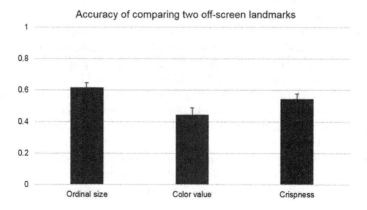

Fig. 5 Participants' accuracy in comparing two off-screen landmarks

6 Discussion

The discussion here presents four sections. The first section addresses the characteristics of using participants from MTurk. The second section addresses the high accuracy in comparing on-screen landmarks. The third section addresses the differences among scenarios in comparing a pair of on-screen and off-screen landmarks. The fourth section addresses the lower accuracy in comparing off-screen landmarks as well as the finding the furthest landmarks.

The ultimate goal of this research is to benefit general public with the ease and spatial awareness in navigation, so a diverse sample of participants would provide helpful insight on the effectiveness of this design. First, the online platform ensures that everyone with internet access can visit and participate. This resembles the characteristics of an app running on a mobile platform that users with internet can use it. Second, the diversity of participants represents different backgrounds of future user, so that issues appear in the experiment can be similar in actual use. Third, the computers used by participants vary greatly, which could also be the situation when users download and use the complete app. Participant's perception of the designed symbols simulates a realistic situation of using different mobile phones in the future. This approach of collecting user's performance, however, has its disadvantage in investigating influencing factors. Due to the protection of participants at the online platform, personal information was not collected to identify possible individual factors. Although the experiment online has designed features such as qualifying questions for excluding invalid answers and requiring participants to type all their answers instead of simply clicking, participant's responses could still include ones that are guessed. More importantly, participants at this online platform can only view screenshot of this app, which is different from the viewing experience in actual environment. While admitting the advantage of using this online platform to evaluate variables, it is the goal of this study to follow up and carry out experiments in actual environment to further assess participant's perception and performance in navigation in the real world.

It is not surprising to find out that almost all participants have no problem finding the closest landmark as well as comparing distances of on-screen landmarks. As these landmarks are presented exactly in the same way as in existing mobile devices, the distance between a landmark symbol and user's symbol proportionally indicate the actual distance between them in environment. Since there are only two on-screen landmarks, the closer landmark out of these two locations on screen is actually the closest landmark location to a user. The later comparison of two on-screen landmarks actually replicates this task. Participant's performance in both tasks also shows similar results indicating that they have no problem understanding those on-screen landmarks.

Regarding the comparison between one on-screen and one off-screen landmark, participants show some differences among three scenarios. While the majority of participants in the ordinal size and crispness scenarios have high accuracy, more participants from the color value scenario make incorrect selection in this task. As a qualitative comparison, the location of any off-screen landmarks is further than the location of any on-screen landmarks. Some participants in the color value scenario, however, have not perceived this information. One major reason is that these participants still treat symbols of off-screen landmark as indicating their positions on screen. Therefore, participants still employed the distance on the display as the only criterion for comparing distance instead of using the lightness of symbol color. When there is an off-screen landmark projected at the edge of screen whose position is closer to user's symbol than the position of an on-screen landmark, some participants would infer that the off-screen landmark's symbol indicates a closer location. This is also a major reason which causes the low accuracy in comparing distance of off-screen landmarks, which is further discussed below.

In the performance of comparing off-screen landmarks and selecting the furthest landmark, participants have the most challenges. Only slightly more than half participants in ordinal size and crispness scenarios have accurate comparison. Participants in color value scenario achieve the lowest accuracy in both types of tasks involving comparing distance between off-screen landmarks. In the task of selecting the furthest landmark, participants in the ordinal size and color value have very low accuracy. Participants in the crispness scenario have higher accuracy. There are two main reasons to explain these results. The first reason is similar to the one above regarding the mechanism that symbols of off-screen landmarks are designed in each scenario. Participants in each scenario may have not fully understand the meaning embedded in the symbol that the change of size, lightness of color, or vagueness of symbol indicates relative distance. Instead of choosing the symbol of the smallest size, lightest color, or most blurred as the furthest landmark, participants still use the distance between symbol's positions on display as the criterion for selecting furthest location. This is probably why that participants in the crispness scenario made higher accuracy in selecting the furthest landmark as the symbol for this landmark happened to have longest distance to the user's symbol on screen. This also explains that participants in the crispness scenarios achieved lower accuracy in the comparison of two off-screen landmarks than in selecting the furthest landmark. The second main reason is related to the perception of designed symbols. This implies that participants understand the mechanism embedded in symbol design, but the change of size, lightness of color, and crispness of symbol add some difficulty for participants to distinguish. That is to say, participant make incorrect comparison in task because of uncertain differentiation between sizes, lightness, or crispness of symbols. The accuracy in comparing two off-screen landmarks will hence decrease. This reason explains the inconsistency between participant's very low accuracy in finding the furthest location and higher accuracy in comparing off-screen landmarks in the ordinal size scenario. When only a pair of different sized symbols are present to a participant, it reduces the difficulty that participants achieve higher accuracy.

7 Conclusion

Addressing the limitation of small screens on mobile devices that adds difficulty to user's cognitive mapping, research has suggested possible way to overcome this shortcoming by visualizing contextual cues to help users integrate the spatial knowledge of a large area. Many approaches have showed the success of visualizing the distance information in the symbols that indicate those distant contextual cues. These approaches, however, have not considered the great advantage of the identity of those contextual cues which can serve as landmarks to further support spatial orientation and cognitive mapping. Studies have examined the effect of visualizing those off-screen landmarks on mobile phones and suggested that spatial orientation is improved and zooming frequency is reduced. This study selected visual variables from cartographic theory including size, color value, and crispness. At the ordinal level of measurement, this study incorporates the information of direction and distance in the design of symbols for off-screen landmarks. This study then further investigate their roles on the perception of distance.

While results show that participants have no problem understanding those symbols of on-screen landmarks which replicates the way of map representation on existing mobile devices. These visual variables, however, lead participants to different levels of accuracy in tasks. In general, all three visual variables hold challenges for participants to select the furthest landmark that is off screen. Participants in the ordinal size and color value scenarios have very low accuracy while participants in the crispness scenario achieve higher accuracy. As pointed out in the discussion, the higher efficiency may be related to the coincidence that the actually distance between symbols on screen is shorter than others which may have helped participants. A future study addressing the factor of distance between symbols on screen is needed to further clarify the effectiveness of crispness.

The visual variable color value leads to the lowest accuracy in every category of tasks. The ordinal size seems slightly more effective than the variable of crispness in comparing off-screen landmarks and comparing a pair of on-screen and off-screen landmarks. The accuracy of performance in these tasks still remain relatively low. This is possibly related to the design as it first introduces two qualitatively distinctive categories of on-screen and off-screen landmarks, and then quantitative information is embedded in the design of symbols for off-screen landmarks for user to compare or distinguish. This combination of both qualitative categories and quantitative information may have increase the cognitive efforts of users to perform related tasks. In short, a possible follow-up is necessary to consider more efficient ways of visualizing on-screen and off-screen landmarks. For example, shape or color could be considered for the representation of both categories of landmarks. Furthermore, only three visual variables are selected in this study. It is the author's goal to compare the effectiveness of other visual variables like transparency. In addition, this study only addresses the perception of distance based on viewing a screen shot on an online platform. The future direction is to assess user's performance using the design for acquiring spatial knowledge and wayfinding in real-world environment. In future development of this application, the distance will also be considered as a selection criterion for displaying off-screen landmarks. Only those off-screen landmarks within a certain range of distance will be displayed for helping users understand the presented area better.

References

Anacta, V. J. A., A. Schwering, R. Li & S. Muenzer (2016) Orientation information in wayfinding instructions: evidences from human verbal and visual instructions. *GeoJournal*, 1–17.

Baudisch, P. & R. Rosenholtz. 2003. Halo: a technique for visualizing off-screen objects. In *CHI 2003*, 481–488. Ft. Lauderdale, Florida, USA.

Burigat, S. & L. Chittaro (2011) Visualizing references to off-screen content on mobile devices: A comparison of Arrows, Wedge, and Overview + Detail. *Interacting with Computers*, 23, 156–166.

Burigat, S., L. Chittaro & S. Gabrielli. 2006. Visualizing locations of off-screen objects on mobile devices: a comparative evaluation of three approaches. In *Proceedings of the 8th conference on Human-computer interaction with mobile devices and services*, 239–246. ACM.

Dillemuth, J. (2005) Map design evaluation for mobile display. *Cartography and Geographic Information Science*, 32, 285–301.

Gollenstede, A. & M. Weisensee. 2014. Animated cartographic visualisation of networks on mobile devices. In *11th International Symposium on Location-Based Services*, 26–28. Vienna, Austria.

Gustafson, S., P. Baudisch, C. Gutwin & P. Irani. 2008. Wedge: clutter-free visualization of off-screen locations. In *CHI 2008*, 787–796. Florence, Italy.

Gustafson, S. & P. Irani. 2007. Comparing visualization for tracking off-screen moving targets. In *CHI 2007*, 2399–2404. San Jose, California, USA: ACM Press.

Harrower, M. & C. A. Brewer (2003) ColorBrewer. org: an online tool for selecting colour schemes for maps. *The Cartographic Journal*, 40, 27–37.

Irani, P., C. Gutwin & X. D. Yang. 2006. Improving selection of off-screen targets with hopping. In *CHI 2006*, 299–308. Montreal, Quebec, Canada: ACM Press.

Ishikawa, T. & M. Kiyomoto. 2008. Turn to the left or to the west: Verbal navigational directions in relative and absolute frames of reference. In *GIScience 2008*, 119–132. Springer.

Li, R., A. Korda, M. Radtke & A. Schwering (2014) Visualising distant off-screen landmarks on mobile devices to support spatial orientation. *Journal of Location Based Services*, 8, 166–178.

Li, R. & J. Zhao (2016) Resizing off-screen landmarks on mobile devices: Levels of measurement and the perception of distance. *Manuscript submitted for publication*.

Lovelace, K. L., M. Hegarty & D. R. Montello. 1999. Elements of good directions in familiar and unfamiliar environments. In *Spatial Information Theory: Cognitive and Computational Foundations of Geographic Information Science 1661*, 65–82.

MacEachren, A. M. 1995. *How Maps Work: Representation, Visualization, and Design*. New York, NY, USA: Guilford Press.

Michon, P.-E. & M. Denis. 2001. When and Why Are Visual Landmarks Used in Giving Directions? In *Spatial Information Theory*, ed. D. R. Montello, 292–305. Heidelberg: Springer.

Münzer, S., Zimmer, H. D., Schwalm, M., Baus, J., & Ilhan, A. (2006). Computer-assisted navigation and the acquisition of route and survey knowledge. *Journal of Environmental Psychology*, 26, 300–308.

Parush, A., S. Ahuvia & I. Erev. 2007. Degradation in spatial knowledge acquisition when using automatic navigation systems. In *Spatial information theory*, 238–254. Springer.

Plaisant, C., Carr, D., Shneiderman, B., 1995. Image-browser taxonomy and guidelines for designers. *IEEE Software*, 12, 21–32.

Raper, J., G. Gartner, H. Karimi & C. Rizos (2007) A critical evaluation of location based services and their potential. *Journal of Location Based Services,* 1, 5–45.

Raubal, M. & S. Winter. 2002. Enriching wayfinding instructions with local landmarks. In *Geographic Information Science,* eds. M. J. Egenhofer & D. M. Mark, 243–259. Springer.

Robinson, A. H., J. L. Morrison, P. C. Muehrcke, A. J. Kimerling & S. C. Guptill. 1995. Elements of Cartography Sixth Edition John Wiley & Sons. Inc.

Schmid, F. (2008) Knowledge based wayfinding maps for small display cartography. *Journal of Location Based Services,* 2, 57–83.

Speake, J. & S. Axon (2012) "I Never Use 'Maps' Anymore": Engaging with Sat Nav Technologies and the Implications for Cartographic Literacy and Spatial Awareness. *The Cartographic Journal,* 49, 326–336.

Waters, W. & S. Winter (2011) A wayfinding aid to increase navigator independence. *Journal of Spatial Information Science,* 2011, 103–122.

Willis, K. S., C. Hölscher, G. Wilbertz & C. Li (2009) A comparison of spatial knowledge acquisition with maps and mobile maps. *Computers, Environment and Urban Systems,* 33, 100–110.

Winter, S., M. Tomko, B. Elias & M. Sester (2008) Landmark hierarchies in context. *Environment and Planning B: Planning and Design,* 35, 381–398.

Investigation of Landmark-Based Pedestrian Navigation Processes with a Mobile Eye Tracking System

Conrad Franke and Jürgen Schweikart

Abstract Eye movements provide information on the mental processing of landmark objects while navigating. The present study investigates landmark-based navigation by pedestrians in real world environments using mobile eye tracking technology. The goal of the study is to identify whether landmarks on maps optimize the navigation procedure and the usage of a map, and imprint the cognitive map sustainably. Two independent test groups navigated through unfamiliar urban environment and were subsequently interviewed. One group had landmark visualized on a map as an additional aid, the control group did not. The results show that objects that are focused longer and more frequently transfer onto the mental map. Upon recalling objects present in the surroundings, on average 8.3 landmarks were named per interview by the landmark group, compared to 7.0 for the control group. For the control group, the usage and duration of observation of the map was thereby approximately 1.7 times greater than for the landmark group. Following the memory test, the participants in the landmark group remembered significantly more objects and located these correctly as compared to the control group. In summary, the results show that the visualization of landmarks on maps optimizes the use of maps for navigation, whereby more landmark objects transfer to long-term memory and the mental map.

Keywords Landmarks · Eye tracking · Navigation · Cognitive processing

C. Franke (✉) · J. Schweikart
Department of Civil Engineering and Geoinformation,
Beuth University of Applied Sciences Berlin, Berlin, Germany
e-mail: cfranke@beuth-hochschule.de

J. Schweikart
e-mail: schweikart@beuth-hochschule.de

1 Introduction

Wayfinding and orientation in unfamiliar surroundings is a complex mental process. During a navigation process the environment is represented in the cognitive map, where salient objects, routes and their topographical relationships to one another are stored mentally (Golledge 1999). The construction of the mental map is organized hierarchically (Barkowsky 2001). Accordingly, the mental map comprises stages of knowledge that vary in their complexity. Landmarks, routes, and survey knowledge are the elementary components (Siegel and White 1975). For landmark knowledge, places and objects are perceived, however are not mentally set in spatial relationship to one another. Route knowledge develops thereupon, whereby objects and spatial patterns are interconnected along an established route. Survey knowledge is the most complex form of spatial perception and makes free spatial navigation possible without going astray in the process (Kettunen 2014:16f).

The focus of the present study is on landmark knowledge. This makes the recognition of objects in the surroundings possible and is the core of spatial knowledge (Ishikawa and Montello 2006). Although landmark and route knowledge develops nearly in parallel, landmark knowledge forms the first step in acquiring spatial knowledge and is thereby the foundation for all further forms of spatial knowledge (McNamara et al. 2008:158f). Landmark knowledge that is stored in long-term memory additionally activates and improves spatial navigation behavior (Gillner and Mallot 1998; Liverance and Scholl 2015).

A distinction is made between global and local landmarks, whereby global objects serve for rough directional orientation and local landmarks make an exact direction change possible at, for example, a route intersection (Michon and Denis 2001; Ross et al. 2004; May et al. 2003). Landmarks stand out due to their visual, semantic, and structural features (Raubal and Winter 2002; Klippel and Winter 2005), which contrast with the immediate environment and are stored on the mental map (Richter and Winter 2014). They characterize geographic locations, construct pathways when there is a change in direction to a destination and are therefore identified as landmarks (Duckham et al. 2010:2). The acquisition of landmark knowledge during navigation has been investigated in numerous studies. It thereby became clear that navigational errors decrease through the use of landmarks and navigation performance is optimized for pedestrians (Lee et al. 2002; Rehri et al. 2010; Kettunen 2014:17ff).

Mobile eye tracking systems are used in order to understand navigation behavior from people's perspective (Kiefer et al. 2012; Ohm et al. 2014). Eye movements provide insights into cognitive processes while solving problems in a spatial context (Liversedge et al. 2011). The perception of landmarks is to be afforded particular attention in order to understand their use in navigation (Kettunen 2014:88). With regard to spatial cognition research, the use of eye tracking should be of central importance in order to create user friendly maps (Kettunen et al. 2015:12). The current technical advances in eye tracking technology now make it possible to investigate map elements and contents in detail (Kettunen et al. 2015). To efficiently visualize the contents of a map, the analysis of cognitive processing is essential. This provides knowledge concerning perception and efficiency when using cartographic products (Dickmann et al. 2015:273). The following research questions are addressed in this study: (1) Are objects fixated longer and more frequently when there is a change in direction? (2) Are objects that are focused more frequently and for longer time transferred to the mental map? (3) How often is the map used as a navigational aid when finding the way? (4) What influence does the presentation of landmarks on a map have on eye movements and thereby on the mental representation of the environment? (5) Is the sustainability of the mental map improved by representing landmarks on maps?

2 Visual Perception and the Mental Representation of Landmarks

Two aspects are relevant to the visual and mental representation of points of spatial orientation. Starting with the viewer perspective, a landmark object (1) is perceived visually using the eyes and is subsequently (2) processed to mind. The information is then recorded in short-term memory and, based on its significance, is stored into long-term memory (Brady et al. 2008). Information is stored mentally in short-term memory for a few seconds up to a couple of hours. The mental storage and cognitive processing for objects in short-term memory is thereby limited (Brady et al. 2011). According to Miller (1956), the storage capacity is limited to seven objects ($\pm$ two). More recent contributions investigated and discussed Miller's results critically and came to the conclusion that the mental storage capacity of short-term memory is reached after an average of approximately four objects ($\pm$ one) (see among others Luck and Vogel 1997; Cowan 2001; Alvarez and Cavanagh 2004; Xu and Chun 2009; Brady et al. 2011). Investigations of the visual working memory go beyond simple numbers of objects and items remembered (Brady et al. 2011). When storing visually recorded objects, information is combined for storage in short-term memory (chunking) (Luck and Vogel 1997). Thereby, the capacity of the visual working-/short-term memory is determined by the integration of objects and their properties (Luck and Vogel 1997). In this way, for example, the number of represented objects decreases as the information content of the respective object increases (Alvarez and Cavanagh 2004). For spatial objects, first their position is stored and then the details of the object are recorded (Xu and Chun 2009).

Over a longer period of time—a couple of days, weeks up to years—information is stored in long-term memory, which has far greater storage capacities than short-term memory (Brady et al. 2008). Landmark studies, for example, have shown that of 10,000 image scenes that were contemplated for just a few seconds, test persons were able to determine, with 83 % certainty, which of two scenes had already been seen (Standing 1973). More recent studies by Brady et al. (2008) additionally show that numerous details from image scenes are stored in long-term memory and recognized at an accuracy of approximately 90 %.

The mental storage capacities in short- and long-term memory are relevant to perception and the representation of landmark objects in real environments. The selection of the information takes place on the basis of its significance to the navigation procedure and the description of the route. Thereby, the task as set is decisive for the mental representation of objects (Brady et al. 2011). Because landmark objects that contrasts with the environment have a higher content of information, they are identified as salient for task achievement, are stored mentally, and used for navigation and routing instructions (Raubal and Winter 2002; Caduff and Timpf 2008; Richter and Winter 2014:60).

Eye movements deliver a wealth of information for comprehending perception and cognition (Borji and Itti 2014:1). Eye tracking technologies are used to record and analyze mental processes. Buswell (1935) and Yarbus (1967) laid the corner stone in order to analyze the relationship between eye movements and cognitive processes. Their investigations showed the relationship between eye movements and a precise task (Borji and Itti 2014). Eyes search for selective information in order to accomplish a task. Number of fixations and their duration are indicators for understanding and analyzing mental processes while accomplishing a task (Rayner 2009). The Eye-Mind Assumption serves as a bridge between visual perception and the mental processing of landmarks. This assumption states that objects that are fixated longer are also processed and stored mentally (Lust and Carpenter 1980).

The Eye-Mind Assumption thereby forms the foundation in order to analyze cognitive processing from the perspective of the observer (Koesling 2003:10ff). Extended fixation on an object corresponds to intense mental processing (Holmqvist et al. 2011). Although the approach of the Eye-Mind Assumption is open to discussion (Viviani 1990; Andersen et al. 2004), the connection between eye movements and visual attention are unequivocal and offer an opportunity to understand mental processes (Salvucci 1999:19; Borji et al. 2015). The Eye-Mind Assumption in particular is therefore quoted as a theoretical foundation (Fabrikant et al. 2010:6) and is used empirically (Fabrikant 2005; Raubal 2009; Servatius 2009; Ooms et al. 2012; Brus et al. 2015) for the investigation of spatial cognition in geography, geo-information and related fields. Mobile eye tracking technologies have been used for a considerable time in order to empirically investigate pathway-finding processes and navigation behavior (Kiefer et al. 2012, 2014). Based on the work of Wiener et al. (2012), Kiefer et al. (2012) investigated the path-finding behavior of persons in real environments with a mobile system. In their pilot study, the use of maps and landmarks was investigated in order to find a pathway in the inner city of Zurich. It was clear that the participants in the study perceived and made use of landmarks to find the path, despite using various path-finding strategies.

Taking into consideration aspects of neuroscience, visual attention and thereby eye movements are directed at prominent objects that, based on their features, are differentiated from the close environment and are thereby salient (Raubal and Winter 2002; Klippel et al. 2005; Caduff and Timpf 2008). Accordingly, based on object salience and a precise task noteworthy objects are focused (Ballard and Hayhoe 2009:2).

Hamid et al. (2010) investigated the usage of landmarks in virtual navigation processes with a table mounted eye tracker. The results show that test persons maintain a view pattern while recording landmarks and observe salient objects longer at intersections than between two intersections. Accordingly, the removal of objects that are observed only briefly had no influence on navigation performance, while the removal of objects with a longer duration of observation had a negative effect on path-finding (Hamid et al. 2010:7). This same approach was followed by Andersen et al. (2012). A table mounted eye tracking system was used in order to investigate the use of landmarks in path-finding processes in virtual surroundings. They studied gender-related differences in navigational behavior and in path-finding strategies. Differences arose in environments without landmarks, but were fewer with an increasing number of landmarks in the environments.

Furthermore, Liao et al. (2016) explored the effect of digital 2D and 3D maps on spatial knowledge acquisition and decision making of pedestrians in a virtual environment using eye tracking technology. In their study 20 participants were instructed to fulfil three tasks: self-localization, map reading and memorizing and navigation along a predesigned route. The results show that landmarks in the 3D geo-browser are significantly fixated more often, and therefore supports the acquisition and recall of spatial knowledge in an unfamiliar urban environment.

In indoor settings, Ohm et al. (2014) investigated landmark-based navigation behavior with mobile eye tracking devices. They discovered that landmarks that are attributed functionality (e.g. doors, stairs etc.) have a greater significance to successful path-finding than visual attributes. Viaene et al. (2014) further investigated landmark-based navigation behavior indoors and discovered that landmarks are identifiable based on fixations recorded with eye tracking systems. They also showed that, in comparison to Think-Aloud protocols, eye tracking data are more helpful for the identification of landmarks (Viaene et al. 2014).

A mobile eye tracking system was used by Kiefer et al. (2014) to investigate the process of self-localization. They discovered that test persons directed their attention longer on landmarks on the map and aligned them with objects in the surroundings during successful localization of their own position. Equally, more jumping in fixation between maps and the surroundings were determined to be present in successful localization.

3 Methods

3.1 Study Area

The Spandau old town of Berlin was selected as an investigational area. The area at the border of the city of Berlin is suitable for an investigation because it has features similar to a laboratory. It has an insular character and is delimited by structural surroundings. Landmarks outside the area of investigation are not relevant to successful navigation. The entire area is a pedestrian zone, whereby street traffic can be excluded as distracting or as a safety concern component during the navigation (Kiefer et al. 2012). Additionally, the Spandau old town features façades that are salient and easy to name in order to describe pathways. For directional decisions, the test persons thus have many potential landmarks from which to select.

3.2 Materials

Two test groups with respectively ten participants in the investigation had the task of destination-oriented navigation through the pedestrian zone of the Spandau old town with the aid of a map with street names and six landmarks (see Fig. 1). This group is described in the following as the landmark group. The control group navigated with the help of a map that offered only street names without landmarks (see Fig. 2).

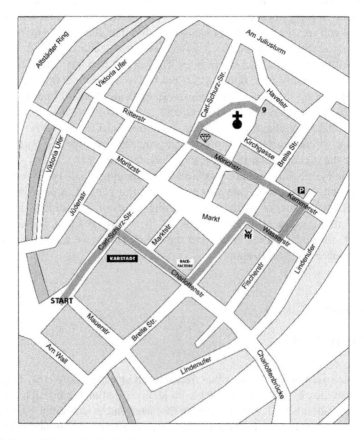

Fig. 1 Map with landmarks for landmark group

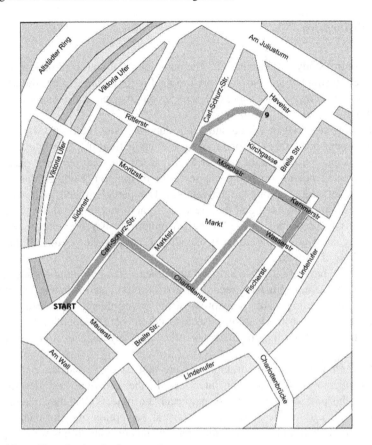

Fig. 2 Map without landmarks for control group

The selection of landmark icons as an additional aid for navigation were based on studies from Hamid et al. (2010) and Janzen and Jansen (2010). They found that landmark at decision point are used preferably during a navigation tasks comparing to landmarks between decision points. Furthermore, Lovelace et al. (1999) showed that global landmarks do not play an important role when describing a route. The localization of the landmarks on the map is based on the studies of Röser et al. (2012). They found that, for decision points, landmarks that appear in the inner angle of a route are used preferably.

The cartographic design is oriented to the results of studies by Elias and Paelke (2008) and Tom and Tversky (2012). Elias and Paelke (2008) showed that an icon-like presentation of landmarks is preferred. The results from Tom and Tversky (2012) make clear that the most associative and lively representational forms are useful in order to support memory of the objects.

3.3 Procedure

A case-control study was carried out. The participants of both landmark and control group were instructed to follow the predesigned route and to explore the area. The length of the pathway was approx. 1 km. Seven changes in direction were to be carried out.

The experiment began with a short introduction about the research project and the experiment procedure. Before starting the navigation, the mobile eye tracking device was put on and calibrated for each participant. Furthermore, participants were instructed not to touch the glasses after calibration. Both groups received identically set tasks, before the navigation. The predetermined path was to be walked destination-oriented and then recalled in an interview. In order to have the test persons focus on the objects in the surroundings, they were instructed at the preliminary stage not to use street names during the interview, but only landmark objects along the route.

During the navigation procedure eye movements (fixation duration and fixation counts) of the participants were recorded. It was investigated, whether various map contents influence the visual perception of the environment, and whether objects that are visualized on the map are also focused on more frequently. Additionally, the duration of observing the map and the frequency of use of the map were investigated.

After the navigation, the participants were requested to recall the pathway and the remembered landmarks in an interview. In the interview they described the extent of the path on a featureless map and the objects that they remembered. Because the focus of the study was on visual perception and the mental representation of landmark objects, the study participants of both groups were requested not to use street names to describe the path.

In order to check whether significant differences existed in the mental representation of the surroundings and the landmark objects in long-term memory, a memory test was carried out with the test persons two weeks after the navigation procedure. Twenty-three objects from the Spandau old town were recorded at day time frontally with a camera. The objects could be seen along the route (see Fig. 3). Because a brief, visual impulse is adequate to stimulate memories in the long-term memory (Brady et al. 2008), every image was shown to the participants in the investigation for 20 s on a video projector. The test persons were requested to draw the pathway walked onto a featureless map from memory and then to correctly locate the shown objects. Alternatively, there was an opportunity to remark that they remembered the object but could not locate it, or to indicate that there was no memory of the object.

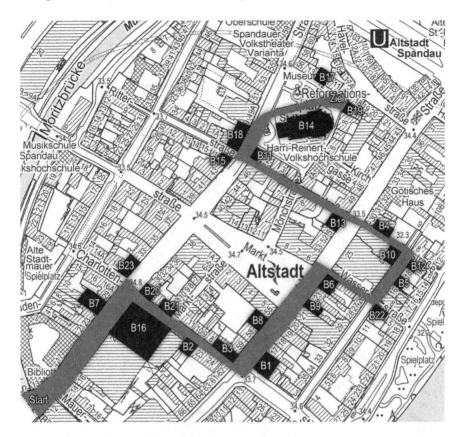

Fig. 3 Objects, which were shown during the memory test

3.4 Apparatus

The Tobii Glasses 2 eye tracking system was used for this study. The eye tracking sensors have a recording frequency of 50 Hz, which is an appropriate sample frequency to record both fixation counts and fixation duration (Andersson et al. 2010). The binocular recording technique of the device records the corneal reflection generated by an infrared light source in relation to the dark pupils. Depending on the direction of view, the relative position of the center of the pupils and the corneal reflection both change, whereby the direction of viewing is determined.

To evaluate the eye tracking data, all objects that were visually acquired by the test persons were defined as Areas of Interest (AOI) along the pathway (see Fig. 4). Every AOI corresponds to a building object. Additionally, the duration and frequency with which the map was observed as a navigational aid were evaluated. The

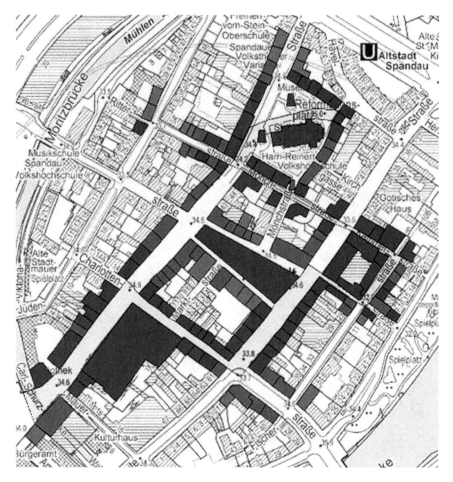

Fig. 4 Area of Interests (AOI) to allocate fixation counts and fixation duration

fixation results (fixation duration and fixation counts) on an object in the environments were allocated to the defined AOI with the aid of analysis software and were subsequently provided as a data table.

In order to identify the duration and numbers of fixations, the I-VT fixation filter (velocity threshold identification) for "Attention" on the Tobii Pro Glasses analyzer software was used. The filter determines the speed of the eyes (velocity) in order to get from one point to another. One hundred degrees per second was established as the limit value for the speed of movement. Eye movements that display a greater speed of rotation are identified as saccades (Salvucci and Goldberg 2000:73). A duration of at least 100 ms was defined (Hornof and Halverson 2002) as the limit value for fixations. The parameters are summarized in the following Fig. 5.

Fig. 5 Settings of I-VT filter

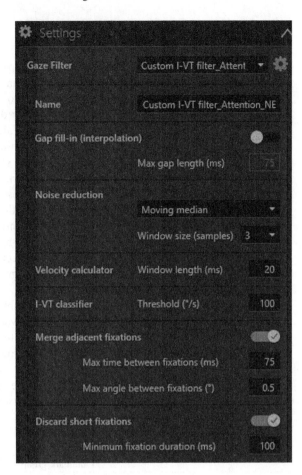

4 Results

The 20 participants in the investigation were an average of 24.5 years old, with an age range of 21–31 years of age. Sixty percent of the participants were male. In the control group were 3 women and 7 men. The landmark group consisted of 5 women and 5 men. The test persons walked the predetermined route through the Spandau old town of Berlin on average in 11:20 min. The ten participants in the landmark group required an average of 10:57 min and thereby almost a minute less than the control group, at 11:42 min. All 20 participants in the investigation took part in the subsequent interview.

4.1 Results: Interview

According to the task, the test persons in the two groups navigated independently of one another along the route from the start to the destination. The test persons were interviewed after the navigation procedure in order to answer questions about the landmarks represented in short- term memory. They described the path that was walked based on landmark objects and without the aid of street names. In total, during 20 interviews, 37 objects were named 153 times. Thereby, 27 from the landmark group and 24 from the control group were stored in short-term memory (see Table 1). Per test person, on average 7.7 landmarks were named per interview. These objects are to be identified as landmarks represented in short-term memory and thereby as components of the cognitive map of the area of investigation.

The comparison of both groups shows that the number of recalled landmark objects are dispersed differently. While the control group shows a bigger interquartile range (IQR) of number of recalled object ($Q_1 = 4.5$, $Q_3 = 9.5$), the interquartile ranges from 7-8 objects in the landmark group. However, the landmark group has outliers in both directions (see Fig. 6).

Because only the data of the control group are normally distributed, the Kruskal-Wallis test for non-parametric, independent variables was used in order to test the differences between the landmark and the control group. The result showed no statistically significant difference between both groups.

Figure 7, in the grey boxes, shows the frequency distribution of the landmark recalls along the extent of the pathway, as well as the seven decision points. Eight peaks in the diagram are notable, showing more than two recalls. From Fig. 7 it can

Table 1 Landmark recall during the interview

	Landmark group	Control group
Number of different landmarks	27	24
Landmark recalls per interview	8.3	7.0

Fig. 6 Dispersion of recalled landmarks during the interview

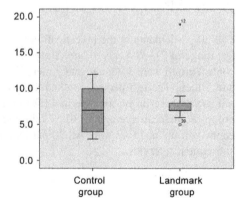

Investigation of Landmark-Based Pedestrian Navigation ... 117

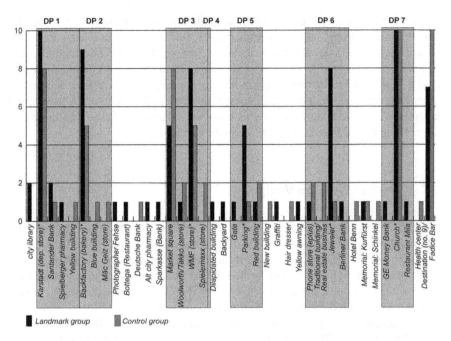

Fig. 7 Landmarks recalls along the route (*objects are visualized as icons on the map)

also be derived that the control group named the market place and the destination location more frequently in the interview than the landmark group. The diagram additionally shows that the landmark group primarily named the six landmarks that are represented on the map.

At decision points, the landmark group named approx. 77 % of a total of 83 landmarks that were named in the interview. For the control group, approx. 76 % of all named landmarks are located at a decision point. The following Fig. 8 shows the seven decision points. It becomes clear that, with both groups, primarily landmarks that were in the inner angle of the route were mentally represented. For the landmark group, it was 70 % of the recalls, for the control group approx. 57 %.

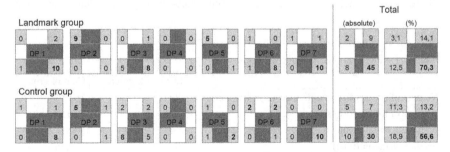

Fig. 8 Landmark recalls at decision points (DP). The route is visualized in *dark grey*

At almost all changes in direction, the landmark group named more landmarks than the control group. Only at decision point 3 were more objects named in total by the control group. Aside from that, it can be observed that, at decision point 6, the landmark group named twice as many as compared to the control group. Additionally, it is noticeable that neither group had a representation of a landmark object at decision point 4. Both data series are distributed normally. As a consequence, the analysis of variance (One Way ANOVA) was applied and established that the groups did not differ significantly ($F(1.12) = 0.326$; $p = 0.579$).

4.2 Results: Eye Tracking

The objects that were recorded visually were derived from the eye tracking data. The number of fixation counts and the fixation duration of contemplating an object along the extent of the pathway were recorded and evaluated for the analysis. As a result of varying light conditions in the infrared frequency range due to daily time and weather conditions, the sample rate for the eye tracking recordings varied between 47 % and 77 % and was, on average, 65.8 %.

In sum, the 20 test persons fixated the building objects along the predetermined route for 10.1 min. The remaining fixations were directed at passers-by, parked cars, the map or the route in front of them. In total, 118 objects were fixated at least one time. In sum, 99 objects were fixated 1,885 times by the landmark group. The control group fixated 105 objects 2,015 times. The participants of the landmark group fixated the objects in the environment in total approximately 4:45 min. The test persons in the control group fixated the surrounding objects longer, at 5:20 min (see Table 2).

The Shapiro-Wilk test shows that the fixation count and the fixation duration on 104 of 118 facades are not normally distributed. In order to determine whether the two groups are different in viewing behavior, a test for non-parametric, independent variables was used. The statistical comparison of the fixation counts and fixation durations by means of the Kruskal-Wallis test show that the viewing behavior of the test groups was similar. Statistically significant differences were only determined in three visually recorded objects. At the decision points, no statistically notable differences were determined (see Fig. 9). The results thereby show that various map

Table 2 Fixations on objects along the route

	Landmark group	Control group
Number of objects that people focused at least once	99	105
Number of fixations (total)	1,885	2,015
Average fixation counts per object per participant	3.48	5.38
Fixation duration (total)	284.72 s	319.46 s
Average fixation duration per object per participant	524 ms	850 ms

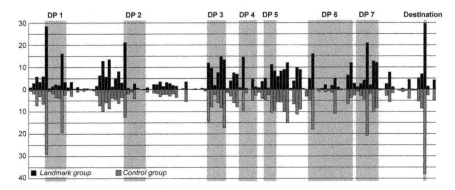

Fig. 9 Fixation counts of both groups per focused object along the route

contents did not influence observational behavior on the surrounding geographic objects or the visual perception of the landmarks in the environment.

In total, barely half of all fixation counts (42.8 %) were directed onto the 37 landmark objects that were named in the interview: 735 (39.0 %) for the landmark group and 933 (= 46.3 %) for the control group.

Upon distinguishing between the objects named in the interview and all the other fixated objects, it was shown that fixation count and fixation duration are distributed differently. Both investigational groups fixated the objects that they named in the interview on average clearly more frequently than all other objects. On average, the represented landmarks were focused by the participants between 1.5 and 3.0 times as often compared to all other objects (see Table 3).

The box plots below show that fixation duration is similar on both, objects that were recalled (IQR Landmark group = 5.6) (IQR Control group = 7.0), and objects which were visually perceived but not named in the interview after navigation (IQR Landmark group = 2.7) (IQR Control group = 2.8). Additionally, the figures confirm that the fixation duration on recalled objects is longer than on other objects which are not mentally represented (see Figs. 10 and 11).

The map fixation result was evaluated in order to determine how often and for what duration the map was used as a navigation aid. The use of the map clearly shows the difference between both test groups. While the landmark groups contemplated the map on average 16.8 s for the entire navigation procedure, the control

Table 3 Average fixation events per participant

	Landmark group		Control group	
	Fixation counts	Fixation duration (ms)	Fixation count	Fixation duration (ms)
On a Landmark*	2.45	409	3.89	599
On other objects with at least one fixation	1.66	236	1.29	209

*Objects named in the interview

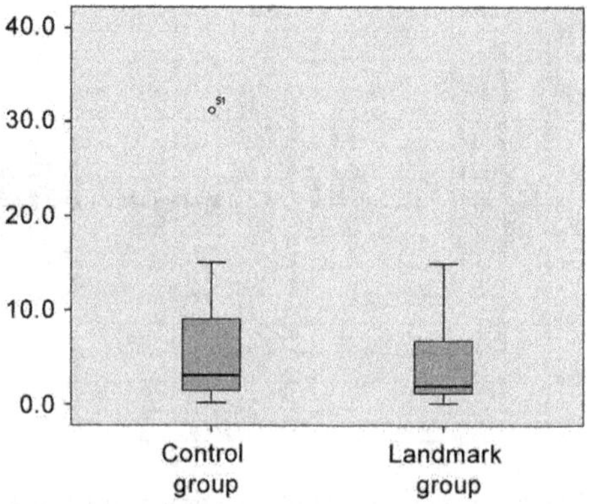

Fig. 10 Fixation duration (in sec.) on a landmarks (objects named in the interview)

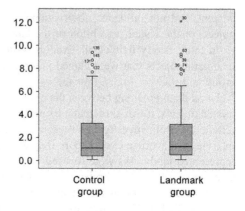

Fig. 11 Fixation duration (in sec.) on objects with at least one fixation, and which were not named in the interview

group used the map almost twice as much, at 29.5 s. The statistical comparison of both groups, however, shows that the fixation results (counts and duration) on the map are not normally distributed and, according to the Kruskal-Wallis test, are not significantly different from one another ($p_{counts} = 0.364$ and $p_{duration} = 0.257$).

As Fig. 12 shows, the control group used the map more often during navigation task than the landmark group. The map was fixated longer by the control group (IQR = 29.1 s), while the landmark group used the map more efficient as a navigational aid (IQR = 16.5 s). Additionally, the control group used the map more often as an aid for navigation than the landmark group (see Fig. 13). Comparing groups statistically, the visit counts on the map are normally distributed (Shapiro-Wilk test) and, according to a One-Way ANOVA, however, are not significantly different from one another ($F_{(1, 18)} = 1.817$, $p = 0.194$).

Alternatively, the heat map analysis is helpful in the visual interpretation of the duration of fixation (Dickmann et al. 2015). In the present study, differences in the fixation pattern have been determined, as the following figures show. Hereby, the fixations for all test persons in a group in a radius of 75 pixels (approx. 2 cm) around the fixation point are summed. Red and orange shades show high fixation

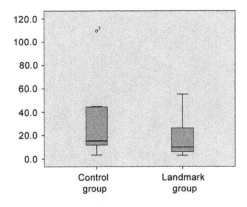

Fig. 12 Fixation duration (in sec.) on map during navigation

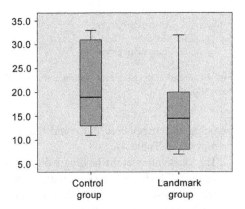

Fig. 13 Number of map visits during navigation

densities and green shades are low values. The landmark group primarily fixated on the presented landmark objects. The control group, in contrast, directed focus on street names. Additionally, it becomes clear that more fixations occurred off of the presented pathway (see Fig. 14).

4.3 Results: Memory Test

In order to investigate the sustainability of the landmark knowledge, a memory test was carried out two weeks after the navigation task. Sixteen of 20 test persons took part: nine from the landmark group and seven from the control group. As the following table shows, the test persons in the landmark group located approx. 6.2 of 23 objects correctly in a featureless map. In the control group it was 3.3 objects, approx. 47 % fewer. Additionally, the error rate for the control group was clearly higher. While approx. 4.7 objects were incorrectly located, for the landmark group it was only 1.4. The differences between both groups are slight for objects that the

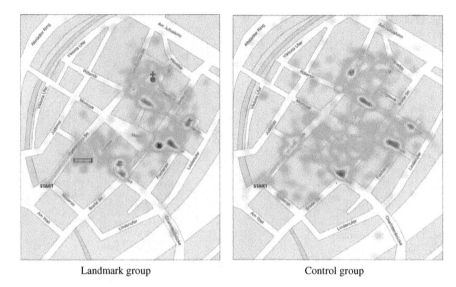

| | Landmark group | Control group |

Fig. 14 Fixation distribution (fixation counts) of all participants of each group on the map during navigation

participants remembered but could not locate and objects for which there was no memory (see Table 4).

The localization of the landmark objects from long-term memory is different in the two groups. The landmark group remembers more landmark objects, and located them correctly on the map (see Fig. 15). The number of correctly localized landmarks is normally distributed in both groups ($p_{OL} = 0.087$, $p_{ML} = 0.837$). The analysis of variance (One-Way ANOVA) showed that both groups were different significantly statistically ($F_{(1,14)} = 5.593$; $p = 0.03$). Because the number of falsely located landmarks are not distributed parametrically ($p_{OL} = 0.275$; $p_{ML} = 0.007$), the Kruskal-Wallis test was used in order to determine whether a statistically significant difference existed. The asymptotic significance is 0.05 ($p = 0.046$) and is thus conspicuous.

Table 4 Results of memory test (per participant)

	Located correct	Located wrong	Memory without location	No memory
Landmark group (9 participants)	6.2	1.4	4.8	10.8
Control group (7 participants)	3.3	4.7	5.3	10.4

Fig. 15 Results of memory test: correct located landmarks

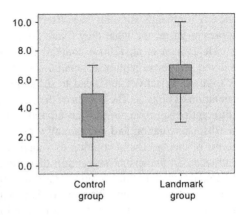

Table 5 Spearman's rank correlation analysis between landmark recalls and eye movements

	Landmark-recall/Fixation-duration (ms)	Landmark-recall/Fixation-count
Landmark group	0.670**	0.674**
Control group	0.584**	0.601**

(** $p < 0.01$)

4.4 Statistical Relationship Between Fixations and Landmark Recall

In order to verify the Eye-Mind Assumption (see among others Just and Carpenter 1980; Bollmann et al. 1997), a statistical relationship between fixation duration, fixation counts and landmark recalls was investigated. The hypothesis that objects focused longer are mentally processed more frequently was tested. Neither landmark recalls nor fixation counts or fixation duration is normally distributed ($p > 0.05$). Fixation frequency for a landmark object and landmark recognition correlate positively. Objects that are observed more frequently or longer are named more frequently in the interview. The longer an object is focused, the greater the probability that it will be recorded mentally (see Table 5).

5 Discussion

The present work is further evidence that eye movements and fixations reflect mental perception and processing of spatial information while solving a spatially related task. To conclude the nature of mental perception processes directly from eye movements is open to discussion because it depends, among other things, on the environmental stimulus, the observer and the actual question being asked, where a person gazes and what they fixate (Borji and Itti 2014:14).

McNamara et al. (2008) contend that the observer perspectives is to be considered in the perception of spatial objects and their relationship to one another. Various perspectives thus lead to different perceptions and thus a different representation of objects. The results of the present study, however, show that despite the different perspectives of the participants, on average identical landmark objects are used for navigation and are mentally represented. The significance of an object in terms of the successful solution to a navigation task thus appears to have a greater influence on the spatial perception than the individual perspective of the person.

Fixation counts and fixation duration represent foveal vision. Peripheral perception is not recorded. The work of Strasburger and Malania (2013) from the field of reading research shows that the perception and representation of objects in the peripheral visual field influence information uptake. Contents in the periphery can be acquired distorted or represented incorrectly. It is further known that intense color contrasts and rapid movements in the peripheral visual field can be acquired (Duchowski 2007). Landmarks are objects that are visually salient and contrasts with the close environment and are thereby also recorded by the peripheral visual field. The actual influence of the periphery on the spatial orientation and the perception of objects during landmark-based navigation, however, would be an area for analysis in future studies.

The acquisition of data by eye tracking systems under real conditions varies depending on the incident light. Ordinarily, the sample rate is lower than in a laboratory environment (Andersson et al. 2010). The range in the present study is between 47 and 77 %, with an average value of approx. 66 %. This results in the fact that not all eye movements can be evaluated statistically. According to Andersson et al. (2010), the recording values are dependent with the recording frequency of the eye tracking system. In their study, they investigated fixation and saccades during a reading procedure. The recording frequencies that were compared were between 1250 and 50 Hz. It was thereby clear that the identification of saccades and fixations that are between 1250 and 250 Hz are hardly different. Below 50 Hz, the recording of the saccade duration is critical, the recording of the fixation duration, in contrast, is not problematic (Andersson et al. 2010:9). In the present study, fixations at 50 Hz were recorded and evaluated. Saccades were not an object of investigation.

The Think-Aloud method and interviews are available in order to confirm what eye movements suggest. A combination of eye tracking and Think-Aloud protocols is recommended in order to determine additional and supplemental information on mental processes (Ruckpaul et al. 2015). This method, however, has the disadvantage that the quality of the Think-Aloud protocol depends on the abilities of the participants, and additionally only slightly provides more information for landmark identification (Viaene et al. 2014). Therefore, the present study made use of eye tracking technology in connection with a subsequent interview.

The investigational environment are a real, unknown urban setting with numerous, influencing factors. Identical conditions for each test person are only approximately repeatable, and hardly correspond to a lab environment. However, there are limitations in a lab environment in the analysis of patterns of fixation. Avoiding obstacles or the avoidance of collisions with other pedestrians are tasks that are difficult to simulate in a lab (Fotios et al. 2015). With empirical investigations in a real environment, the behavior of test persons is investigated under real conditions. Hereby, for example, test persons tend to assume their natural gait and react instinctively to unexpected events or barriers (Foulsham et al. 2011).

Based on a literature review, landmark objects were selected and presented for the landmark group in the fundamentals of the map. The selection of landmarks is in line with the study results of Lovelace et al. (1999), Janzen and Jansen (2010) and Röser et al. (2012), according to which, when there is a turning procedure, landmarks in the inner angle of the route are used preferably. Von Stülpnagel and Frankenstein (2015) supported these results. They show that the striking structural feature of a landmark has greater significance than visual salience. The present study confirm those results. Because the control group tended to represent the same landmarks and use them to describe the route as the landmark group, it is clear that the structural properties of landmarks assume a central significance in navigation.

To investigate the long-term memory of landmark knowledge a selection of images along the route was shown during a memory test. However, the experiment could be further improved by showing all objects along the route. According to Brady et al. (2008) thousands of object can be stored in visual long-term memory with remarkable fidelity. In future research, the number and the sequence of objects shown during a long-term memory test need to be investigated in further detail.

Furthermore, participants of the landmark group were able to recall landmarks with higher fidelity during the long-term memory test. Particularly those objects which were visualized as icons on the map during the navigation task were located correctly. However, the visualization of the icons on the map were different from the images shown during the memory test. According to Brady et al. (2008), the representation of individual item in long-term memory is highly structured on multiple levels. In addition, items are mentally stored by chunks of information (Cowan 2008). Accordingly, it can be assumed that objects are stored by a combination of functionality for a navigation task, their visual saliency in the environment and their presentation on a map. Hence, it should be investigated, if participants' landmark knowledge is a result of the provided map, the experience during the navigation process or a combination of both. Regarding the visualization methods of landmark icons on maps, the study results of Franke and Schweikart (2016) indicated that the learning process of landmarks along a route depends more likely on the task than on the cartographic presentation on a map. However, future research should address the link between the mental storage of landmark item in long-term memory, their salience in the environment and the visualization on the map.

Sex differences were not observed and go beyond the scope of the present study. However, previous studies investigated gender differences in detail (see Rahman et al. 2003; Gallagher et al. 2006; Voyer et al. 2007). Gallagher et al. (2006) discussed methodological issues when investigating sex differences in object location memory, and found no overall sex differences. Edler et al. (2014)

investigated the influence of map features on the recall of learned object locations. Their results did not reveal significant differences in spatial orientation. On the other hand, Voyer et al. (2007) investigated gender and age differences in object location memory tasks. They observed small to intermediate gender differences in object identity and object location memory under most circumstances, where object location memory often show tendencies in favor of females (Wolbers and Hegarty 2010). Since various researchers came to different results, further research should address gender differences in the context of object location memory.

The present study further shows that, by visualizing landmarks on a map, objects from the environment are more frequently stored in short-term and in long-term memory. Statistically significant differences between both groups, however, were determined only with representation in long-term memory. It thereby becomes clear that cartographic landmark representation has little positive influence on landmark knowledge stored short-term. The sustainability of the mental map through the presentation of landmarks is supported by evidence. The recognition of objects as well as spatial localization on a map are supported by landmark presentation in maps. Equally, erroneous localizations are significantly decreased. The mutual influence of localization and objects represented by visual properties should be investigated in greater depth in future studies, because numerous studies show that objects in association with spatial information are encoded (Brady et al. 2011). The results of the present study suggest that the visual properties of an object allow recognition and are associated with the location at which the object is found. Alternatively, whether visual properties of an object can be concluded based on the location should be tested.

6 Conclusion and Outlook

The representation of landmarks on a map leads to a simplification of a navigation process, because a map is required less often as an aid to orient spatially. At the same time, landmark presentation in maps allows more environmental objects to incorporate sustainably into the mental map. The differences between the two groups in terms of the mental storage and correct localizations from long-term memory are statistically significant. The landmark groups remembered more objects that were correctly localized and thereby made fewer errors compared to the control group. As a central component of the cognitive map, cartographic contents supports the expression of landmark knowledge. In future studies, an investigation should be carried out in more depth as to what influence the time span between the navigation procedure and the memory test has. A comparison of the object salience and the representation in long-term memory should be investigated.

The study further shows that frequently focused and longer fixated objects are represented mentally at a higher probability. When landmarks were represented on maps, the representation of the objects on the mental map takes place at a higher probability. The present study therefore contributes to the understanding of landmark-based pedestrian navigation from the perspective of the individual. The use of a mobile eye tracking system provides insight into navigational behavior and the mental representation of landmarks. The study thereby ties the study together with the work of Hamid et al. (2010) and Kiefer et al. (2014) and shows that, when perceiving landmarks, the duration of fixation and the fixation count follow a clear pattern.

Gaze behavior relative to surrounding objects during the navigation procedure and the visual acquisition of landmarks, however, is not influenced by the two different contents of the maps. Both test groups have similar fixation behavior onto buildings, whereas the cartographic presentation of landmarks supports the mental representation of the environment more strongly. The presented landmark objects are visually recognized on the map, aligned relative to the environment and identified. The two test groups thereby used the map differently. The test persons in the landmark group used the map less frequently and registered the map contents more rapidly as compared to the control group. The period of observation and the frequency with which the map was used by the control group are approximately 1.7 times higher than for the landmark group. In order to walk the route, the landmark group used the map on average 16.6 s/test person. The control group clearly fixated on the map longer (29.9 s/test person). It thereby becomes clear that the representation of landmarks simplified the navigation process and arranges the use of the map more effectively. In order to investigate the navigational performance in more depth, further parameters, such as the number, duration and reasons for stopping before and after a decision point should be investigated in future studies.

The goal of the study, the identification of the contents of maps that provide for a sustainable expression of the mental map, was realized. At the same time it was determined that the use of a map is more efficient if landmarks are visualized on a map. The navigation performance of a pedestrian in unfamiliar environments is improved by landmark presentation, and the time spent using navigational assistance is optimized.

References

Alvarez G A, Cavanagh P (2004) The capacity of visual short-term memory is set both by visual information load and by number of objects. Psychological Science, 15, 106–111.

Andersen N E, Dahmani L, Konishi K, Bohbot V D (2012) Eye tracking, strategies, and sex differences in virtual navigation. Neurobiology of learning and memory, 97(1), 81–89.

Andersson R, Nyström M, Holmqvist K (2010) Sampling frequency and eye-tracking measures: how speed affects durations, latencies, and more. Journal of Eye Movement Research, 3(3), 6.

Ballard D H, Hayhoe M M (2009) Modelling the role of task in the control of gaze. Visual cognition, 17(6–7), 1185–1204.

Barkowsky T (2001) Mental processing of geographic knowledge. In Spatial Information Theory, 371–386. Springer Berlin Heidelberg.

Bollmann J, Heidmann F, Johann M (1997) Kartographische Bildschirmkommunikation: Methodische Ansätze zur empirischen Untersuchung raumbezogener Informationsprozesse. Aktuelle Forschungen aus dem Fachbereich VI. Geographie/Geowissenschaften, 267–284.

Borji A, Itti L (2014) Defending Yarbus: Eye movements reveal observers' task. Journal of vision, 14(3), 29.
Borji A, Lennartz A, Pomplun M (2015) What do eyes reveal about the mind?: Algorithmic inference of search targets from fixations. Neurocomputing, 149, 788–799.
Brady T F, Konkle T, Alvarez G A (2011) A review of visual memory capacity: Beyond individual items and toward structured representations. In: Journal of vision, 11(5), 4.
Brady T F, Konkle T, Alvarez G A, Oliva, A (2008) Visual long-term memory has a massive storage capacity for object details. Proceedings of the National Academy of Sciences, 105, 14325–14329.
Brus J, Vondrakova A, Vozenilek V (Eds.) (2015) Modern Trends in Cartography. Selected Papers of CARTOCON 2014. Springer Cham Heidelberg, New York, Dordrecht, London.
Buswell G (1935) How people look at pictures. Chicago: University of Chicago Press.
Caduff D, Timpf S (2008) On the assessment of landmark salience for human navigation. Cognitive processing, 9(4), 249–267.
Cowan N (2001) The magical number 4 in short-term memory: A reconsideration of mental storage capacity. In: Behavioral and Brain Sciences, 24, 87–185.
Cowan N (2008) What are the differences between long-term, short-term, and working memory? In: Progress in brain research, 169, 323–338.
Dickmann F, Edler D, Bestgen A-K, Kuchinke L (2015) Auswertung von Heatmaps in der Blickbewegungsmessung am Beispiel einer Untersuchung zum Positionsgedächtnis. In: Kartographische Nachrichten, 65 (5), 272–280.
Duchowski A (2007) Eye tracking methodology: Theory and practice (Vol. 373). Springer Science & Business Media.
Duckham M, Winter S, Robinson M (2010) Including Landmarks in Routing Instructions, in: Gartner, G. Journal of Location Based Services, volume 4, issue 1, 28–52. Taylor & Francis, London.
Edler D, Bestgen A K, Kuchinke L, Dickmann F (2014) Grids in topographic maps reduce distortions in the recall of learned object locations. PloS one, 9(5).
Elias B, Paelke V (2008) User-centered design of landmark visualizations. In Map-based mobile services, 33–56. Springer Berlin Heidelberg.
Fabrikant S I (2005) Towards an Understanding of Geovisualization with Dynamic Displays: Issues and Prospects. In AAAI Spring Symposium: Reasoning with Mental and External Diagrams: Computational Modeling and Spatial Assistance, 6–11.
Fabrikant S I, Hespanha S R, Hegarty M (2010) Cognitively inspired and perceptually salient graphic displays for efficient spatial inference making. Annals of the Association of American Geographers, 100(1), 13–29.
Fotios S, Uttley J, Yang B (2015) Using eye-tracking to identify pedestrians' critical visual tasks. Part 2. Fixation on pedestrians. Lighting Research and Technology, 47(2), 149–160.
Foulsham T, Walker E, Kingstone A (2011) The where, what and when of gaze allocation in the lab and the natural environment. Vision research, 51(17), 1920–1931.
Franke C, Schweikart J (2016) Mental Representation of Landmarks on Maps – Investigating Cartographic Visualization Methods with Eye Tracking Technology. In: Special Issue on Eye Tracking for Spatial Research in Spatial Cognition and Computation: An Interdisciplinary Journal. 1–19.
Gallagher P, Neave N, Hamilton C, Gray J M (2006) Sex differences in object location memory: Some further methodological considerations. Learning and Individual Differences, 16(4), 277–290.
Gillner S, Mallot H A (1998) Navigation and Acquisition of Spatial Knowledge in a Virtual Maze. Journal of Cognitive Neuroscience, 10(4), 445–463.
Golledge R (1999) Human Wayfinding and Cognitive Maps, in: Wayfinding Behavior, John Hopkins Press, 5–46.
Hamid S N, Stankiewicz B, Hayhoe M (2010) Gaze patterns in navigation: Encoding information in large-scale environments. Journal of Vision, 10, 1–11.
Holmqvist K, Nyström M, Andersson R, Dewhurst R, Jarodzka H, Van de Weijer J (2011) Eye tracking: A comprehensive guide to methods and measures. Oxford University Press.

Hornof A J, Halverson T (2002) Cleaning up systematic error in eye-tracking data by using required fixation locations. In: Behavior Research Methods, Instruments, & Computers, 34(4), 592–604.

Ishikawa T, Montello D R (2006) Spatial knowledge acquisition from direct experience in the environment: Individual differences in the development of metric knowledge and the integration of separately learned places. Cognitive Psychology 52(2): 93–129.

Janzen G, Jansen C (2010) A neural wayfinding mechanism adjusts for ambiguous landmark information. NeuroImage, 52(1), 364–370.

Just M A, Carpenter P A (1980) A theory of reading: From eye fixations to comprehension. Psychological review, 87, 329–354.

Kettunen P (2014) Analysing landmarks in nature and elements of geospatial images to support wayfinding. Academic Dissertation in Geoinformatics at Aalto University School of Engineering.

Kettunen P, Sarjakoski T, Sarjakoski L T (2015) Elements of Geospatial Images to Support Cognitive Tasks in Wayfinding. 27th International Cartographic Conference, ICC 2015.

Kiefer P, Giannopoulos I, Raubal M (2014) Where am I? Investigating map matching during self-localization with mobile eye tracking in an urban environment. Transactions in GIS, 18(5), 660–686.

Kiefer P, Straub F, Raubal M (2012) Towards location-aware mobile eye tracking. In Proceedings of the Symposium on Eye Tracking Research and Applications, 313–316. ACM.

Klippel A, Richter K F, Hansen S (2005) Structural salience as a landmark. In Workshop mobile maps.

Klippel A, Winter S (2005) Structural salience of landmarks for route directions. In Spatial information theory 347–362. Springer Berlin Heidelberg.

Koesling H (2003) Visual perception of location, orientation and length: an eye-movement approach. Dissertation. – Bielefeld, Germany.

Lee P, Tappe H, Klippel A (2002) Acquisition of landmark knowledge from static and dynamic presentation of route maps. KI, 16(4), 32–34.

Liao H, Dong W, Peng C, Liu H (2016) Exploring differences of visual attention in pedestrian navigation when using 2D maps and 3D geo-browsers. Cartography and Geographic Information Science, 1–17.

Liverance B M, Scholl B J (2015) Object persistence enhances spatial navigation: A case study in smartphone vision science. Psychological Science, 26, 955–963

Liversedge S P, Meadmore K, Corck-Adelman D, Shih S I, Pollatsek A (2011) Eye movements and memory for objects and their locations. Studies of Psychology and Behavior, 9(1), 7–14.

Lovelace K L, Hegarty M, Montello, D R (1999) Elements of good route directions in familiar and unfamiliar environments. In International Conference on Spatial Information Theory (65–82). Springer Berlin Heidelberg.

Luck S J, Vogel E K (1997) The capacity of visual working memory for features and conjunctions. Nature, 390, 279–281.

May A J, Ross T, Bayer, S H, Tarkiainen M J (2003) Pedestrian navigation aids: information requirements and design implications. Personal and Ubiquitous Computing, 7, 331–338.

McNamara T P, Sluzenski J, Rump B (2008) Human spatial memory and navigation. Cognitive psychology of memory, 2, 157–178.

Michon P-E, Denis M (2001) When and why are visual landmarks used in giving directions? In Montello, D. R., editor, Spatial Information Theory, volume 2205 of Lecture Notes in Computer Science, pages 292–305. Springer, Berlin.

Miller G A (1956) The magical number seven, plus or minus two: Some limits on our capacity for processing information. In: Psychological Review 63 (2), 81–97.

Ohm C, Müller M, Ludwig B, Bienk S (2014) Where is the Landmark? Eye Tracking Studies in Large-Scale Indoor Environments. In: 2nd International Workshop on Eye Tracking for Spatial Research co-located with the 8th International Conference on Geographic Information Science (GIScience 2014), September 23, 2014, Vienna, Austria.

Ooms K, De Maeyer P, Fack V, Van Assche E, Witlox F (2012) Interpreting maps through the eyes of expert and novice users. International Journal of Geographical Information Science, 26, 1773–1788.

Rahman Q, Wilson GD, Abrahams S (2003) Sexual orientation related differences in spatial memory. J Int Neuropsychol Soc 9: 376–383.

Raubal M (2009) Cognitive engineering for geographic information science. Geography Compass, 3(3), 1087–1104.

Raubal M, Winter S (2002) Enriching wayfinding instructions with local landmarks (243–259). Springer Berlin Heidelberg.

Rayner K (2009) Eye movements and attention in reading, scene perception, and visual search, The Quarterly Journal of Experimental Psychology 62:1457–1506.

Rehri K, Häusler E, Leitinger S (2010) Comparing the effectiveness of GPS-enhanced voice guidance for pedestrians with metric- and landmark-based instruction sets. Geographic information science, 189–203. Springer Berlin Heidelberg

Richter K F, Winter S (2014) Landmarks. GIScience for Intelligent Services. - Springer, Cham, Heidelberg, New York, Dordrecht, London.

Röser F, Krumnack A, Hamburger K, Knauff M (2012) A four factor model of landmark sali-ence-A new approach. In Proceedings of the 11th International Conference on Cognitive Modeling (ICCM), 82–87.

Ross T, May A, Thompson S (2004) The use of landmarks in pedestrian navigation instruc-tions and the effects of context. In: Mobile Human-Computer Interaction-MobileHCI 2004, 300–304. - Springer, Berlin, Heidelberg.

Ruckpaul A, Fürstenhöfer T, Matthiesen S (2015) Combination of Eye Tracking and Think-Aloud Methods in Engineering Design Research. In Design Computing and Cognition'14 (pp. 81–97). Springer International Publishing.

Salvucci D D (1999) Mapping eye movements to cognitive processes (Doctoral dissertation, Carnegie Mellon University).

Salvucci D D, Goldberg J H (2000) Identifying fixations and saccades in eye-tracking protocols. In: Proceedings of the 2000 symposium on Eye tracking research and applications; Santa Barbara, CA, USA, 71–78.

Servatius K (2009) Dynamische Unterstützungsformen in kartographischen Medien. Selbstverl. d. Geograph. Ges. Trier, 41.

Siegel A W, White S H (1975) The Development of spatial representations of large-scale environments. Advances in Child Development and Behavior, 10, 9–55.

Standing L (1973) Learning 10,000 pictures. Q J Exp Psychol 25:207–222.

Strasburger H, Malania M (2013) Source confusion is a major cause of crowding. Journal of Vision, 13(1), 24.

Tom A C, Tversky B (2012) Remembering routes: streets and landmarks. Applied cognitive psychology, 26(2), 182–193.

Viaene P, Ooms K, Vansteenkiste P, Lenoir M, De Maeyer P (2014) The Use of Eye Tracking in Search of Indoor Landmarks.

Viviani P (1990) Eye movements in visual search: Cognitive, perceptual, and motor control aspects. In E. Kowler (Ed.), Eye Movements and their Role in Visual and Cognitive Processes (353–393). New York: Elsevier Science Publishing.

von Stülpnagel R, Frankenstein J (2015) Configurational salience of landmarks: an analysis of sketch maps using Space Syntax. Cognitive processing, 16(1), 437–441.

Voyer D, Postma A, Brake B, Imperato-McGinley J (2007) Gender differences in object location memory: a metaanalysis. Psychon Bull Rev 14: 23–38.

Wiener J M, Hölscher C, Büchner S, Konieczny L (2012) Gaze behaviour during space perception and spatial decision making. Psychological research, 76(6), 713–729.

Wolbers T, Hegarty M (2010) What determines our navigational abilities? Trends in cognitive sciences, 14(3), 138–146.

Xu Y, Chun M M (2009) Selecting and perceiving multiple visual objects. Trends in cognitive sciences, 13(4), 167–174.

Yarbus A L (1967) Eye movements and vision. New York: Plenum.

Increasing the Density of Local Landmarks in Wayfinding Instructions for the Visually Impaired

Rajchandar Padmanaban and Jakub Krukar

Abstract Multiple approaches to support non-visual navigation have been proposed, of which traditional auditory turn-by-turn navigational systems achieved high popularity. Despite being modified according to the needs of visually impaired users, the underlying dataset communicated to the wayfinder is sourced primarily from traditional POI databases which are of limited use to blind navigators. This work proposes the use of environmental features spontaneously detected by blind navigators during their everyday locomotion as 'local landmarks' for enriching auditory navigational instructions. We report results of a survey which served to identify such environmental features. Consequently, we propose a list of potential local landmarks for the blind. Next, in a usability study, we demonstrate that enriching traditional turn-by-turn auditory instructions with local landmarks can improve the subjective satisfaction and confidence in navigation. Results indicate that the improvements seem to be achieved even without increasing the subjective complexity of the instructions. Finally we discuss how using local landmarks to enrich auditory navigational instructions can benefit visually impaired users.

Keywords Local landmarks · Wayfinding · Visually impaired

1 Introduction

The visually impaired population face a range of navigational problems different from those experienced by the sighted. Gathering information from the immediate environment without vision is a much slower process and it is limited to one's closest proximity; for this reason, traveling through an unfamiliar area can be a greatly distressing activity. In the context of wayfinder's confidence, one topic

R. Padmanaban · J. Krukar (✉)
Institute for Geoinformatics, University of Münster, Münster, Germany
e-mail: krukar@uni-muenster.de

R. Padmanaban
e-mail: charaj7@gmail.com

© Springer International Publishing AG 2017
G. Gartner and H. Huang (eds.), *Progress in Location-Based Services 2016*, Lecture Notes in Geoinformation and Cartography, DOI 10.1007/978-3-319-47289-8_7

receiving a lot of attention in the studies of sighted individuals is the inclusion of local landmarks—distinct features of the environment communicated along the route or at decision points. Although the benefit of including landmarks in a route description is clear, relatively little attention has been given to understanding what constitutes as appropriate landmarks for blind travelers. It thus remains unclear whether local landmarks for the visually impaired might similarly increase subjective confidence during travel through unknown areas. This is despite the proliferation of electronic travel aids which assist in interpreting and communicating information from the surrounding environment while providing the blind navigator with environmental details unavailable through the visual sense. This work focuses on (a) identifying suitable spatial elements for selection as local landmarks for the visually impaired individuals, and (b) measuring the impact of including local landmarks in auditory instructions on the subjective satisfaction and confidence of the blind navigators. We hypothesize that increasing the density of local landmarks communicated to the visually impaired navigators *en route* will result in increasing subjectively reported measures of confidence and satisfaction. In the following sections, we review existing work on navigational aids for the blind and point out the research's over reliance on performance measures. Next, we present results of a questionnaire that identified a number of commonly encountered environmental features used by the visually impaired in their everyday locomotion. Lastly, we construct and evaluate a set of navigational instructions enriched by the presence of the newly identified local landmarks.

2 Related Work

2.1 *Wayfinding, Wayfinding Assistance and Landmarks*

Navigation is an activity traditionally considered to consist of two components: *locomotion* and *wayfinding* (Montello 2001, 2005). Locomotion is the process of moving one's body across space without the need to consider spatial information other than what is immediately available to us. It involves movement around obstacles, selection of surfaces to stand on, passing through openings or moving towards perceivable landmarks. Wayfinding involves the activity of planning efficient routes in a goal-oriented manner, using the cognitive model of space, often based on information that is not directly accessible at the starting location (ibid.). Navigating through space usually (but not always) consists partially of each of these two components, although the proportions of their importance might vary greatly. Montello (2005) gives the examples of a bus passenger moving their body without much planning (other than at the moment of getting on and off the bus) and of a blind person, who might locomote efficiently through their direct surrounding but may experience troubles orienting themselves with respect to distal landmarks.

The process of wayfinding can be supported by providing users with navigational information at each decision point, which–together with the development of GPS technology and location based services–has become the dominant approach in turn-by-turn navigation assistance systems. This approach, while being widely used, has been shown to bear many disadvantages, such as decreasing the user's ability to build a cognitive representation of their traveled area (Burnett and Lee 2005). Common examples include car navigation systems as well as foot navigation instructions provided by commercial smartphone applications.

Humans naturally use landmarks in their mental representations of space (Montello 1998) and in communicating routes to each other (May, Ross, Bayer, nd Tarkiainen 2003; Michon and Denis 2001). Enriching wayfinding information with landmarks has therefore been shown to benefit navigators (Deakin 1996) and has since been adapted in some commercially available GPS-based systems.

While a method for predicting what constitutes a (good) landmark is a topic open to intensive investigation (Caduff and Timpf 2008; Miller and Carlson 2011; Richter and Winter 2014), the underlying assumption is that correct identification and inclusion of landmarks in wayfinding instructions is beneficial to the wayfinder's performance (Deakin 1996), route memory (Denis, Mores, Gras, Gyselinck, and Daniel 2014) and the construction of one's cognitive map (Schwering, Li, and Anacta 2013). Highlighting landmarks located along the route in wayfinding instructions can also help navigators to build a mental representation of the presently unfamiliar part of the route. This serves to prepare them for challenging stretches of the travel (Michon and Denis 2001). Therefore, while landmarks can be beneficial to the spatial memory and performance of the navigator, a significant part of the benefit lies in the users' subjectively perceived satisfaction and confidence *en route* through an unfamiliar environment. The subset of landmarks located along a specific route, i.e. along straight stretches of road and at decision points, has been termed *local landmarks* (Raubal and Winter 2002).

2.2 Navigational Aids for the Blind

One of the key challenges faced by the blind population is independent travel (Giudice and Legge 2008). As studies have shown, the reason behind this, however, is unlikely due to deficiencies in spatial skills (Loomis et al. 1993) but rather in the informational needs which cannot be easily fulfilled without vision (Passini and Proulx 1988). Multiple technological approaches have been suggested to aid the visually impaired population in independent navigation, but none so far have gained or maintained a dominant popularity among users. Giudice and Legge (2008) provide a historical overview of these diverse approaches and outline the differences between them. As they note, the loss of sight often occurs at an older age, when adapting to a new technology might present a serious barrier to users (Giudice and Legge 2008). Perhaps for this reason, GPS-based navigation devices including auditory support remain the most popular. Since their accuracy is limited, they typically are meant to complement, and not substitute, the use of a white cane. This is also dictated by the fact that the use of the cane is ubiquitous among blind navigators and its utility extends beyond the detection of objects within reach. For instance, it also provides echolocation cues through tapping (Giudice and Legge, 2008).

A number of commercial companies, open source initiatives, and research projects have developed navigational systems supporting visually impaired users in wayfinding. To date, the Wikipedia article devoted to the issue lists 23 such GPS-based systems.[1] They typically run on regular smartphones, and can be controlled with a purpose-built interface, based on audio input and output. An important feature is that they attempt to increase the density of communicated Points-of-Interest (POIs). AriadneGPS,[2] for instance, allows the user to add and manage personal POI favorites which can trigger automatic alerts during wayfinding. Another popular application, BlindSquare,[3] integrates the online social network Foursquare, sourcing a larger number of POIs.

The above approaches, however, depend largely on POIs that are contributed to the database *by* the sighted population for purposes considered relevant *to* the sighted population. Without a careful selection of information, using these datasets might create a cluttered set of instructions that are difficult to interpret for blind users (Giudice and Legge 2008). Such POIs are also unlikely to gain the function of local landmarks in the course of non-visual navigation. Traditionally, the process of enriching wayfinding instructions with local landmarks consists of attaching a textual label to a perceivable object, which is unique enough to be conceptually and semantically distinguished from its neighbors (Raubal and Winter 2002). This implies that the landmark is perceptible in the proximity of other perceptible objects of a similar type (i.e. 'non-landmarks'). To a blind navigator, a typical POI (e.g. 'Ben's Cafe') can still be uniquely identified with a textual label, but without access to visual information (and in a busy urban soundscape) the means of matching that label with any actually perceived environmental object is limited. As a result, a blind user might not have the chance to compare it with other environmental objects of a potentially competing level of salience. A solution to this issue could be the inclusion of a denser set of landmark candidates perceivable by blind navigators within their wayfinding instructions; some of which candidates can be then identified using textual labels associated with standard POIs.

While OpenStreetMap (OSM) supports mapping of two types of features driven by the needs of the blind[4] (tactile paving and distinct types of road crossings), the density of those features in the actual urban environment (even assuming they were all mapped) might always remain disproportionally skewed towards most dangerous urban locations. This potentially limits the utility of enhanced systems for

---

[1]https://en.wikipedia.org/wiki/GPS_for_the_visually_impaired.
[2]http://www.ariadnegps.eu.
[3]http://blindsquare.com.
[4]http://wiki.openstreetmap.org/wiki/OSM_for_the_blind.

'raising red flags' over the course of non-visual navigation instead of supporting cognitive mapping of the travelled area. For the same reason, communicating the location of such real-world features can possibly increase the subjective confidence of the navigator. This confidence, however, would be specifically related to *locomotion* safety and not to *wayfinding* in a way similar to how local landmarks communicated by humans to humans increase such confidence among sighted individuals.

This is a neglected research area. As Giudice and Legge (2008) note, creators of navigational aids for the blind are limited in their understanding of true user needs and wishes. Literature is dominated by evaluations based on the measures of performance, and subjective satisfaction measures are often left behind (but see e.g. Bradley and Dunlop 2005; Golledge et al. 2004). While such performance metrics can be used to estimate the efficiency of the system (e.g. Kutiyanawala et al. 2010), they might not contribute to the understanding of why many similar aids do not gain popularity among visually impaired navigators.

2.3 Wayfinding Instructions for the Blind

With respect to subjective measures, Bradley and Dunlop (2005) studied how blind (compared to sighted) navigators could judge their workload while navigating to four distinct landmarks with the use of different types of instructions. The authors observed that blind participants experienced lower subjective workload when navigating with instructions which contained a larger proportion of cues *alternative to* textual descriptions of street names and regular Points-of-Interest.

The distribution of workload is not trivial. As Giudice and Legge (2008) note, the balance of effort required to conduct efficient locomotion (even without considering wayfinding) is significantly larger for the visually impaired. It therefore could be favorable to communicate wayfinding instructions by building on the effort already required for successful locomotion. The concept of a local landmark that can be perceived by the visually impaired bears the potential for such a role.

Serrão et al. (2012) considered a wide of range of permanent objects inside a building, and created a system with the ability to recognize landmarks potentially useful to blind navigators. In doing so, however, the authors did not focus on the verification of how easy it might be for the visually impaired to detect these objects.

The CrossingGuard system developed by Guy and Truong (2012) provides detailed auditory information about pedestrian crossings as they are being approached by visually impaired wayfinders. Users of the system expressed reduced stress as well as increased comfort and confidence when presented with details about the layout of the tested intersections. The authors used a crowdsourced approach based on StreetView for coding relevant details of the intersection, although the system did not extend beyond the context of a road crossing.

Swobodzinski and Raubal (2009) proposed an indoor routing system based on Orientation and Mobility training, a process which is typically undertaken by individuals after losing sight (Blasch et al. 1997). As part of their algorithm, the authors specify the distinction between four classes of objects known in Orientation and Mobility training as *obstacles*, *hazards*, *cues* and *landmarks*. Obstacles and hazards are objects which need to be avoided while locomoting. Cues and landmarks, on the contrary, are environmental features which can support navigation. While both must be potentially detectable by a blind navigator, the distinction lies in their unique properties. A cue is an environmental feature which–due to its standard and repeatable property—cannot be uniquely identified, while a landmark has distinct properties that make it a salient element of the environment, potentially beneficial for wayfinding. The authors give the examples of a ceiling fan being a cue (if it's one of multiple such fans in the given environment), and of a small fountain in a hotel lobby serving as a landmark. As the authors note, however, the distinction between what constitutes as a cue or a landmark can be ambiguous, since the concept of a landmark could be perceived as any object that has significance to the blind navigator (given their strategies, skills and preferences). This note is supported by landmark research with sighted navigators, which demonstrates that different objects can gain or lose landmark status depending on the circumstances. For instance, even a perceptually non-salient object can become a landmark due to its salient location in space useful for the task at hand (Miller and Carlson 2011). Knowing that visually impaired navigators need to make wayfinding decisions more often and at shorter intervals (Passini and Proulx 1988), the potential for a landmark to become salient due to its salient location in space is much greater in non-visual navigation.

For this reason, communicating the presence of a larger number of perceivable spatial features in wayfinding instructions for the visually impaired introduces the chance of raising the status of an object spontaneously detected through *locomotion* to the role of a landmark involved in *wayfinding*. This could translate to increasing subjective confidence and satisfaction of the blind navigator—an outcome similar to using local landmarks in instructions for sighted navigators, when traversing unfamiliar environments.

This paper therefore aims to build upon the fact that spatial elements without a distinct, visual meaning, can gain salience due to their utility in the everyday locomotion of visually impaired individuals. This work focuses on (a) identifying suitable local landmarks for the blind, and (b) measuring the impact of their inclusion in auditory wayfinding instructions on the wayfinder's subjective satisfaction and confidence.

3 Identifying Local Landmarks for Visually Impaired Wayfinders

In order to verify the relevance of potential environmental features used to enrich wayfinding instructions for the blind, visually impaired navigators were interviewed and asked about their wayfinding strategies. Particular care was given to communicating their abilities to spontaneously detect distinct elements of the environment during everyday locomotion.

3.1 Interview

Ten blind participants (7 men and 3 women) aged between 35 and 68 were recruited with the help of a local association. Seven of them were congenitally blind, two lost sight during mature adulthood and one during childhood. The majority of the participants spoke English; for the two who did not, German translations of all subsequent materials were provided. Standard ethical procedures were applied throughout the study, with participants giving their informed consent prior to the study.

In a semi-structured interview, participants were requested to give an account of their wayfinding strategies. All ten participants reported that they typically used a cane to support their navigation, eight used dedicated smartphone applications, and one participant relied on help from a guide dog. When asked about their use of auditory navigational software, eight participants declared that they used commercially available solutions on an everyday basis. Participants also expressed the preference to loudspeakers over headphones in order to preserve the ability to hear other environmental noise while navigating.

Participants expressed their strong dependence on Audible Traffic Signal (ATS) when crossing the road. In order to estimate the metric range of ATS as a potential landmark, participants were asked to estimate their hearing range of ATS under typical traffic conditions. The estimated distance fell in the range of 5–20 m, with a reported mean of 12.5 m.

A list of potentially relevant local landmarks (i.e. objects or environmental features which can be spontaneously detected and are ubiquitous in the urban environment) has been composed based on previous informal conversations and the existing literature (see e.g. short summary by Bradley and Dunlop (2005) and a list by Strothotte et al. 1995). Participants reported an ability to sense a wide range of materials and objects with their white cane. When asked to select exact objects they considered possible to detect while traveling, their answers were relatively homogenous (see Table 1).

Table 1 Number of participants (out of 10) who answered 'yes' to the question: "Can you detect the following object when traveling on a road?"

Landmark	No. of participants
Access and exit areas	10/10
Traffic lights (ATS)	10/10
Surface materials	10/10
Tactile areas and tactile strips	10/10
Railings	8/10
Walls	8/10
Bus stops with a shelter	8/10
Tree pits	8/10
Staircases	8/10
Bus stops without a shelter	0/10
Others (please specify)	2 mentions of *street gutters*

3.2 Local Landmarks for Visually Impaired Wayfinders

Below we provide a list of local landmarks established from the interview. The potential usability of some listed landmarks is supported by literature. In some cases, however, despite being described as easily detectable, their usefulness for navigation is limited due to associated spatial characteristics. For example, following street gutters would often force the user to move dangerously close to motorized traffic.

- **Access/Exit Areas** are sections of sidewalks allowing motorized traffic to access properties across the sidewalk. Typically constructed from a distinct surface material and having a distinct slant (Fig. 1a), there are instances when their utility for non-visual navigation is limited, e.g. when the surface material is uniform (Fig. 2b) or the arrangement is temporary (Fig. 3c).
- **Tactile Areas** (Fig. 2a) are larger planes constructed of tangibly distinct surfaces and of contrasting colors in order to aid navigation of the blind and partially sighted individuals. Typically constructed near junctions, waiting areas (e.g. bus stops), and public entrances, they are purpose-built, and thus of clear utility to visually impaired wayfinders.
- **Tactile Strips** often extend from Tactile Areas in order to aid faster movement along safe pathways (Fig. 2b). Tactile paving is a feature supported by the OpenStreetMap database, though mappers seldom specify its existence. Some commercial navigational applications make use of this database.
- **Tree Pits** are the constructions built around trees for their protection in an urban environment. Often aligned on pathways at regular intervals, they present potentially helpful local landmarks, given that their location lies within the range of a white cane from the safe pathway (Fig. 3a, b). Often, however, tree pits might be located on the outer edge of the pedestrian traffic area (Fig. 3c), in a non-linear manner potentially confusing to the navigator (Fig. 3d), or behind a cycling path (Fig. 3e).

Increasing the Density of Local Landmarks in Wayfinding ...

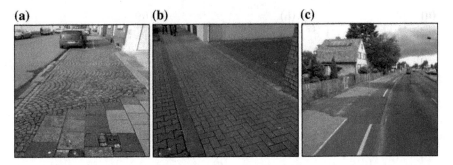

Fig. 1 **a** Slanted, **b** uniform, and **c** temporary access and exit areas

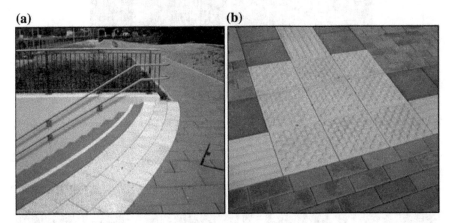

Fig. 2 **a** Tactile area/paving and **b** tactile area connected with tactile strips

- **Staircases** can be easily detected with the use of a white cane. The additional landmark-related advantage of staircases is that they can be direction-specific when the downstairs/upstairs distinction is made. In order to avoid confusion, this work will not use the term 'staircase' when stairs lead directly to a building entrance.
- **Railings and fencing** can be useful non-visual local landmarks when adjacent to the pedestrian area (Fig. 4a) if they are not temporary (Fig. 4b). Due to potential difficulties with the detection of short stretches of railings along long straight stretches of a route, railings are particularly useful when they lead to, or originate from, a turn.
- **Walls,** similar to railings and fencing, can help blind navigators to establish a straight traveling direction (Swobodzinski and Raubal 2009) Moreover, walls help individuals identify turns when arriving at, or departing from a turning point.
- **Audible Traffic Signals** (ATS) are standardized, and in many countries ubiquitous, devices that use sound to indicate the presence of a pedestrian crossing

Fig. 3 Tree pits, potentially useful as local landmarks (**a** and **b**) and those of limited utility (**c**, **d** and **e**)

Fig. 4 **a** Permanent and **b** temporary fencing

and vibrations to indicate whether the crossing has islands. As presented in the previous sections, participants differ with regard to their self-reported abilities to locate the ATS signal, with the mean range of hearing reported as 12.5 m. Similarly to Tactile Paving, pedestrian crossings are a feature supported by OSM, with a distinction into multiple sub-categories.

- **Surface Materials and transitions between them** are another environmental feature participants feel confident at detecting. Since these transitions often coexist with other local landmarks (such as Access and Exit Areas), they were not considered separately in the current study.
- **Bus/Tram Stops** were reported to be only easy to detect if a shelter around the stop was present. Therefore, stops without the shelter were not considered as local landmarks.

With regard to the results of the interview, it is worth noting that humans are generally more likely to remember pre-listed objects (Craik and McDowd 1987). Therefore, the frequency of objects not explicitly asked for in the interview is likely to remain underestimated. For instance, two participants mentioned street gutters when asked for 'other possibly detectable objects'. We thus do not claim this list to be exhaustive. Further research is also needed into the factual accuracy with which blind navigators can detect potentially salient spatial elements with regard to 'false positive' and 'false negative' types of error. This can vary especially under the presence of heterogeneous wayfinding strategies and spatial abilities within the visually impaired population (Schinazi et al. 2015).

4 Evaluating Landmark-Enriched Navigational Instructions

4.1 Procedure

A user study was conducted in the city of Münster, Germany with the same pool of participants who took part in the aforementioned interview. The study took place in an urban area selected due to its non-trivial spatial arrangement as well as its accessibility to a location that was known to the participants. Participants were asked to walk the route with the support of audio instructions (played or read by the accompanying researcher at the corresponding location in a 'wizard-of-oz' arrangement; see Fig. 5).[5]

All participants walked the same route, in the same direction which took approximately 12 min. Participants were randomly divided into two groups. Each group traveled half of the route with Landmark-Enhanced Instructions (*LE*) and the other half with regular instructions, without additional landmarks mentioned (*non-LE*). Depending on the group, the order at which the given set of instructions was played varied: the first group was assisted by *LE* for the first half of the route, and *non-LE* for the second. The second group was guided by the *non-LE* instructions for the first half and *LE* for the second half of the journey.

---

[5]It bears noting that the technique of communicating auditory information is, in itself, a research problem central to separate studies; however, the current work focuses on the content, and not the means of transmitting the landmark-enhanced wayfinding instructions.

Fig. 5 The 'Wizzard-of-Oz' experimental procedure

Each participant was asked twice to respond to a usability questionnaire. This survey was provided to participants half-way through the route as well as at its end. It was made clear that all questions asked were directly related to the recently completed stretch of the route. In this repeated-measure study design, each participant contributed two sets of responses: one for the *LE* and one for the *non-LE* set of verbal wayfinding instructions. The potential effect of order and the resulting linkage between the given route stretch and the instruction type was counterbalanced.

4.2 *Materials*

The instructions were played to participants over loudspeakers from a text-to-speech software (or read aloud by a researcher[6]) as the participants approached the relevant location. The location of local landmarks was encoded earlier from 3D video data in an effort to verify whether objects identified in Sect. 3.2 could be detected and digitized remotely. A sample set of instructions is provided in Table 2. Note that some landmarks identified in Sect. 3.2 are not used, as the study area did not contain all types of landmarks that would be usable in this wayfinding scenario. For instance, existing tree pits cannot be located too far away from the main line of locomotion.

The *LE* set consisted of 20 commands in total (10 of which mentioned a local landmark), and the *non-LE* set contained 13 commands (4 of which contained a standard POI-like label).

---

[6]Two participants asked for the instructions to be read in German, instead of the English recordings.

Table 2 Sample comparison of Landmark-Enhanced (LE) and non-Landmark-Enhanced (non-LE) instructions. The non-LE instructions were generated based on existing wayfinding applications for the blind (BlindSquare and AriadneGPS)

Landmark-enhanced	Non-landmark-enhanced
Walk 10 m	Walk 10 m
Turn right and go downstairs	Turn right
Turn right onto access and Exit area for Platten-Peter Fliesenzentrum	Turn right for Platten-Peter Fliesenzentrum
Walk 50 m and pass by access and Exit area	Walk 200 m
Walk 150 m	
Walk 25 m	Walk 200 m
Follow right side small wall	

A custom-built questionnaire was constructed based on the System Usability Scale (Brooke 1996), with additional questions relating to specific landmark properties. Table 3 presents all questions used. Questions were presented together with their German translations. A five-step 'strongly agree - strongly disagree' scale was used for collecting responses.

4.3 Results

The main motivation for evaluating local landmarks in this study was to increase subjective satisfaction of the blind participants. For this reason, further analysis does not concern performance measures but focuses on the subjective estimates of the instruction's usability.

Linear Mixed-Effect Models (Baayen et al. 2008) were used to statistically verify the influence of *condition* (*LE* vs *non-LE*) on questionnaire responses using R and the lme4 package (Bates et al. 2014). Questionnaire items were recoded so that the higher number would always correspond to a more favorable answer. These responses were entered as the dependent variable of the model. Intercepts for *Participant ID* and *Question ID* were entered as random effects, together with the random slope for the *by-participants* and *by-question* effect of *condition*. The *p-value* for the fixed effect of *condition* was estimated using lmerTest package (Kuznetsova et al. 2014) and was equal to $p = 0.0026$. Marginal R^2 estimated by the MuMIN package (Bartoń 2014) was 0.18 and Conditional R^2 was 0.47.

Landmark-Enhanced instructions significantly increased subjective questionnaire responses of the participants, although the magnitude of improved satisfaction varied across participants, and for individual questionnaire items. After accounting for this by-participant and by-question variability, approximately 18 % of the variance in the responses can be associated with the influence of the instruction type, i.e. *condition* (*LE* vs *non-LE*).

Table 3 Questions used in the evaluation study

	Question
Q1	I would like to use this system frequently for navigation
Q2	I thought the system has made the navigation more complex
Q3	I found this system has more detailed instruction
Q4	I think I need practice to use this system
Q5	It feels easy to handle this system
Q6	I found this system helps me to identify turns and curves easily
Q7	I found it was harder to find streets and routes with this system
Q8	I thought this system has irrelevant landmarks for guidance
Q9	I found this system leads me to correct path
Q10	I thought the system aids me in identifying the landmarks
Q11	I found the system helps me to travel faster
Q12	I found the system guides me to identify the crossings
Q13	I think I need technical support before using this system
Q14	I could reach the destination precisely
Q15	I felt the verbal command was inconsistent
Q16	I felt very confident using this navigation system

Figure 6 illustrates the mean responses to all questionnaire items together with 95 % Confidence Intervals.

As the sample size was relatively small and the questionnaire was custom-built, in what follows we do not apply dimension-reduction techniques on the questionnaire items but offer an exploratory discussion of individual question-by-question differences based on Fig. 6.

Participants were more satisfied (Q1) with the *Landmark-Enhanced* set of instructions. This was accompanied by higher subjectively perceived confidence (Q16) during navigation, as well as higher perceived efficiency of the journey; navigators felt they were able to travel faster (Q11), and with a higher degree of precision (Q14). The system appeared to have made it easier to identify turns (Q6), pathways (Q7, Q9), and road crossings (Q12). The *Landmark-Enhanced* set of instructions was particularly useful in providing relevant (and subjectively perceived as useful) landmark-related information (Q3, Q10). It seems that the *Landmark-Enhanced* set of instructions provided relevant information (Q8) in comparison to the traditional instruction set. It also seems that the benefits of the new system were not associated with its increased perceived complexity (Q2, Q4, Q5, Q13, Q15).

In short, the *Landmark-Enhanced* set of instructions had a significant influence on the participants' subjective satisfaction and confidence, while it was not seen as subjectively more complex.

Increasing the Density of Local Landmarks in Wayfinding ...

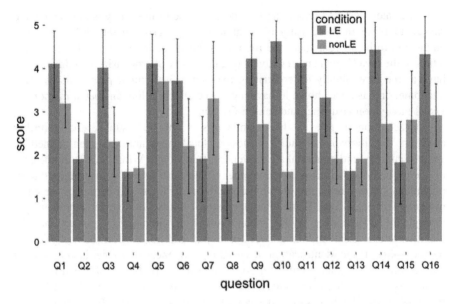

Fig. 6 Mean responses to questionnaire items with 95 % CIs (note that here the values are not recoded: refer to Table 3 for the meaning of a higher/lower response within each question)

4.4 Discussion

Results suggest that traditional turn-by-turn navigational systems for the blind can be improved by the simple means of increasing the density of relevant local landmark information they communicate. For an urban area selected for this study, the density of the newly introduced landmarks did not result in increased subjective complexity. This could become an issue if the communicated information was irrelevant or trivial—imposing an increased load with little benefit to the navigator.

While it is not surprising that the system communicated landmark-related information better than the traditional set of instructions, it is worth noting that subjective satisfaction and confidence were increased by these means. This effect is similar to the effect that local landmarks have on visual navigation, where one of their main advantages lies in increasing the confidence of the navigator by ensuring they stays on the correct trajectory.

An alternative explanation must consider the fact that participants could simply feel more confident when provided with more information making their locomotion easier. This is possible, however the list of communicated landmarks did not contain elements generally considered to be 'hazards' or 'obstacles' in non-visual navigation. Access and exit areas, walls, tactile paving and staircases are all easily detectable by wayfinders after Orientation and Mobility training, and even form part of their everyday mobility strategies. Nevertheless, having these features mentioned in the navigational process increased subjective satisfaction and confidence.

One related question remaining outside the scope of this study is to what degree can an increase in the density of local landmarks result in the ability of blind navigators to better remember the route or to build a survey knowledge representation of the area? The increased diversity and density of emphasized landmarks can have a role in explicitly differentiating two otherwise similar path sections. This phenomenon has been shown to have a positive effect on human spatial memory in visual navigation (Buchner and Jansen-Osmann 2008).

The selection of relevant local landmarks is another issue deserving further attention. Traditionally, the attractiveness of local landmarks has been defined by visual, semantic and structural attractiveness (Raubal and Winter 2002). In the case of the visually impaired population, the visual aspect clearly loses importance, potentially becoming overwritten by the auditory or haptic properties of the environmental object, as well as the blind navigator's ability to spontaneously identify it through the course of everyday locomotion. It is, therefore, important that the identification of local landmarks does not require additional effort (be it physical or mental)—just as the identification of visual attractiveness of a local landmark does not require 'other than regular' effort in a visual exploration of the urban surroundings by a sighted individual. This property of landmarks might be largely responsible for the low subjective complexity of the tested LE instructions—a relation which is not self-evident. Presenting an increased volume of information in a serial manner (imposed by the auditory medium) is generally associated with increased cognitive load. It therefore appears that the key to useful local landmark selection is in understanding which spatial objects would be detected by the blind navigator spontaneously over the course of regular locomotion. This once again puts the emphasis back on the heterogeneous abilities and preferences of the visually impaired user (Schinazi et al. 2015).

The structural salience of selected landmarks is of no lesser importance. Given a dense set of subsequent local landmarks, all detectable over the course of everyday locomotion by an individual blind user, it might be more beneficial for one's spatial memory to hear the information about some landmarks over others. Environmental elements linked to larger structures (e.g. access areas to large public buildings) can be of particular importance, as they afford the understanding of one's location and bodily orientation at multiple points, with respect to a single feature (e.g. two sides of the same building). At the same time, this creates the potential for confusion, if these spatial relations are not communicated in sufficient detail; knowing that one passes three separate access areas to the same public building over the course of a complex travel might be not beneficial if the relation between these points is not explained. This is an aspect not tackled in this work.

Another view on the structural salience of non-visual local landmarks can consider travel as a movement along a 1-dimensional sequence of auditory landmarks. Then, the concept of their structural salience can be viewed as the landmark's location with respect to other landmarks along that route. The distance between subsequently communicated landmarks, as well as the contrast between the salience of sequentially encountered objects, might provide dominant cues for structural attractiveness.

It is important to note that the current work considers landmarks in relation to environmental elements that do not necessarily have any uniquely identifying properties. Just as 'any' doorway is not a landmark to a sighted individual, 'any' bus shelter, or 'any' access area might not become a landmark to a visually impaired navigator. However, at least two known cases suggest that this potential repeatability and 'non-landmarkness' can be understood differently in the context of non-visual navigation. Firstly, even non-unique structures can have an important status in wayfinding, so long as their location and function make them an important point for understanding one's orientation in relation to the surrounding structures - staircases in complex buildings being one example (Hölscher et al. 2006). Secondly, as locomotion is a more–physically and mentally–demanding activity for the blind than it is for the sighted (Giudice and Legge 2008), the energy spent on an individual's movement along and through these repeatable everyday environmental elements is likely to be greater. The level of one's energy spent on the process of locomotion is linked to spatial learning processes driven by attentional states and heuristic judgment of traveled distances (Montello 2005). A person locomoting with greater effort is, therefore, likely to pay greater attention to one's environment which might translate to a greater ability to distinguish seemingly similar objects and spatial relations between them. In such an approach to non-visual local landmarks, the complexity of the urban environment must also be considered (Gaunet 2004).

The effort has already been made towards creating computational models with hierarchical data structures adequate for the techniques blind navigators use to learn their spatial environment (Gaunet and Briffault 2005; Yaagoubi et al. 2012). A denser set of local landmarks, such as what has been proposed in the present work, can increase the subjective satisfaction and confidence of the visually impaired travelers using systems based on such databases.

5 Limitations and Conclusion

In the current work, the location of local landmarks used in the instructions was digitalized remotely using a 3D video dataset provided by the Hansa Luftbild company. It remains to be evaluated whether all local landmarks identified in Sect. 3.2 can be distinguished for digitization, especially under more heterogeneous urban conditions. Potential alternative sources of delivering this information are worth noting. For instance, Guy and Truong (2012) demonstrated how StreetView could be used for a similar task in a crowdsourced approach. Infrastructural elements which can serve as local landmarks to the visually impaired population are rarely distinguished in authoritative databases at a sufficient level of detail while automated or crowdsourced approaches offer a promising alternative.

It is also worth noting that subjective measures of satisfaction and complexity–while offering a worthwhile perspective on the problem–are not always completely reliable. People have been shown to both under and overestimate their subjective perception of various metrics in usability studies. Further work is required to compare these subjective estimates to alternative measures of cognitive effort. The proliferation of psychophysiological techniques nowadays offers such a possibility in the context of naturalistic navigation (Mavros et al. 2013; Tröndle et al. 2014).

To conclude, subjective measures of user satisfaction and perceived complexity rarely form a part of the evaluation of navigational systems for the blind. The current work proposes that increasing the density of local landmarks in navigational instructions can yield a positive increase over many measures of a user's subjective satisfaction. Future work needs to further investigate the suitability of proposed environmental features for their local landmark status, as well as seek to explore the cognitive benefit that enriched instructions can have on the visually impaired wayfinders.

Acknowledgments This paper is based on Rajchandar Padmanaban's Master Thesis carried out at the Institute for Geoinformatics of the University of Münster. The authors wish to thank Reinhard Silvers, Kerstin Gonschorrek and the Hansa Luftbild company for their support with data collection. We also thank all participants involved in the study for their time, effort, and many helpful advices as well as to Emily Harason for proofreading.

References

Baayen, R. H., Davidson, D. J., & Bates, D. M. (2008). Mixed-effects modeling with crossed random effects for subjects and items. *Journal of Memory and Language, 59*(4), 390–412.

Bartoń, K. (2014). MuMIn: Multi-model inference. R package version 1.15.1. Retrieved from http://cran.r-project.org/package=MuMIn.

Bates, D., Maechler, M., Bolker, B., & Walker, S. (2014). lme4: Linear mixed-effects models using Eigen and S4. R package version 1.1–7. Retrieved from http://cran.r-project.org/package=lme4.

Blasch, B. B., Wiener, W. R., & Welsh, R. L. (1997). *Foundations of orientation and mobility*. Amer Foundation for the Blind.

Bradley, N. A., & Dunlop, M. D. (2005). An Experimental Investigation into Wayfinding Directions for Visually Impaired People. *Personal and Ubiquitous Computing, 9*(6), 395–403. http://doi.org/10.1007/s00779-005-0350-y.

Brooke, J. (1996). SUS-A quick and dirty usability scale. *Usability Evaluation in Industry, 189* (194), 4–7.

Buchner, A., & Jansen-Osmann, P. (2008). Is Route Learning More Than Serial Learning? *Spatial Cognition & Computation, 8*(4), 289–305.

Burnett, G. E., & Lee, K. (2005). The effect of vehicle navigation systems on the formation of cognitive maps. In *International Conference of Traffic and Transport Psychology*.

Caduff, D., & Timpf, S. (2008). On the assessment of landmark salience for human navigation. *Cognitive Processing, 9*(4), 249–67. http://doi.org/10.1007/s10339-007-0199-2.

Craik, F. I., & McDowd, J. M. (1987). Age differences in recall and recognition. *Journal of Experimental Psychology: Learning, Memory, and Cognition, 13*(3), 474.

Deakin, A. K. (1996). Landmarks as navigational aids on street maps. *Cartography and Geographic Information Systems, 23*(1), 21–36.

Denis, M., Mores, C., Gras, D., Gyselinck, V., & Daniel, M.-P. (2014). Is Memory for Routes Enhanced by an Environment's Richness in Visual Landmarks? *Spatial Cognition & Computation, 5868*(May 2016), 1–39. http://doi.org/10.1080/13875868.2014.945586.

Gaunet, F. (2004). Verbal guidance rules for a localized wayfinding aid intended for blind-pedestrians in urban areas. *Universal Access in the Information Society, 4*(4), 328–343. http://doi.org/10.1007/s10209-004-0109-7.

Gaunet, F., & Briffault, X. (2005). Exploring the Functional Specifications of a Localized Wayfinding Verbal Aid for Blind Pedestrians: Simple and Structured Urban Areas. *Human–Computer Interaction, 20*, 267–314. Retrieved from http://gsc.up.univ-mrs.fr/gsite/Local/lpc/dir/gaunet/articles/Exploring the functional specifications of a localized wayfinding verbal aid.pdf.

Giudice, N. A., & Legge, G. E. (2008). Blind Navigation and the Role of Technology. *The Engineering Handbook of Smart Technology for Aging, Disability, and Independence*, 479–500.

Golledge, R. G., Marston, J. R., Loomis, J. M., & Klatzky, R. L. (2004). Stated preferences for components of a personal guidance system for nonvisual navigation. *Journal of Visual Impairment & Blindness, 98*(3), 135–147.

Guy, R., & Truong, K. (2012). CrossingGuard: exploring information content in navigation aids for visually impaired pedestrians. *Proceedings of the 2012 ACM Annual Conference on Human Factors in Computing Systems - CHI '12*, 405–414. http://doi.org/10.1145/2207676.2207733.

Hölscher, C., Meilinger, T., Vrachliotis, G., Brösamle, M., & Knauff, M. (2006). Up the down staircase: Wayfinding strategies in multi-level buildings. *Journal of Environmental Psychology, 26*(4), 284–299.

Kutiyanawala, A., Kulyukin, V., & Bryce, D. (2010). On Self-sufficiency of verbal route directions for blind navigation with prior exposure. In *Proceedings of the 25th Annual International Technology and Persons with Disabilities Conference (CSUN 2010), San Diego, CA*.

Kuznetsova, A., Bruun Brockhoff, P., & Haubo Bojesen Christensen, R. (2014). lmerTest: Tests for random and fixed effects for linear mixed effect models. R package version 2.0-11.

Loomis, J. M., Klatzky, R. L., Golledge, R. G., Cicinelli, J. G., Pellegrino, J. W., & Fry, P. A. (1993). Nonvisual navigation by blind and sighted: assessment of path integration ability. *Journal of Experimental Psychology. General, 122*(1), 73–91. http://doi.org/10.1037/0096-3445.122.1.73.

Mavros, P., Austwick, M. Z., & Hudson Smith, A. (2013). Geo-EEG: Towards the use of EEG in the study of urban behaviour, (April), 1–6. http://doi.org/10.1007/s12061-015-9181-z.

May, A. J., Ross, T., Bayer, S. H., & Tarkiainen, M. J. (2003). Pedestrian navigation aids: Information requirements and design implications. *Personal and Ubiquitous Computing, 7*(6), 331–338. http://doi.org/10.1007/s00779-003-0248-5.

Michon, P.-E., & Denis, M. (2001). When and why are visual landmarks used in giving directions? *Spatial Information Theory, 2205*, 292–305. http://doi.org/10.1007/3-540-45424-1_20.

Miller, J., & Carlson, L. (2011). Selecting landmarks in novel environments. *Psychonomic Bulletin & Review, 18*(1), 184–191. http://doi.org/10.3758/s13423-010-0038-9.

Montello, D. R. (1998). A new framework for understanding the acquisition of spatial knowledge in large-scale environments. In M. J. Egenhofer (Ed.), *Spatial and temporal reasoning in geographic information systems* (pp. 143–154). New York: Oxford University Press.

Montello, D. R. (2001). Spatial Cognition. In *International Encyclopedia of Social and Behavioral Sciences* (pp. 14771–14775).

Montello, D. R. (2005). Navigation. In P. Shah & A. Miyake (Eds.), *The Cambridge Handbook of Visuospatial Thinking* (pp. 257–294). Cambridge: Cambridge University Press.

Passini, R., & Proulx, G. (1988). Wayfinding without Vision: An Experiment with Congenitally Totally Blind People. *Environment and Behavior, 20*(2), 227–252. http://doi.org/10.1177/0013916588202006.

Raubal, M., & Winter, S. (2002). Enriching Wayfinding Instructions with Local Landmarks. *Proceedings of the Second International Conference on Geographic Information Science, 2478*(1), 243–259. http://doi.org/10.1007/3-540-45799-2_17.

Richter, K.-F., & Winter, S. (2014). *Landmarks - GIScience for Intelligent Services*. Springer.

Schinazi, V. R., Thrash, T., & Chebat, D.-R. (2015). Spatial navigation by congenitally blind individuals. *Wiley Interdisciplinary Reviews: Cognitive Science*, *7*(February), n/a–n/a. http://doi.org/10.1002/wcs.1375.

Schwering, A., Li, R., & Anacta, V. J. A. (2013). Orientation information in different forms of route instructions. In *Short Paper Proceedings of the 16th AGILE Conference on Geographic Information Science, Leuven, Belgium*.

Serrão, M., Rodrigues, J. M. F., Rodrigues, J. I., & Du Buf, J. M. H. (2012). Indoor localization and navigation for blind persons using visual landmarks and a GIS. *Procedia Computer Science*, *14*(Dsai), 65–73. http://doi.org/10.1016/j.procs.2012.10.008.

Strothotte, T., Petrie, H., Johnson, V., & Reichert, L. (1995). MoBIC : user needs and preliminary design for a mobility aid for blind and elderly travellers. *Computer*, *1*, 348–351.

Swobodzinski, M., & Raubal, M. (2009). An Indoor Routing Algorithm for the Blind : Development and Comparison to a Routing Algorithm for the Sighted. *International Journal of Geographical Information Science*, *23*(10), 1315–1343. http://doi.org/10.1080/13658810802421115.

Tröndle, M., Greenwood, S., Kirchberg, V., & Tschacher, W. (2014). An Integrative and Comprehensive Methodology for Studying Aesthetic Experience in the Field Merging Movement Tracking, Physiology, and Psychological Data. *Environment and Behavior*, *46*(1), 102–135.

Yaagoubi, R., Edwards, G., Badard, T., & Mostafavi, M. A. (2012). Enhancing the mental representations of space used by blind pedestrians, based on an image schemata model. *Cognitive Processing*, *13*(4), 333–347. http://doi.org/10.1007/s10339-012-0523-3.

Identifying Divergent Building Structures Using Fuzzy Clustering of Isovist Features

Sebastian Feld, Hao Lyu and Andreas Keler

Abstract Nowadays indoor navigation and the understanding of indoor maps and floor plans are becoming increasingly important fields of research and application. This paper introduces clustering of floor plan areas of buildings according to different characteristics. These characteristics consist of computed human perception of space, namely isovist features. Based on the calculated isovist features of floorplans we can show the possible existence of greatly varying alternative routes inside and around buildings. These routes are archetypes, since they are products of archetypal analysis, a fuzzy clustering method that allows the identification of observations with extreme values. Besides archetypal routes in a building we derive floor plan area archetypes. This has the intention of gaining more knowledge on how parts of selected indoor environments are perceived by humans. Finally, our approach helps to find a connection between subjective human perceptions and defined functional spaces in indoor environments.

Keywords Indoor navigation · Floor plans · Archetypal analysis · Isovist features

---

S. Feld (✉)
Mobile and Distributed Systems Group, LMU Munich, Munich, Germany
e-mail: sebastian.feld@ifi.lmu.de

H. Lyu
Department of Cartography, Technische Universität München, Munich, Germany
e-mail: hao.lyu@tum.de

A. Keler
Applied Geoinformatics, University of Augsburg, Augsburg, Germany
e-mail: andreas.keler@geo.uni-augsburg.de

1 Introduction

Navigation is one of the most popular use cases for Location-Based Services (LBS) (Gartner et al. 2007; Gartner and Rehrl 2009; Gartner and Ortag 2011; Krisp 2013; Krisp and Meng 2013). In particular, navigation inside buildings, also referred to as indoor navigation, has gained increasing attention in recent years. Estimations of Shekhar et al. (2016) state that people currently spend 10–20 % of their lifetime using LBSs and around 80–90 % of their lifetime in indoor environments. One area of LBS applications is the calculation of routes for visitors of large buildings, e.g., hospitals (Hughes et al. 2015), fairs, or airports (Ruppel et al. 2009). Also, non-human entities like autonomous mobile robots in store houses or non-player characters in computer games utilize such geospatial trajectories (Zheng and Zhou 2011).

Besides finding a shortest path between two given points, the calculation of alternative routes is an important task. There are several definitions for quality metrics for alternative routes in street networks, see for example Camvit (2013), Delling and Wagner (2009) and Kobitzsch et al. (2013). The first definition of alternative routes in indoor navigation scenarios is given by Werner and Feld (2014), where algorithms for creating said routes have been proposed as well. Subsequently, the concept of archetypal routes has been defined in Feld et al. (2015) using archetypal analysis (Cutler and Breiman 1994), a multivariate data analysis and clustering mechanism to find the most extreme observations or pure types inside a given dataset. In that work, the authors have used simple features like the area of the convex hull of a route or the overlength regarding the shortest path.

How humans visually perceive the environment has got a large influence on both navigation performance and emotional response. Franz and Wiener (2008) illustrated that isovists and visibility graph measures are able to capture behaviorally relevant properties of space, allowing the prediction of affective responses and navigation behavior. Emo et al. (2012) found connections between streets' spatial geometry and spatial decisions with space syntax as an interpreter. Unlike outdoor road networks, indoor environments are restricted by walls and other obstacles such as furniture and installations. They also provide navigable space with more degrees of moving freedom.

However, it is still unclear how to incorporate visibility properties of an environment into LBS applications (e.g., navigation systems) to provide users reasonable choices. Different from psychological and behavioral researches, we use perceptual properties in a reversed way. We first inspect perceptual differences among different locations of an environment. Then we use these results to create alternative routes to enable users to have different perceptual experiences when they traverse the environment. Basically, the main idea is to use extreme types of indoor perception to cluster freely walkable space into functional areas. To achieve this goal, we combine the concepts of archetypal analysis (Cutler and Breiman 1994) and isovist features (Benedikt 1979).

Our first contribution creates insights about a building's structure by using just the plain floorplan itself. For this, we calculate isovist features for all accessible points on the map and cluster them via archetypal analysis. The key benefits of using archetypal analysis are the identification of the most extreme observations and the use of multiple features at the same time. This results in the classification of indoor areas and its surroundings into values like entrance areas, corridors, halls or streets.

Our second contribution expands the idea of archetypal routes (Feld et al. 2015) and analyzes the effect of isovist features on the clustering of a set of routes traversing a building. Thus, we identify a small set of proper alternative routes between two points based on perception features only, i.e. isovist measures, and not geometric measures regarding the routes themselves.

2 Background

2.1 Archetypal Analysis

Cutler and Breiman (1994) define archetypal analysis as a statistical data analysis technique. The results of archetypal analysis are defined clusters, which are comparable to the results of other data clustering methods (Kaufman and Rousseeuw 1990; Jain and Dubes 1988) as, for instance, k-means clustering (Hartigan 1975). In contrast to the last mentioned examples of hard clustering, archetypal analysis is a fuzzy clustering technique.

The general goal of clustering is to organize data in feature space into useful partitions. This organization of a collection of patterns into clusters is often based on similarity (Jain et al. 1999). Archetypal analysis partitions certain amounts of not necessarily equally spaced data. In contrast to traditional clustering techniques, archetypal analysis searches for the points on the outer rim of data space. It approximates the convex hull of the data mainly by looking for data points that are maximally distinct from each other. This characteristic renders archetypal analysis fundamentally different to other clustering techniques, as the aim is to find "pure types" within specific data sets (Eugster and Leisch 2009).

The original algorithm is described in Cutler and Breiman (1994). Seiler and Wohlrabe (2012) propose an iterative version of archetypal analysis which alternates between two steps. The idea is to find a convex hull approximation (in data space) using relatively few points, which results in solving a linear optimization problem.

Referring to the practical approach in this work, we consider a data set with N observations, which in our case are routes consisting of pixel values, and m attributes. This results in a $N \times m$ matrix X. Afterwards, the number k of archetypes to be extracted needs to be defined. These archetypes are then specified by the $k \times m$-

dimensional matrix Z, which is computed by minimizing the residual sum of squares (*RSS*):

$$RSS = \left\| X - \alpha Z^T \right\|_2$$

Consequently, we compare matrix X with the product of the $N \times k$-dimensional coefficient α and the matrix of archetypes Z. In the formula of the *RSS* the part $\| \cdot \|_2$ represents a fitting matrix norm, which in our case is the L_2-norm. Subsequently, α is the coefficient matrix, which is needed to generate X from a given set of archetypes Z. Further mathematical and computational details of the algorithm are explained in Eugster and Leisch (2009). Archetypal analysis by Cutler and Breiman (1994) is also called alternating least square algorithm since it alternates between calculating the best coefficient α for given archetypes Z and calculating the best archetypes Z for given coefficient α. Archetypal analysis iterates until it finds a minimum. It always terminates, but does not necessarily find the global minimum of the *RSS*. It can find a local minimum instead of, for example, the best approximation of the convex hull of the data using k points.

There is no universal rule for determining the initial number of archetypes k. The common approach for its determination is the so called "elbow criterion": a flattening of the *RSS* scree plot is indicating a potentially good value of k.

Our literature review on archetypal analysis includes publications that focus on details like numerical issues, stability, computational complexity and robustness. These issues are based on concrete applications and mentioned in Cutler and Breiman (1994), Eugster and Leisch (2009, 2011) and Seiler and Wohlrabe (2012). Recently, applying archetypal analysis is becoming popular in economics (Eugster and Leisch 2009). Relatively unexplored is the use of archetypal analysis with geodata.

2.2 Qualitative Perceptual Analysis of Space

Human beings experience surrounding environments through senses including seeing, hearing, and smelling. The objects that are seen usually shape the most basic and important part of our experiences. Much effort has been made by behavioral researchers and environmental psychologists to explore how visual properties of an environment affect human's subjective feelings and behaviors in it.

The term *isovist* has been introduced by Tandy (1967) as a visible space obtained at a specific place. Benedikt (1979) then provided a formal definition of isovists with a set of analytic measurements to enable quantitative descriptions of spatial environments. Isovist fields characterize the whole environment with recorded measurements. Using a discretized representation—either some selected locations or evenly distributed locations—is practical to approximate isovist fields (see e.g. Peponis et al. 1997; Batty 2001). To the best of our knowledge, the selection of representative locations and granularity of the discrete representation still have no universal answers. There is always a compromise between coverage (Davis and Benedikt 1979) and computing costs. In most of the mentioned researches, isovist analysis is performed on 2D representations, such as maps and floor plans. Though Emo (2015) advocates isovist analysis in 3D space, she admits such egocentric isovist analysis is still far from mature.

Space syntax (Hillier and Hanson 1984) and visibility graph (Turner et al. 2001) are also frequently used tools that focus on visual perception. Space syntax captures mainly topologic structures (or inter-visibility connections) of an environment, and defines no explicit geometric measurements in Euclidean space, which isovists are capable of, while visibility graph counts the inter-visibility between locations.

In the context of walkable area and alternative routes, we assume that any accessible location can potentially be traversed. We focus on properties of each single location in an environment. Since the generation of visibility graph shares the same idea as generating isovist fields, measurements on visibility graph can easily be integrated in future studies.

3 State of the Art

Orientation in indoor scenarios is still a problem to solve, since street names or other characteristic landmarks are missing, unlike in outdoor environments (Yang and Worboys 2011; Viaene et al. 2014; Ohm et al. 2015). Since the focus of this work is on indoor environments we want to review research on indoor wayfinding, perceptual analysis and the representation of indoor space.

3.1 Indoor Wayfinding and Navigation in Complex Buildings

Early research on indoor navigation has been initiated by Best (1970) where challenges of wayfinding in complex indoor environments have been formulated. The work of Best (1970) deems reasoning on choice points, distances and changes in direction to be relevant for wayfinding in road networks or buildings (Hölschner et al. 2005).

Hölschner et al. (2005) stated that the difficulties of wayfinding in complex buildings are connected with individual spatio-cognitive abilities and the architecture of the building. Therefore it is obvious to link research on architectural design and human spatial cognition to gain knowledge in the field of indoor wayfinding.

Wayfinding behavior in complex buildings has already been investigated by studies coming from the community of environmental psychology (Hölschner et al.

2005). Typical buildings are hospitals (Haq and Zimring 2003), shopping malls (Dogu and Erkip 2000) and airports (Raubal 2002). In general, spatial knowledge and wayfinding acquisition are complex tasks (Li et al. 2011).

3.2 Perceptual Analysis and Wayfinding Behavior

Isovist features and axial lines have shown high predictive power in previous research. Wiener and Franz (2004) studied the interrelations between isovist measures and performance on navigation related tasks. They derived isovist measures from visibility graphs (Turner et al. 2001). Their experiment illustrated the meaningfulness of isovist measures, however, the most important wayfinding behavior—orientation—was not considered. Davies et al. (2006) proposed to use isovist features and build prediction models for spatial orientation and implied that correlations among isovist measures have important implications. By respecting the spatial configuration in indoor environments, we can derive measures for describing the imagined spatial perception of moving persons. Important for these cases are fields of vision (Schwab 2016), which are isovists that are connected with individual vantage points (Turner et al. 2001). This is strongly connected with the previously mentioned wayfinding strategies, mainly due to the fact that isovist measures are information sources for individual decision making. The base for these findings comes from previous studies by Conroy (2001), Wiener et al. (2011) and Schneider and König (2012), who evaluated the potential of isovist measures by providing different case studies in indoor environments.

3.3 Representation of Navigable Space

Shekhar et al. (2016) state that a major research question on indoor localization includes the conversion of indoor floor plans (e.g. CAD drawings) into navigable maps. Lorenz et al. (2013) argue that the map design of indoor floor plans is more complicated to realize than outdoor environments since people usually walk across different elevation levels of buildings. This could be solved by including 3D indoor space visualization (Brown et al. 2013) or by a combination of 2D floor plans for different elevation levels. Cartographic considerations on floor plans show that most examples are inappropriate as they are non-generalized and too detailed for simple orientation (Lorenz et al. 2013).

In general, there are still just few publications on cartographic design guidelines for indoor maps (Lorenz et al. 2013). Nevertheless, Puikkonen et al. (2009), May et al. (2003) and Vinson (1999) deliver some results from provided user studies on created indoor environment maps.

Other questions connected with floorplans and navigable maps consist of estimating reliabilities of indoor positioning and the handling of missing indoor building information (Shekhar et al. 2016).

Richter et al. (2009) solve the question of how to represent indoor space by introducing a hierarchical representation of indoor spaces. In general we can state that it is possible to describe indoor environments similar to outdoor environments as an arrangement of elements in space, which is also referred to as spatial configuration (Schwab 2016). This spatial configuration consists of topologic space information together with indoor geometry and the arrangement of objects (Frankenstein et al. 2010), which is the key information for indoor navigation in many approaches (Brown et al. 2013).

4 Methodology

Our concept combines two different approaches for classifying indoor environments, namely isovist analysis and archetypal analysis. The former focuses on estimated human senses for optical perception of inner building structures. The latter makes use of given features, in our case isovist measures, and derives extrema of their appearance, namely archetypes. Like Krisp et al. (2010, 2012a, b) we have used the main building of the Technische Universität München (TUM) in Munich for our case study, which is characterized by its complexity and the high number of entrances. The two mentioned aspects are mainly caused by the possibility to enter varying elevation levels differently and by the diversity of options to traverse the building by using more than twenty entrance points. There is a high number of alternative routes to be expected, which will be one of the leading aspects for our case study. The difference between our idea and previous visibility analysis approaches that use isovist measures in indoor environments is the way of inspecting the calculated isovist measures: all measures are utilized simultaneously for calculating the archetypes. The aim of this approach is to design a functional segmentation method for indoor spaces. This includes the classification of indoor areas and its surroundings into values like entrance areas, corridors, halls or streets.

The actual investigation of the maps and routes represented in a multidimensional feature space will be realized using an extension of the archetypal analysis framework proposed by Feld et al. (2015). See Fig. 1 for an overview of the algorithm's workflow.

4.1 Input Requirements

As mentioned in the previous section, we adopt a 2D representation to depict the study environment. The representation can be generated from floor plans or street maps or the mix of both. 2D polygons represent walkable areas, their boundaries

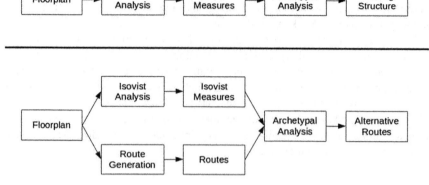

Fig. 1 Workflow for analyzing a given floorplan (*top*) and a set of routes (*bottom*)

and inner holes represent restrictions such as walls, installations, and other obstacles. This representation is also reasonable for multi-level indoor environments. Each level of a building is represented in 2D separately. It does no harm to isovist analysis when the calculation is based on a single level since ceiling and floor are nontransparent. To calculate isovist fields for the full description of the environment, we use regular grid tessellation and select the geometric center of each cell as the representative location. The appropriate grid resolution should be determined to capture meaningful properties of the environment and fit the application scenario. A too coarse resolution may fail to reveal changes of isovist features in transitional areas like doorways or turnings. A 0.5 m resolution is enough for our case study. A finer resolution will reveal more details, however leads to more computing costs (time and storage). Besides, 0.5 m is also a reasonable approximation of body-size and normal walking step length for human navigation and wayfinding scenarios.

A set of routes between a given start and goal will be created using the penalty algorithm as proposed in Werner and Feld (2014). The algorithm works on plain bitmaps and creates a node for each white pixel and an edge for all neighboring white pixels. The edge weight is set to 1 for horizontal or vertical edges and accordingly to $\sqrt{2}$ for diagonal edges. The algorithm iterates between performing shortest path routing using Dijkstra's algorithm (Dijkstra 1959) and increasing the edge weights of the shortest path just found. Thus, by iteratively increasing the edge weights the resulting shortest path will change with time.

4.2 Feature Extraction

In this paper we extract isovist features by following Benedikt's definition and measurements (Benedikt 1979). A single isovist is the set of all points in space that

Fig. 2 Definition of an isovist by line-of-sight and its corresponding features

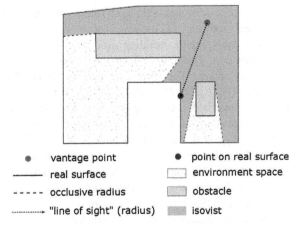

are visible from a given point (vantage point) in the space. Figure 2 illustrates an exemplary isovist with occlusive radius and line-of-sight.

The six measurements are:

(a) A_x, the area of the isovist;
(b) P_x, the real-surface perimeter of the isovist that indicates the amount of obstacle surface visible from a vantage point;
(c) Q_x, the occlusivity of the isovist measures the length of the occluding radial boundary;
(d) $M_{2,x}$, the variance of the radius measures the distribution of the radials length;
(e) $M_{3,x}$, the skewness of the radius measures the asymmetry of the distribution of the radius length;
(f) N_x, the circularity of the isovist which can be calculated by the following formula,

$$N_x = |\partial V_x|^2 / 4\pi A_x$$

where $|\partial V_x|$ is the perimeter of the isovist.

We calculate the six isovist measurements using the given floorplan in vector format and rasterize the calculated values afterwards for further processing.

4.3 Clustering Building Structures and Routes

As stated in the paper's introduction, we focus on the extraction of insights about a building's structure as well as on the identification of a small set of proper alternative routes between two given points. For doing so, we need to transform the given map, isovist features, and routes into a form that can be processed by archetypal analysis.

When analyzing the building structure we consider each pixel of the floorplan as an observation. Thus, we have $N = r \times c$ observations, whereas r is the number of pixel rows and c is the number of pixel columns. Each observation has got $m = 6$ attributes $\{A_x, P_x, Q_x, M_{2.x}, M_{3.x}, N_x\}$. Thus, the input of the archetypal analysis is the $N \times m$ matrix X consisting of all the floorplan's pixels each having six features.

The clustering of a (huge) set of routes each having the same start and goal behaves slightly different. Each route consists of multiple pixel coordinates indicating the trajectory's course. We define each route to be an observation and for each route pixel we determine the corresponding isovist measures from the previously calculated matrix. For each route and for each of the six isovist measures we calculate the minimum, maximum, mean, median and variance. Thus, the input of the archetypal analysis is the $N \times m$ matrix X consisting of all the given routes each having $5 \times 6 = 30$ attributes.

5 Results and Discussion

We focus our discussion on a real world scenario, namely the main building of the Technische Universität München (TUM) and its surrounding area (Theresienstraße, Arcisstraße, Gabelsbergerstraße, Luisenstraße). We have simplified the original floorplan (e.g. removal of doors) in order to be compatible with the route generation algorithm in use.

5.1 Clustering the Map

Our first contribution is about identifying functional space inside a given, unlabeled floorplan. Thus, we perform archetypal analysis on the floorplan's isovist measures. We utilized each walkable pixel as an observation and calculated the corresponding six isovist measures as the attributes.

Archetypal analysis works by approximating the convex hull of the observations in the multidimensional feature space. A common way to estimate a suitable value for the number of archetypes k is to inspect the resulting approximation errors, i.e. the residual sum of squares (*RSS*). A flattening of the curve indicates that a further addition of an archetype would not help to improve the accuracy of the approximation. For our evaluation we have performed archetypal analysis with numbers of archetypes ranging from $k = 1$ to $k = 10$ and repeated the calculations multiple times to prevent local minima. The scree plot of the resulting *RSS* as shown in Fig. 3 suggests to choose a value of $k = 4$, $k = 5$, or even $k = 7$. From an application point of view a high number of archetypes like $k = 7$ harbors the danger that the

Fig. 3 Scree plot showing the value of k and the resulting residual sum of squares (RSS)

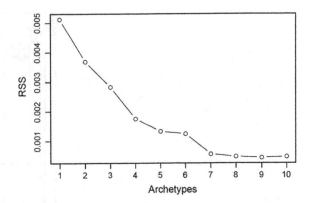

interpretation gets confusing. Thus, for the further course of this section we will focus on up to five archetypes.

Archetypal analysis is a so-called fuzzy clustering method meaning that each observation is represented by a convex combination of the identified clusters, i.e., the archetypes. We now discuss the points in the map that have a certain level of assignment to an archetype. Graphically spoken, we highlight and discuss the observations (pixels) that are some kind of "near" to an archetype in the feature space. Thus, we assign a pixel to an archetype if the corresponding value in the coefficient matrix α is higher than a defined threshold. Of course, the threshold for the definition of an "archetypal pixel" can be varied, but for the sake of clarity and visualization we restrict ourselves to the representative threshold of $\alpha > 0.5$.

The case $k = 2$ is often hard to interpret: If the analyzed data set contains several natural clusters, it happens that they get uncontrollably combined. The left-hand side of Fig. 4 shows the parallel coordinates plot for $k = 2$. The abscissa shows the six isovist features, the gray lines represent all observations (i.e. all walkable pixels) and the colored lines the identified archetypes. Basically, the red archetype can be summarized as having high isovist measures while the green archetype has got low values. Just the skewness of the radials $M_{3,x}$ seems quite similar. Roughly speaking, high isovist measures indicate that the point of view is a good lookout (large isovist area A_x) and that in turn involves much obstacle surface to be seen (high real surface P_x). In the case at hand, the green archetype has got very low values. In particular the low radial's variance $M_{2,x}$ indicates that the area of view (the isovist itself) has got a more regular shape. The right-hand side of Fig. 4 shows the corresponding coloring of pixels; it accompanies with our preliminary interpretation. Basically, the red pixels represent places in the floorplan where large parts of the space can be seen. Green pixels are the opposite of that, since the view is strongly restricted.

The left-hand side of Fig. 5 focusses on $k = 3$. It shows that the green archetype stayed nearly the same. In fact, all the values are lower than when using $k = 2$. The red archetype, however, still has got high levels of values, but they are somehow more "radical". The isovist's area A_x is extremely high and the moderate real-surface P_x, the high occlusivity Q_x, and the high variance $M_{2,x}$ indicate a more "diversified" line-of-sight. Again, this archetype will be something like open space.

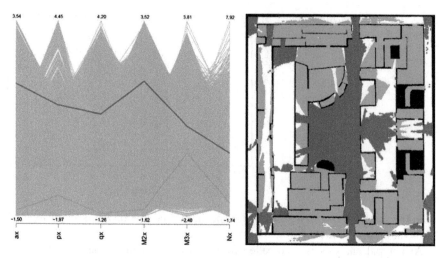

Fig. 4 Parallel coordinates plot for $k=2$ (*left*) and correspondingly colored pixels with a threshold of $\alpha > 0.5$ (*right*)

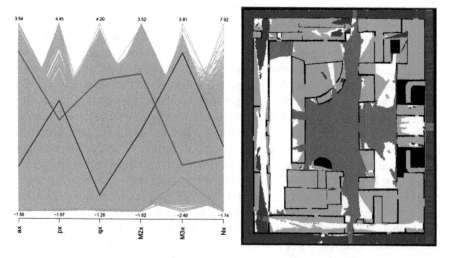

Fig. 5 Parallel coordinates plot for $k=3$ (*left*) and correspondingly colored pixels with a threshold of $\alpha > 0.5$ (*right*)

The new blue archetype is interesting, since the moderate area of visibility A_x, plenty of walls to be seen (high P_x) and the low values of occlusivity Q_x suggest a very uniform and simplistic structure. If we follow the initial interpretation by looking at the right-hand side of Fig. 5, we see red pixels representing free and open space, but with one obvious difference to the results of $k=2$: the structure of the red area is more compact and just pixels having a "large outlook" are selected.

The green archetype can again be interpreted as areas where the view is restricted, such as rooms or smaller halls. The new blue archetype can be interpreted as points in the map where the field of vision is very restricted in two directions and very wide in the other two directions, like in narrow hallways—or as in our case—the streets around the building.

Please note the white pixels of the floorplan indicating that the corresponding α values are all below the threshold. This means that the observations are somehow poor to be described using the identified archetypes. When looking at the white area at the left-hand side of the floor plan in Fig. 5 it is noticeable that there are properties from each archetype: Somehow the area is room-like due to the regular shape and the restricted view, somehow it is open space due to its large area and somehow it is like a hall or a street due to the length.

Figure 6 shows the top pixels for $k=4$ (left) and $k=5$ (right), the number of archetypes that are most suitable at least when inspecting the scree plot in Fig. 3. The corresponding parallel coordinate plots (not shown) and a visual inspection of the pixels indicate that the green archetype ("rooms") and blue archetype ("halls or streets") stayed nearly the same, just the interpretation of "open space" is split. We still have the red archetype indicating places with a large view, but additionally there is the turquoise archetype that has got deep views into entrances (very high occlusivity Q_x, i.e. the length of the occluding radial boundary). The right-hand side of Fig. 6 refines that setup further with the pink archetype where a spectator would see much area while having a wall behind his or her back.

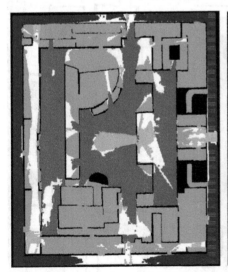 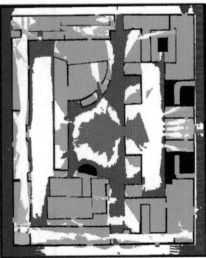

Fig. 6 Colored pixels for $k=4$ (*left*) and $k=5$ (*right*) for a threshold of $\alpha > 0.5$

5.2 Identifying Alternative Routes

Our second contribution is about analyzing the effect of isovist features on the clustering of a set of routes between two given points in order to generate a small set of alternative routes. To this end, we created sets of routes between multiple pairs of starts and goals inside floorplans. In the work at hand we used a representative result for a set of 400 routes going from the main entrance of the building to another entrance at the opposite side of the building. We used the penalty algorithm as proposed in Werner and Feld (2014) to create the routes.

Figure 7 shows a plot of the resulting routes that traverse the map in multiple ways ranging from completely detouring the building, going with variations through several rooms, transiting the patio, etc. Through visual analysis of the routes created, one can identify a strong correlation with real movement flows of students during semesters on working days. Like stated before, our second contribution is about extracting different archetypes of alternative routes, thus we try to reduce the size of the set of routes from 400 to a low single-digit number. In terms of archetypal analysis we consider each route as an observation. For each route, we calculate the minimum, maximum, mean, median, and variance of the isovist measures. Thus, for each of the 400 routes we have got $5 \times 6 = 30$ attributes.

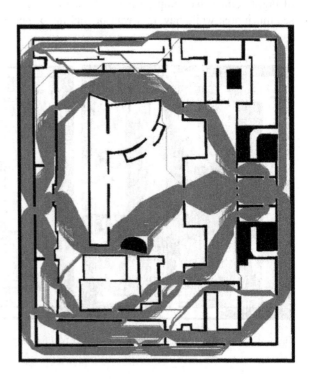

Fig. 7 Set of routes used for the discussion

Fig. 8 Scree plot showing the value of k and the resulting *RSS*

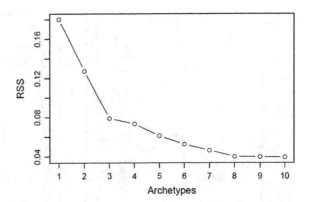

As in the previous section we consult the scree plot of the *RSS* to check for an indication of a proper value of k (see Fig. 8). There is a strong bend of the decrease at $k = 3$ visible, what also holds as an appropriate value from an application point of view.

The calculated archetypes, i.e., the configurations of features, do not necessarily exist or can be observed. Thus, and according to Feld et al. (2015), we call the archetypes' nearest neighbor in the feature space the "realized archetypal route". Figure 9 shows the corresponding archetypal routes for $2 \geq k \geq 5$.

The top-left part of Fig. 9 shows the case for $k = 2$. The red route traverses the map through the center while having a large view in the patio and proceeds quite twisting and winding to the goal. The green archetype is, at least to this end, the complete opposite. It is a straight path with regular turns and a more restricted view. It follows the street around the building and traverses some rooms in the northern and western part of the building.

The result for $k = 3$ is shown in the top-right part of Fig. 9. The red archetype is again a windy route going through the open space. The green archetype has changed a bit, since it is more "extreme" in being very straight with a minimum number of turns. The new blue archetype is quite complicated; it traverses the map in a diverse way and has got narrow parts that go through different rooms, doors, and hallways.

When calculating $k = 4$ archetypes (bottom-left part of Fig. 9) we basically get the same three routes as before, although archetypal analysis is not nested per se. Additionally, there is the turquoise archetype being a kind of mix between the green and blue archetype. This route is quite straight with few turns, but mostly uses rooms and doors instead of the more extreme, i.e., even straighter, green route which uses the road.

The bottom-right part of Fig. 9 shows the results for $k = 5$. The routes are identical to the ones before, but additionally, the new pink route is—at least at a first sight—very similar to the red one. It also traverses the patio but runs around the isolated building to the top.

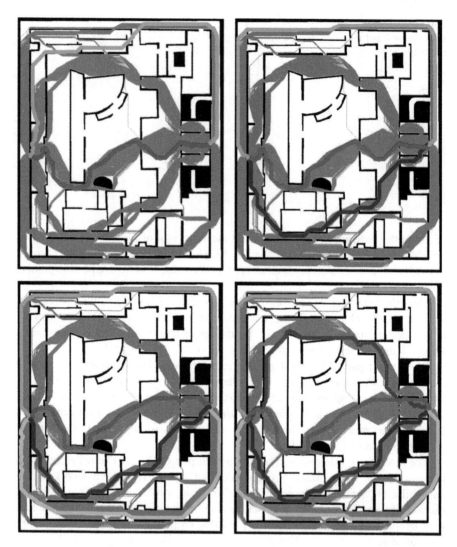

Fig. 9 Archetypal routes for $k=2$ (*top left*), $k=3$ (*top right*), $k=4$ (*bottom left*), and $k=5$ (*bottom right*)

All five routes are proper alternatives with very different characteristics. Beginning from $k=6$ the resulting routes start to "collapse". That means that at least two archetypal routes lie over another (note: in the observation space, not in the feature space).

5.3 Sets of Archetypal Routes

Just focusing on a single representation of an archetype can sometimes bias the interpretation. Just like with the clustering of the map, we now focus not only on the nearest neighbor of each archetype, but on the observations that reach a certain threshold regarding the coefficient matrix α.

Figure 10 shows this for $k = 3$ and a threshold of $\alpha > 0.8$. It can be clearly seen that the identified archetypes are based on the impression while traversing the building and not on their geographic location. The green archetype consistently traverses the patio having several variations at the start and the end of the route. The blue archetype characterized by going very straight through the streets or long and narrow halls can be found not only at the top of the floorplan, but also in variations at the bottom of the map. The variations, be it shortcuts or detours, are located in rather narrow spaces. Finally, also the red archetype is variable and consistently follows variations of rooms and doors, but all the time in a quite complicated fashion.

Summarized, the resulting sets of routes are useful in application scenarios where variety is desirable. Think of computer games where non-player characters choose a set of archetypal routes based on their strategy and out of this set a random concrete route to surprise the player. Another use case is a navigation system for pedestrians that proactively prevents bottlenecks via non-deterministic route suggestions.

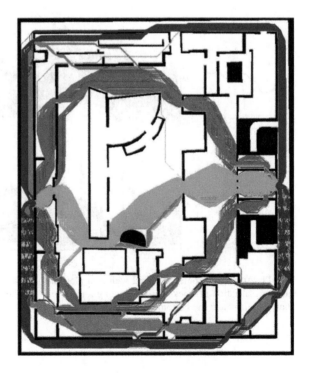

Fig. 10 *Top* routes for $k = 3$, i.e. routes with a corresponding coefficient value of $\alpha > 0.8$

5.4 Archetypal Routes: Features Versus Time

To further interpret the properties of archetypal routes, we need to take a closer look at the isovists of each of the selected locations along the route. To keep the explanation clear and simple we have chosen the archetypal routes for $k=2$ as an example. We also adopt a visual analytics way rather than using statistical analysis as in many isovist related works already done.

Figure 11 depicts the case of archetypal routes for $k=2$. The top row corresponds to the red route and the bottom row to the green route in Fig. 9, top-left. The six measurements are normalized and plotted in a stacked area chart (left-hand side). Isovists for each location are centered to have the same vantage point in order to form a radar-like view (right-hand side). This visualization is related to the Minkowski Model (Benedikt 1979), however, the original model is difficult to perceive clearly when hundreds of isovists are stacked along a route. Our visualization loses time dimension information, but it is sufficient to capture the pattern changes of isovists along different routes. In this example, we can already find obvious differences between the two archetypal routes indicating that our method does not only find routes that are geometrically different, but also with different perceptual properties. There are still details to be extracted from this visualization in future work.

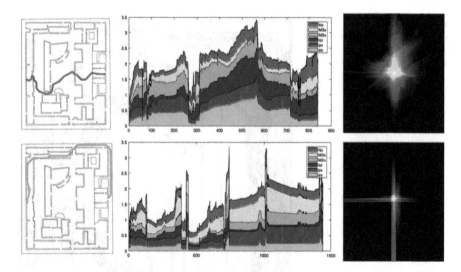

Fig. 11 Archetypal routes for $k=2$ with area plot for the normalized isovist features (*mid*) and radar visualization of stacked isovists along a route (*right*). The top row corresponds to the red route (*top left*) and the bottom row to the green route (*bottom right*) in *Fig.* 9 top left

6 Conclusion

This paper proposed to use archetypal analysis, a fuzzy clustering method, to gain insights of an environment based on modelled perceptual properties (visibility). Our approach is able to find archetypes, i.e., extreme types, of perceptual properties of the given environment. By applying this information we can create proper alternative routes going through a building that enable different user experiences when traversing the environment. Previous psychological and behavioral research tried to establish correlations between perceptual properties and wayfinding behavior. Our method enables the creation of routes with extremely different visibility properties. Additionally, we segment 2D Euclidean space into areas with fundamental differences depending on their visibility (based on the isovist feature values).

In future work we want to investigate other isovist measurements for extracting environment information and for creating alternative routes. Also we try to focus on the measurements along the routes using the stacked isovist visualization and visual analytics. Finally, different route generation algorithms together with routes having different start and goals need to be examined.

References

Batty M (2001) Exploring isovist fields: space and shape in architectural and urban morphology. Environment and planning B: Planning and Design, 28(1), 123–150.

Benedikt M (1979) To take hold of space: isovist and isovist fields. Environment and Planning B 6 (1): 47–65

Best G (1970) Direction finding in large buildings. In: Canter DV (ed) Architectural Psychology - Proceedings of the conference at Dalandhui. London: RIBA, pp. 72–75

Brown G, Nagel C, Zlatanova S, Kolbe TH (2013) Modelling 3D Topographic Space Against Indoor Navigation Requirements. In: Pouliot J, Daniel S, Hubert F, Zamyadi A (eds) Progress and New Trends in 3D Geoinformation Sciences. Springer, Berlin, pp. 1–22

Camvit (2013): Cambridge Vehicle Information Technology Ltd. - Choice Routing. Online, 2013. http://www.camvit.com/camvit-technical-english/Camvit-Choice-Routing-Explanation-english.pdf.

Conroy R (2001) Spatial Navigation in Immersive Virtual Environments. The faculty of the built environment, University College London, London, U.K., Dissertation

Cutler A, Breiman L (1994) Archetypal analysis. Technometrics 36(4): 338–347

Davies C, Mora R, Peebles D (2006) Isovists for Orientation: Can space syntax help us predict directional confusion? In: Proceedings of the 'Space Syntax and Spatial Cognition' workshop, Spatial Cognition 2006, Bremen, Germany, 24 September 2006

Davis L, Benedikt M (1979) Computational models of space: Isovists and isovist fields. Computer graphics and image processing, 11(1), 49–72.

Delling D, Wagner D (2009) Pareto paths with sharc. In: Experimental Algorithms Springer, Berlin, pp. 125–136

Dijkstra EW (1959) A note on two problems in connexion with graphs. Numerische mathematic 1 (1): 269–271

Dogu U, Erkip F (2000) Spatial factors affecting wayfinding and orientation: A case study in a shopping mall. Environment and Behavior 32(6): 731–755

Emo B (2015) Exploring isovists - the egocentric perspective. In: Proceedings of the 10th International Space Syntax Symposium, London: Space Syntax Laboratory, 2015.

Emo B, Hölscher C, Wiener J, Dalton RC (2012) Wayfinding and spatial configuration:evidence from street corners. In: Proceedings of the 8th Space Syntax Symposion, Santiago de Chile, 2012, pp. 1–16

Eugster MJA, Leisch F (2009) From spider-man to hero – archetypal analysis in r. Journal of Statistical Software 30(8): 1–23

Eugster MJA, Leisch F (2011) Weighted and robust archetypal analysis. Computational Statistics & Data Analysis 55(3): 1215–1225

Feld S, Werner M, Schönfeld M, Hasler S (2015) Archetypes of Alternative Routes in Buildings. In: 6th International Conference on Indoor Positioning and Indoor Navigation (IPIN 2015), pp. 1–10

Frankenstein J, Büchner S, Tenbrink T, Hölscher C (2010) Influence of geometry and objects on local route choices during wayfinding. In: Hölscher C, Shipley T, Belardinelli M, Bateman M, Newcombe N (eds) Spatial cognition VII, LNAI 6222, pp. 41–53

Franz G, Wiener JM (2008) From space syntax to space semantics: a behaviorally and perceptually oriented methodology for the efficient description of the geometry and topology of environments. Environment and Planning B: Planning and Design 35(4): 574–592

Gartner G, Cartwright W, Peterson P (2007) Location based services and telecartography. Springer, Berlin

Gartner G, Ortag F (2011) Advances in location-based services. Springer, Berlin

Gartner G, Rehrl K (2009) Location based services and telecartography II. Springer, Berlin

Haq S, Zimring C (2003) Just down the road a piece: The development of topological knowledge of building layouts. Environment and Behavior 35(1): 132–160

Hartigan JA (1975) Clustering Algorithms. John Wiley & Sons, Inc., 1975.

Hillier B, Hanson J (1984). The social logic of space. Cambridge university press.

Hölscher C, Meilinger T, Vrachliotis G, Brösamle M, Knauff M (2005) Finding the way inside: Linking architectural design analysis and cognitive processes. Spatial Cognition IV - Reasoning, Action, Interaction, pp. 1–23

Hughes N, Pinchin J, Brown M, Shaw D (2015) Navigating In Large Hospitals. In: 6th International Conference on Indoor Positioning and Indoor Navigation (IPIN 2015), pp. 1–9

Jain AK, Dubes RC (1988) Algorithms for Clustering Data. Prentice-Hall advanced reference series. Upper Saddle River, NJ: Prentice-Hall, Inc.

Jain AK, Murty MN, Flynn PJ (1999) Data clustering: a review. ACM computing surveys (CSUR) 31(3): 264–323

Kaufman L, Rousseeuw PJ (1990) Finding Groups in Data: an Introduction to Cluster Analysis. John Wiley & Sons.

Kobitzsch M, Radermacher M, Schieferdecker D (2013) Evolution and evaluation of the penalty method for alternative graphs. In: Proceedings of the 13th Workshop on Algorithmic Approaches for Transportation Modelling, Optimization, and Systems (ATMOS'13), pp. 94–107

Krisp JM (2013) Progress in location based services (LBS). Springer, Berlin

Krisp JM, Ding L, Jin Y, Peer P (2012b) Indoor Routing - Is a centrality measure for an indoor routing network useful? Mobile Tartu 2012, 22.-25. August 2012, Tartu, Estonia

Krisp JM, Liu L, Berger T (2010) Goal Directed Visibility Polygon Routing for Pedestrian Navigation. In: Proceedings 7th International Symposium on LBS & TeleCartography, 20–22.. September, Guangzhou, China, pages pending

Krisp JM, Meng L (2013) Going local – evolution of location based services. In: Krisp JM (ed) Progress in location based services. Springer, Berlin, pp. 11–15

Krisp JM, Peer P, Ding L (2012a) Classification of an indoor routing network based on graph theory. GeoInformatics 2012, 15–17 June, Hong Kong, China

Li R, Klippel A, Liben LS, Christensen AE (2011) The Impact of Environmental Qualities and Individual Differences on Spatial Orientation in a Mobile Context. In: Proceedings of the

Workshop of Cognitive Engineering for Mobile GIS in conjunction with COSIT 2011, Belfast, ME, September 12, 2011, Vol. 780, pp. 9–16. CEUR-WS.org, ISSN 1613-0073

Lorenz A, Thierbach C, Baur N, Kolbe TH (2013) App-Free Zone: Paper Maps as Alternative to Electronic Indoor Navigation Aids and Their Empirical Evaluation with Large User Bases. In: Krisp JM (ed) Progress in Location-Based Services, Lecture Notes in Geoinformation and Cartography. Springer, Berlin, pp. 319–338

May AJ, Ross T, Bayer SH, Tarkiainen MJ (2003) Pedestrian navigation aids: information requirements and design implications. Personal Ubiquitous Computing 7: 331–338

Ohm C, Müller M, Ludwig B (2015) Displaying landmarks and the user's surroundings in indoor pedestrian navigation systems. Journal of Ambient Intelligence and Smart Environments 7(5): 635–657

Peponis J, Wineman J, Rashid M, Kim SH, Bafna S (1997). On the description of shape and spatial configuration inside buildings: convex partitions and their local properties. Environment and Planning B: planning and design, 24(5), 761–781.

Puikkonen A, Sarjanoja A-H, Haveri M, Huhtala J, Häkkilä J (2009) Towards Designing Better Maps for Indoor Navigation: Experiences from a Case Study. In: Proceedings of the 8th International Conference on Mobile and Ubiquitous Multimedia. ACM, New York, NY, USA, pp. 16:1–16:4

Raubal M (2002) Wayfinding in Built Environments: The Case of Airports. In: IfGIprints (Vol. 14). Institut für Geoinformatik, WWU Münster, Münster

Richter K-F, Winter S, Ruetschi U-J (2009) Constructing Hierarchical Representations of Indoor Spaces. In: Proceedings of the 2009 Tenth International Conference on Mobile Data Management: Systems, Services and Middleware (MDM 2009). IEEE, Los Alamitos, pp. 686–691

Ruppel P, Gschwandtner F, Schindhelm CK, Linnhoff-Popien C (2009) Indoor Navigation on Distributed Stationary Display Systems. In: 33rd Annual IEEE International Computer Software and Applications Conference (COMPSAC 2009), pp. 37–44

Schneider S, König R (2012) Exploring the Generative Potential of Isovist Fields. In: Achten H, Pavlicek J, Hulin J, Matejovska D (eds) Digital Physicality - Proceedings of the 30th eCAADe Conference - Volume 1, pp. 355–364

Schwab C (2016) Visibility-Based Obstacle Placing - Automated Obstacle Placing based on Circularity. Faculty of Informatics, Vienna University of Technology, Diploma Thesis

Seiler C, Wohlrabe K (2012) Archetypal Scientists. CESifo Working Paper Series 3990, CESifo Group Munich, 2012.

Shekhar S, Feiner SK, Aref WG (2016) Spatial Computing. Communications of the ACM 59(1): 72–81

Tandy CRV (1967) The isovist method of landscape survey. In: Methods of Landscape Analysis.

Turner A, Doxa M, O'Sullivan D, Penn A (2001) From isovists to visibility graphs: a methodology for the analysis of architectural space. Environment and Planning B: Planning and Design 28: 103–121

Viaene P, Vanclooster A, Ooms K, De Maeyer P (2014) Thinking Aloud in Search of Landmark Characteristics in an Indoor Environment. In: Ubiquitous Positioning Indoor Navigation and Location Based Service (UPINLBS 2014), pp. 103–110

Vinson NG (1999) Design guidelines for landmarks to support navigation in virtual environments. In: Williams MG (ed) Human factors in computing systems: Chi 99 conference proceedings the CHI is the limit; conference on human factors in computing systems, association for computing machinery. Addison-Wesley, New York, Harlow, pp. 278–285.

Werner M, Feld S (2014) Homotopy and alternative routes in indoor navigation scenarios. In: Proceedings of the 5th International Conference on Indoor Positioning and Indoor Navigation (IPIN'14).

Wiener J, Franz G (2004) Isovists as a means to predict spatial experience and behavior. In: Spatial Cognition IV - Reasoning, action and interaction. Proceedings of the International Conference on Spatial Cognition 2004, Lecture Notes in Computer Science. Springer: Berlin, pp. 42–57.

Wiener JM, Hölscher C, Büchner S, Konieczny L (2011) Gaze Behaviour during Space Perception and Spatial Decision Making. Psychological Research 76(6): 713–729

Yang L, Worboys M (2011) A navigation ontology for outdoor-indoor space. In: Proceedings of the 3rd ACM SIGSPATIAL Int'l workshop on indoor spatial awareness, pp. 31–34

Zheng Y, Zhou X (2011) Computing with spatial trajectories. Springer Science & Business Media, 2011

Generation of Meaningful Location References for Referencing Traffic Information to Road Networks Using Qualitative Spatial Concepts

Karl Rehrl, Richard Brunauer, Simon Gröchenig and Eva Lugstein

Abstract Location referencing systems (LRS) are a crucial requisite for referencing traffic information to a road network. In the past, several methods and standards for static or dynamic location referencing have been proposed. All of them support machine-interpretable location references but only some of them include human-interpretable concepts. If included, these references are based on pre-defined locations (e.g. as location catalogue) and often miss meaningful interlinking with road network models (e.g. locations being simply mapped to geographic coordinates). In a parallel research strand, ontological concepts for structuring road networks based on human conceptualizations of space have been proposed. So far, both methods have not been integrated. The current work closes this gap and proposes a generation process for meaningful location references on top of road networks based on qualitative spatial concepts. A prototypical implementation using OWL, Neo4J graph database and a standardized nationwide road network graph shows the practical applicability of the approach. Results indicate that the proposed approach is able to bridge the gap between existing road network models and human conceptualizations on multiple levels of abstraction.

Keywords Digital road networks · Location referencing · Qualitative spatial concepts

1 Introduction

One of the crucial aspects of digital road information such as traffic events, real-time traffic information or road condition information is concerned with the referencing of information entities to a digital representation of a road network. Usually this is accomplished by using GIS-based location referencing methods (Hackeloeer et al. 2014). In the past, several different methods and data models for

---

K. Rehrl (✉) · R. Brunauer · S. Gröchenig · E. Lugstein
Salzburg Research Forschungsgesellschaft mbH, Salzburg, Austria
e-mail: karl.rehrl@salzburgresearch.at

© Springer International Publishing AG 2017
G. Gartner and H. Huang (eds.), *Progress in Location-Based Services 2016*, Lecture Notes in Geoinformation and Cartography,
DOI 10.1007/978-3-319-47289-8_9

location referencing have been proposed (Hackeloeer et al. 2014; Nyerges 1990; Scarponcini 2002; Vonderohe et al. 1997; Zhou et al. 2000). In general, the different methods and data models can be categorized into three approaches: the first approach uses shared road network models (e.g. each information processing system has to use the same road network model), the second approach is based on dynamic location referencing (e.g. information processing systems code location references in a way that these references can be mapped on different road network models) and the third approach uses shared catalogues of pre-defined locations (e.g. a standard set of well-known locations is defined which is then used by different information processing systems). All approaches have in common that they primarily support machine-interpretable locations. Examples of the first approach are Geographic Data Files (GDF) (ISO 14825 2011) or Austria's National Transport Graph GIP.at (Mandl-Mair 2012). OpenLR[1] or TPEG-ULR (Ernst et al. 2012) are examples of the second approach. The third approach also provides human-interpretable location references. In the past, two standard formats for expressing human-interpretable location references have been proposed, namely TMC location tables (ISO 14819-3 2013) and TPEG-Loc by the Transport Protocol Expert Group (TPEG).[2] These location referencing systems provide human- as well as machine-interpretable location references based on a catalogue of pre-defined locations such as road junctions, road sections or points of interest. However, pre-defined location catalogues are typically not interlinked with digital road network models, but are simply mapped to geographic references such as WGS84. Moreover, the location tables often have to be derived through a time-consuming manual process which hinders frequent updates. In another research strand, authors have proposed qualitative concepts for structuring road networks based on human conceptualizations of space for supporting human-interpretability of road information (Car and Frank 1994; Timpf et al. 1992; Wang and Meng 2009). These approaches model a road network on multiple levels of abstraction and therefore provide valuable contributions for cognitively adequate location references which can be read and interpreted by human beings as well. So far, these approaches have not been integrated into location referencing systems. Furthermore, a process for generating such information automatically on top of digital road network models is missing.

In this work we propose an integrated process for enhancing digital road network models with qualitative spatial concepts on multiple levels of abstraction. Therefore, cognitively adequate abstraction levels and qualitative spatial concepts are derived from natural language traffic messages and formally defined by a multi-level ontology. The main contribution of the work is the description of a process for generating individuals of the ontology (location references) on top of arbitrary road network graphs. A prototypical implementation based on the ontology modelling language OWL and Neo4J graph database shows the practical

---

[1] http://www.openlr.org/.
[2] http://tisa.org/technologies/tpeg/.

applicability of the approach. The proposed ontology and the generation process are tested with data from Austria's National Transport Graph (GIP.at). Results show that the proposed approach is able to enhance digital road networks with qualitative spatial location references on multiple levels of abstraction to enable meaningful location referencing.

The remainder of the work is structured as follows: Sect. 2 discusses related work with respect to location referencing approaches. Section 3 proposes the ontology classes as well as their relationships. In Sect. 4 a prototypical implementation based on OWL and Neo4J graph database is presented. Section 5 discusses results from applying the approach to a nationwide road network graph and Sect. 6 concludes the work.

2 Related Work

GIS-based modelling of road networks for referencing road or traffic information has been addressed by several research strands. One of the first proposals came from Nyerges (1990). Nyerges proposed three different locational reference strategies for highways, namely (1) road name and milage, (2) control sections with equidistant or variable lengths and (3) link (chain) and node. Although the proposal solely addresses highways, it can be adapted to other roads as well. Vonderohe et al. (1997) proposed a generic data model for linear referencing systems which may be considered the foundation of most location referencing systems. The authors proposed anchor points (e.g. intersections) and anchor sections (e.g. links between intersections) as natural segmentations of road networks and proposed road segmentation strategies as linear references. Scarponcini (2002) extended this approach with a generalized model for linear referencing in transport which takes into account different linear referencing methods. Curtin (2007) proposed a comprehensive process for linear referencing based on the generalized model by Scarponcini. In another noticeable work Curtin et al. (2007) took this model into account for discussing general principles of network analysis in geographic information science. Nowadays, most standard formats for describing static or dynamic road network references such as GDF, TMC, TPEG-Loc, TPEG-ULR or OpenLR use similar models with some form of linear referencing. However, most of these models are designed with respect to data exchange between systems, not taking cognitive spatial concepts into account. TMC, TPEG-Loc and TPEG-ULR also provide human-interpretable formats, but have to be either manually maintained and are limited to catalogues of pre-defined locations without providing the possibility to link concepts with existing road network models.

Beside the technical approaches, authors proposed conceptual modelling approaches of road networks taking human conceptualizations of space into account. Timpf et al. (1992) proposed a conceptual model for highway navigation with three levels of abstraction, namely *planning*, *instruction* and *driver* level. For each of the different abstraction levels, spatial concepts for describing a road network are defined. Car and Frank (1994) proposed a hierarchical algorithm for path search in large road networks based on human conceptualizations. More recent works also proposed ontological modelling approaches of road networks: Timpf (2002) presented an ontology of wayfinding from a traveler's perspective. The most recent approach with respect to our work comes from Wang and Meng (2009). The authors proposed a hierarchical ontology for modelling road networks on multiple scales. However, they did not consider the conceptualization of road networks from a traffic information perspective—their concepts are tailored to navigation and routing. Moreover, their approach does not take existing road network models into account.

Our approach of generating meaningful location references on top of digital road network models is placed somewhere in the middle of the presented previous approaches and tries to integrate them: On the one hand it takes into account the perspective of human conceptualizations of road networks, as proposed by Timpf and other authors, and on the other hand, it considers the more technically-oriented approaches, as anchor point theory, static and dynamic linear referencing. The approach bridges the gap between purely static and purely dynamic location referencing systems by proposing a process for automatic generation of meaningful location references as enhancement of digital road network models. Therefore, qualitative spatial concepts modelled with the standardized web ontology language OWL (Bechhofer 2009) are used. This enables the management and query of location references by using standard tools from the semantic web community like graph databases or query languages such as Cypher or SPARQL.

3 Qualitative Spatial Concepts for Road Network Referencing

For generating meaningful location references on top of digital road network models, we firstly define cognitive spatial concepts which are commonly used by road users in their everyday language. As starting point, we analyzed spatial references being used in traffic messages. For example, a natural language traffic message could be composed as follows: "On motorway A1 in travel direction Salzburg between exit Mondsee and Thalgau at kilometer 231 be aware of a broken vehicle". From this and similar examples, we derive the following structure of human-interpretable spatial references in the context of traffic messages: (1) the message uniquely identifies a road by its reference number or a well-known name, (2) the message identifies the driving direction by using qualitative spatial direction concepts, (3) the message identifies the relevant road section (this can be optional if the message is related to the whole road) and (4) if available and necessary, the message identifies a more detailed linear reference of the event (e.g. mileage point). From this example it gets obvious, that human-interpretable descriptions of traffic-related events contain spatial concepts on at least three cognitive levels, namely road, section and optionally a more detailed segmentation such as a road segment or a linear reference such as a single location or a dynamic segment. For separating sections the cognitive spatial concept of a junction (including interchanges and intersections) is used and for identifying driving directions we propose qualitative direction concepts.

Therefore, our conceptual modelling approach resulted in six concepts for representing road networks on different conceptual levels, namely **Road**, **Section**, **Junction**, **Segment**, **LinearReference** and **Direction**. For the definition of the concepts and the relationships we integrated results from previous approaches discussed in the related work section. As proposed by Guarino (1998), we define these concepts and relationships as ontology.

3.1 Spatial Concepts

Road: On the most abstract cognitive level we define **Road** as the basic concept. A road is determined by its (at least locally) **unique name** or **reference number**. The concept of a road is used in different levels of abstraction since a national road, for example, could cover long distances but may also include different local roads. Higher level roads are typically designed to connect prominent places such as cities or villages and local roads are designed to connect city districts. Due to their prominent nature in any road network and due to their specific characteristics (e.g. separate driving directions, on/off ramps, etc.) we define **Motorway** as a sub-concept of Road. Prominent examples of roads in Austria are the federal road "B1—Wiener Bundesstraße" or the motorway "A1—Westautobahn". It is worth to mention that the ontology could contain several other sub-concepts of roads, but in our case we keep it simple and for most traffic messages both concepts are sufficient to distinguish. If necessary, the ontology can be easily extended with additional sub-concepts (Fig. 1).

Section: On the next cognitive level we define the concept **Section**. A **Section** is defined as part of a road with start and end point at prominent junctions with other roads. This concept has been defined in analogy to anchor points and anchor sections as proposed by Vonderohe et al. (1997). For example, the part of "A1—Westautobahn" motorway in Austria between the exits "Mondsee" and "Thalgau" is modelled as a **Section**. In case that a road starts or ends without a junction, than start or end node of the road is selected as section start or end. As sub-concept of **Section** we define the concept **Ramp** for representing sections connecting sections of different roads.

Segment: On the lowest cognitive level we define the concept of a **Segment**. A **Segment** separates a section in more detailed parts, being characterized by equal attributes (e.g. form, lanes, width) or more detailed anchor nodes (e.g. minor roads being connected to the road). For example, the part of a road section between two non-prominent nodes can be characterized as a road segment. This concept may be used for linking the higher-level concepts with any other road network model being

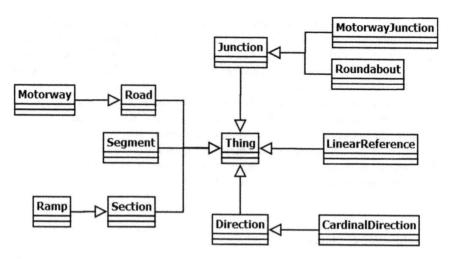

Fig. 1 The basic ontology of spatial concepts and selected sub-concepts

typically modelled on the abstraction level of segments only (sometimes also called links).

Junction: In our conceptual modelling approach, sections are the result of structuring the road network with respect to road interchanges and intersections. Therefore, we define **Junction** as central cognitive concept for structuring road networks. Accordingly, we defined two sub-concepts, namely **MotorwayJunction** (including all ramps as well as the road segments between the ramps) and **Roundabout** (including all segments of a roundabout) since both junctions have special characteristics and thus play a major role in mental models. It is worth to mention that a junction may be composed of several ramps and segments, e.g. in case that the junction is a motorway interchange or any other complex junction.

Direction: Until now we are able to model roads on different cognitive levels, but we are not able to indicate directions in a qualitative way. With respect to traffic information, each information entity has to be related to a unique driving direction. Thus, we complete our road ontology with the concept **Direction**. Similar to the LinearReference concept, the concept may be used on each cognitive level, representing the direction of a road, a section or a segment. From a technical perspective, directions of roads, sections or segments are represented by defining start and end node of the corresponding entity. While this approach is well-suited for systems, human beings are used to express directions with qualitative spatial concepts. One of the simplest qualitative concepts is the **Cardinal Direction** concept (e.g. 4- or 8-sector model) which classifies driving directions as cardinal directions (Frank 1996). However, depending on the cultural background, people more likely use other or additional qualitative direction concepts on different cognitive levels: (1) on a city level inbound or outbound direction in relation to the city center or names of prominent city districts, points of interest or junctions, (2) on a regional level names of prominent towns and villages or junctions, (3) on a

Table 1 The six concepts of the proposed ontology

Concept	Description
Road	A road describes a continuous infrastructure for driving with the same reference number or name
Section	A section splits a road according to well-known anchor points (e.g. junctions). The sub-concept Ramp connects a road or road section with another road or road section
Segment	A segment splits a section into several parts with equidistant or variable length. The segmentation of a road section is usually disjoint and complete
Junction	A junction is used to model the connection of two or more roads. A junction subsumes all sections that are part of the junction. This includes all ramps and deceleration/acceleration lanes but also the parts of the connected roads between the outer ramps
Direction	The directions is used to qualitatively express driving directions in a human-interpretable way
LinearReference	A LinearReference may be used to linearly reference arbitrary parts (single position or part) of a road, road section or road segment with a relative start and end offset relation to the entering point of the corresponding road, road section or road segment. LinearReferences may be point or line references

national level names of prominent cities and prominent junctions and (4) on an international level national borders or prominent cities. Since such qualitative direction concepts need further modelling effort, for now we limit our ontology to the concept of cardinal directions and postpone the definition of more detailed qualitative direction concepts to future work.

LinearReference: For referencing road information to arbitrary road parts (on each cognitive level), the literature proposes linear reference methods (LRM) as commonly agreed approach (Hackeloeer et al. 2014). A linear reference can be either static (e.g. pre-defined mileage points) or dynamic (e.g. dynamically referenced traffic events) and either a point reference (e.g. mileage point, traffic event at a single location) or a line reference (e.g. 100 m segments, traffic event of a certain length). For our road ontology, we propose a generic concept **LinearReference**. It is defined with relative start and end offsets in driving directions on each cognitive level (e.g. as offset to a road, section or segment). We propose to define more detailed linear reference concepts (e.g. MilageLocation, 100MeterLinearReference or EventLocation) as sub-concepts of LinearReference if needed (Table 1).

3.2 Relationships

Until now, we have defined the spatial concepts but we excluded the relationships between these concepts. From a cognitive perspective, road-related spatial references are typically derived from three hierarchical spatial reference frames, namely (1) an international reference frame (e.g. international reference numbers for motorways), (2) a national reference frame (e.g. nationwide names or references for national roads) and (3) a local reference frame (e.g. local names for regional or national roads within a city). For example, for messages within a city center people use local road names, whereas outside the city center nationwide road names are used. Accordingly, on long distance travels people switch to an international reference frame. Since we conceptualize roads according to their name or reference, the same physical road may be conceptualized twice or threefold in our ontology. For example, a road may be conceptualized by its local name (e.g. "Ignaz-Harrer-Straße") and by its national name (e.g. "B1—Wiener Bundesstraße"). Now, one could argue that the relationship between both road conceptualizations has to be modelled as is-part-of-relationship, but this is not generally valid. For example, in the city of Salzburg, only parts of the local road "Ignaz-Harrer-Straße" belong to the national road "B1—Wiener Bundesstraße". Other parts belong to the national road "B155—Münchner Bundesstraße". Therefore, we decided to model this relationship not on the road-level, but on the section-level since both parts of the "Ignaz-Harrer-Straße" are modelled as separate sections anyway. Furthermore, since sections may vary between local and national roads (due to different naming strategies), we model the section-to-section-relationship as hierarchical relationship (*isSubSectionOf*). Consequently, a section may have several relationships, namely being part of different roads (a local and a national road) but also being part of other sections. This modelling strategy allows us to express the cognitive hierarchical relationship of changing spatial reference frames between local and national frames seamlessly.

4 Implementation

4.1 Overall Process

From an implementation perspective, we propose the following 3-step approach for describing the ontology as well as generating the individuals of the ontology on top of a road network model:

1. The proposed qualitative spatial concepts have to be defined using a standardized ontology language. Due to its broad acceptance, we decided to use the Web Ontology Language (OWL). We modelled the OWL structure by using the Protégé editor. Figure 2 shows the resulting OWL ontology as OntoGraf[3] visualization (Protégé plugin). This ontology is the foundation for all further processing steps.
2. We expect an arbitrary road network to be modelled as directed road network graph with road segments as vertices and connections between the segments as

---

[3]http://protegewiki.stanford.edu/wiki/OntoGraf.

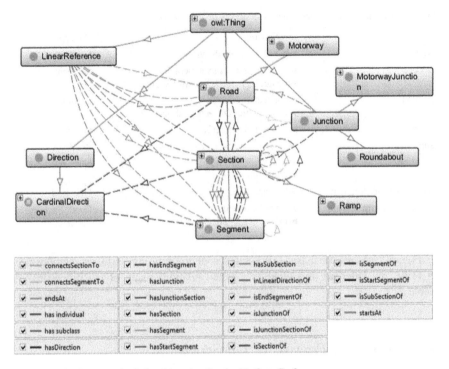

Fig. 2 OWL classes and relationships visualized with OntoGraf

edges. Attributes such as geometries, road names and road class hierarchies have to be attributed to the directed road segments. In our implementation approach we used the standardized format for modelling transport networks in Austria (GIP.at) and imported the road model into a Neo4J graph database.

3. The generation process for individuals of the ontology takes the OWL ontology and the road network graph as input and generates the individuals. The overall process is listed in the next subsection. The process has been implemented as Neo4J Plugin. This plugin accesses the OWL ontology using the OWL API[4] and stores the OWL ontology together with the generated individuals in the Neo4J graph database following the method described in the Neo4J Blog.[5] Storing the ontology in Neo4J instead of a triple store has the benefit that it can be accessed and queried using the graph query language Cypher.[6] Furthermore, it is possible to traverse the individuals of the ontology using standard graph algorithms.

---

[4]http://owlapi.sourceforge.net/.
[5]http://neo4j.com/blog/using-owl-with-neo4j/.
[6]http://neo4j.com/docs/developer-manual/current/#cypher-query-lang.

Table 2 Relationships between the six concepts of the ontology

Concept	Relationship	Concept
Road	►hasSection, ◄isSectionOf	Section
Section	►hasSegment, ◄isSegmentOf, ◄isStartSegmentOf, ◄isEndSegmentOf	Segment
Section	►connectsSectionTo ►isSubSectionOf, ◄hasSubSectionOf	Section
Segment	►connectsSegmentTo	Segment
Junction	◄hasJunction, ►isJunctionOf	Road
Junction	◄hasJunctionSection, ►isJunctionSectionOf	Section
Road, Section, Segment	►hasDirection,	Direction
LinearReference	►startsAt, ►endsAt, ►inLinearDirectionOf	Road, Section, Segment

Figure 2 shows the basic OWL classes of the ontology and its relationships. In OWL the latter are modelled as directed relations using object properties (cf. Table 2).

4.2 Generation Process for Individuals

To generate the individuals of the ontology on top of a road network, the following four steps have to be executed:

1. The road segments and connections have to be read from a file or database. Properties have to include at least the segment id, the road name and/or reference number, the start and end node identification number, the road type and connections to neighbor segments at both ends. In order to derive junctions and sections, all topological connections between segments must be known or determinable. Road graphs often include drivable connections only, which would yield incomplete or unrecognized junctions, e.g., at junctions with one or more one-way streets. In this case, it depends on the data model, if the complete topological connection information can be extracted from the road network model in an efficient way. For the proposed algorithm the missing information is derived by comparing the start and end node ids of the segments.
2. *Sections* and *roads* can now be derived from the chain of road segments. A *section* is composed of an ordered list of segments being located between two road junctions or between a road junction and the end of a road. The proposed extraction algorithm starts at an arbitrary segment (i.e. a segment which occurs first in the database) and proceeds in both directions by following the topological connections to other segments. The section grows as long as a segment has exactly one topological connection to another segment. This strategy is repeated until each segment has been processed. Next, sections are divided into

sub-sections where the road name changes. Each sub-set of segments with a unique road name becomes its own section which is a child of the original section. Complex junctions are composed of *Sections* which have been identified as *Ramps* as well as the road sections of the intersecting roads enclosed by the ramps (compare with Fig. 6). A property is used to differentiate between on- and off-ramps. In case of a motorway junction, all sections of a junction are merged into a *MotorwayJunction,* which inherits from the general *Junction* concept. Similarly, sections of a roundabout belong to one *Roundabout* individual which also inherits from the *Junction* class.

3. *Roads* are modelled as sequences of sections with the same road name. In analogy to segments, road sections are interlinked via the relation *connectsSectionTo*. If a *Segment* contains a local and national road name, individual roads are created. In case of minor road name errors (e.g. small segments with missing road names), the algorithm detects these errors and merges nearby road parts. Due to their special characteristics, motorways are mapped to the *Motorway* class.

4. *Cardinal Directions* are derived by comparing the first and the last coordinate of the start and end node of a segment, the start and end segment of a section or the start and end section of a road. The concept *LinearReference* provides the possibility to extract a part of a road, section or segment (e.g. between 10.4 and 12.3 km of a road). It can be used for dynamic referencing of traffic messages on multiple abstraction levels.

5 Results

5.1 Test Data

For testing our approach we generated individuals of the ontology on top of Austria's National Transport Graph (GIP.at)[7] which is a geo-referenced dataset of a nationwide transport network published under the Creative Commons 3.0 license.[8] We imported the road network data (47,236 km) from the provided CSV-format into a generic road graph model using Neo4J graph database. In the GIP.at format (CSV file), the road network is modelled on the abstraction level of topologically connected road segments (so called links). In our approach, each link from the CSV file is represented as individual of the *Segment* class. We filtered the data using the functional road class (FRC) attribute (which is used to express the hierarchy of roads in the network) in order to apply the ontology generation process to the strategic road network only. The filter range was from FRC 0 (which stands for motorways) to FRC 4 (which stands for roads connecting villages or city districts).

---

[7] https://www.data.gv.at/katalog/dataset/3fefc838-791d-4dde-975b-a4131a54e7c5.
[8] https://creativecommons.org/licenses/by/3.0/at/.

The generation process resulted in a total count of 223.328 road segments. After import of the road segments in a Neo4J database, we executed our generation process (Neo4J Plugin) for deriving the individuals according to the proposed ontology. The generation process finished and resulted in 25.072 roads and 66.987 sections.

5.2 Examples and Queries

To test the ontology and the generated individuals, we implemented several Cypher queries. With the Cypher queries we tried to answer the following questions:

- Q1: Are roads and sections adequately represented on different levels of abstraction?
- Q2: How well does the modelling of junctions work?
- Q3: How can we use junctions as selector for between-sections?
- Q4: Does the qualitative direction concept proof useful?

For answering these four questions we present three example queries and visualize the results. Figure 3 shows the locations of the three examples on a map of the Western provinces of Austria (including parts of Bavaria).

To answer the first **question Q1**, we have chosen the Fernpassstraße (~75 km) which is an important national transit highway in the province of Tyrol and crosses

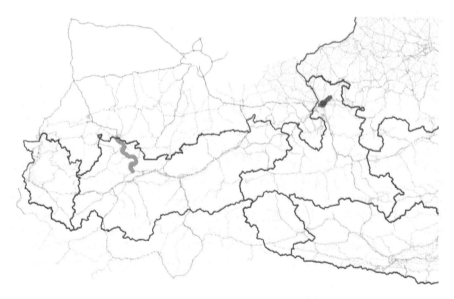

Fig. 3 Location of roads and sections from the examples (*green* B179—Fernpassstraße; *red* B1—Wiener Straße; *blue* motorway junction Salzburg Nord)

several villages with different/additional local names—hence a prominent example of a road with different levels of abstraction, junctions and ramps.

Example 1 Cypher query to retrieve all sections of "B179—Fernpassstraße".

```
MATCH (r:Road {roadRef:'B179'})-[:hasSection]-(sec)
OPTIONAL MATCH (sec)-[:hasSubSection]-(ssec)
RETURN r,sec, ssec
```

Results of the query are shown in Fig. 4. It contains the individuals of the road "B179—Fernpassstraße" (in the center) surrounded by 30 *Sections* (green). The arrows show the *isSectionOf-* and *hasSection*-relations. The remaining 22 (blue) individuals show ramps.

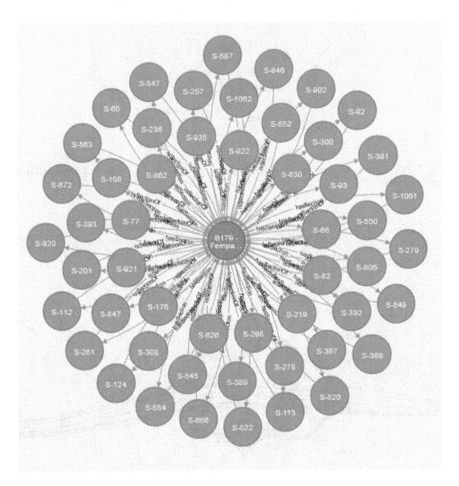

Fig. 4 Automatically derived road sections of the B179—Fernpassstraße in Austria (federal highway); visualized with Neo4j Browser

Another aspect of the ontology is presented in Fig. 5. Figure 5 shows a map of a long road section of 'B179—Fernpassstraße' (blue) and two small sub-sections representing local roads (orange and green). The long section does not contain any intersections with other roads but the two sub-sections are attributed with local

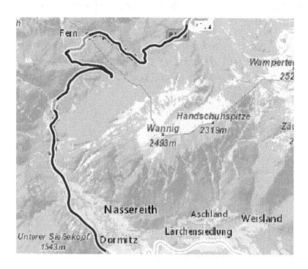

Fig. 5 Local roads 'Fernpaß' and 'Fernstein' as sub-sections (*orange, green*) of a national road section (*blue*)

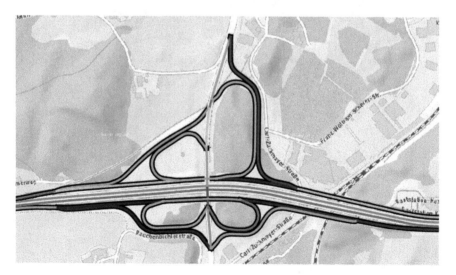

Fig. 6 Sections of junction "A1—Westautobahn—Anschlussstelle Salzburg Nord". Each section is illustrated in a unique colour, and *Ramps* are marked with a *black solid* outline

names. This example nicely visualizes the relationship between national and local abstraction levels. While non-local road users with distant destinations will primarily conceptualize the national road ('B179—Fernpassstraße') and most likely conceptualize this section as one section, local road users will most likely conceptualize the local names ('Fernstein' and 'Fernpaß') and therefore conceptualize the section in a more granular way. With the concept of sub-sections we are able to model each of the granularities accordingly.

To address **question Q2** concerning the modelling of junctions, the next query retrieves a complex motorway junction. As Fig. 6 shows, the junction has on- and off-ramps, acceleration and deceleration lanes.

Example 2 Cypher query to retrieve all sections of junction "Anschlussstelle Salzburg Nord" (federal highway × motorway).

```
MATCH (j: MotorwayJunction
       {junctionName:' Anschlussstelle SalzburgNord'})
       -[:hasJunctionSection]-> (sec)-[:hasSegment]-
       > (seg)
RETURN sec.section_id, seg.segment_id, labels(sec)[0]
```

The Cypher query presented in Example 2 selects all road sections of the *Junction* 'Anschlussstelle Salzburg Nord' connecting the motorway 'A1—Westautobahn' with the roads 'B150' and 'B156'. As visible from Fig. 6, the two motorway sections (in the middle from east to west) and the intersecting roads (from south to north) are also part of the junction, which is intuitively expected. From this example we conclude, that the ontology and the generation process are able to model complex motorway junctions in a cognitively adequate way. Since junctions are an important mental concept, this is a necessary step for human interpretability of traffic messages. Furthermore, the *Sections* which are part of the *Junction* can be used to reference fine-granular traffic information, e.g., problems on the on-ramp or on the side roads.

Question Q3 addresses the challenge of selecting sections between two junctions and **question Q4** deals with driving directions. As example, a 12 km long diverse road section of the federal highway 'B1—Wiener Bundesstraße' has been chosen. Two roundabouts delimit the section, one close to a motorway junction, the other in the city center of Salzburg. This road section has features of an urban main road but also of an arterial road and highway.

Example 3 Cypher query to retrieve all sections sections of "B1—Wiener Straße" (federal highway) between roundabout "Kreisel Eugendorf A1" and roundabout "B1—KV Hans Schmid Platz".

```
MATCH (j:Roundabout {name: 'Kreisel Eugendorf A1'})-
      [:hasJunctionSection]-> (n:Section)
WITH startSec LIMIT 1
MATCH (j:Roundabout {name: 'B1 - KV Hans Schmid Platz'})-
      [:hasJunctionSection]-> (n:Section)
```

```
WITH endSec LIMIT 1
MATCH (r:Road {roadName: 'B1 - Wiener Straße'})-[:hasSection]-
       > (sec)
WITH sec
MATCH p = shortestPath((startSec)-[:connectsSectionTo*0..
       9999]-> (endSec))
WHERE p IN sec
WITH p
MATCH (p)-[:hasDirection]->
       (d:CardinalDirection)
RETURN p, d
```

The query selects the *Road* 'B1—Wiener Straße' and retrieves all sections between the *Junctions* 'Kreisel Eugendorf A1' and 'B1—KV Hans Schmid Platz'. First, the starting *Section* at the first *Junction* and the end *Section* at the last *Junction* are derived (relationships *hasJunctionSection*). The shortest path method finds the sequence of *Sections* connecting both *Junctions* via the *connectsSectionTo*-relationships. The resulting *Sections* and the driving directions as *Cardinal Directions* are visualized in Fig. 5 (Fig. 7).

Fig. 7 Sections of the road 'B1—Wiener Straße' between junction to 'Kreisel Eugendorf A1' and junction to 'B1—KV Hans Schmid Platz'. The different colors represent the different *Sections*. The short *grey lines* reveal the *Segments*. The *black arrows* and letters indicate the driving direction of the sections as *Cardinal Direction*

6 Conclusions

In this work we proposed an integrated process for generating meaningful location references for enhancing road network models based on an ontology with six qualitative spatial concepts (*Road, Section, Segment, Junction, Direction, LinearReference*) and their relationships. This approach is on the one hand intended to bridge the gap between technically-oriented and human-interpretable location referencing systems and on the other hand to bridge the gap between static and dynamic referencing systems. We presented a prototypical implementation of an automatic process for generating the individuals of the ontology on top of a standardized, nationwide road network graph using a standardized ontology description language (OWL) and an open source graph database (Neo4J) for storing the ontology as well as the individuals. From the conceptualization and implementation approach we derived the following conclusions:

1. The proposed approach has demonstrated the use of qualitative spatial concepts for enhancing digital road networks with for the purpose of locational referencing on different levels of abstraction. This may be considered as clear benefit in comparison to previous static (e.g. TMC location table) as well as dynamic (e.g. OpenLR) approaches.
2. The proposed approach was tested with a nationwide road network graph. This goes beyond previous conceptual modelling approaches and demonstrates practical applicability. Although the approach has only been tested with one road network data model, due to the similar modelling concepts (e.g. road segments, topological connections, attributes) it may be easily adapted to other data models as well.
3. The quality of the resulting ontology instances heavily depends on the quality of the underlying road network graph, e.g. correctness of topological connectivity, road names, and hierarchical structure. Before generating the individuals of the ontology, a thorough quality evaluation process by taking into account standardized quality measures (ISO 19157 2013) has to be considered.
4. The use of cardinal directions as qualitative spatial direction concepts satisfies human-readability, applicability and interpretability only partly. The need for an automated process of deriving more expressive spatial direction concepts, which are closer to the everyday language, has been identified. However, the proposed ontology can be taken as foundation for a more detailed ontology by integrating additional qualitative spatial concepts.
5. So far an empirical evaluation of the proposed spatial concepts is missing. We have only demonstrated a plausibilisation for some challenging road network parts. However, since the basic spatial concepts have been derived from natural language and existing human-interpretable location referencing techniques, the overall ontology has a good empirical foundation.

In the future we plan to address the question of extending the ontology with more detailed qualitative direction concepts which are closer to everyday language.

Furthermore, we are planning to apply the ontology for different application scenarios to further evaluate adequacy and applicability empirically.

References

Bechhofer, S. (2009). OWL: Web Ontology Language. In *Encyclopedia of Database Systems* (pp. 2008–2009). Boston, MA: Springer US. http://doi.org/10.1007/978-0-387-39940-9_1073.

Car, A., & Frank, A. U. (1994). Modelling a hierarchy of space applied to large road networks. In *IGIS '94: Geographic Information Systems* (pp. 15–24). Springer Berlin Heidelberg. http://doi.org/10.1007/3-540-58795-0_30.

Curtin, K. M. (2007). Network Analysis in Geographic Information Science: Review, Assessment, and Projections. *Cartography and Geographic Information Science, 34*(2), 103–111. http://doi.org/10.1559/152304007781002163.

Curtin, K. M., Nicoara, G., & Arifin, R. R. (2007). A comprehensive process for linear referencing. *URISA Journal, 19*(2), 41–51.

Ernst, T., Schmidt, M., & Schramm, A. (2012). TPEG-ULR: A New Approach for Dynamic Location Referencing. In *Proceedings of the 19th ITS World Congress*. Vienna.

Frank, A. U. (1996). Qualitative spatial reasoning: cardinal directions as an example. *International Journal of Geographical Information Systems, 10*(3), 269–290. http://doi.org/10.1080/02693799608902079.

Guarino, N. (1998). Formal Ontology and Information Systems. In N. Guarino (Ed.), *Proceedings of FOIS98 Trento Italy 68 June 1998* (Vol. 46, pp. 3–15). IOS Press. http://doi.org/10.1.1.29.1776.

Hackeloeer, A., Klasing, K., Krisp, J. M., & Meng, L. (2014). Georeferencing: a review of methods and applications. *Annals of GIS, 20*(1), 61–69. http://doi.org/10.1080/19475683.2013.868826.

ISO 14819-3. (2013). Intelligent transport systems – Traffic and travel information messages via traffic message coding – Part 3: Location referencing for Radio Data System – Traffic Message Channel (RDS-TMC) using ALERT-C. ISO - International Standards Organisation.

ISO 14825. (2011). Intelligent transport systems – Geographic Data Files (GDF) – GDF5.0. ISO - International Standards Organisation.

ISO 19157. (2013). ISO 19157:2013 Geographic information - Data quality. ISO - International Standards Organisation.

Mandl-Mair, I. (2012). GIP. at: The Basis for a Modern Administration of Austria's Transport Routes. In *Proceedings of the 19th ITS World Congress*. Vienna, Austria.

Nyerges, T. L. (1990). Locational referencing and highway segmentation in a geographic information system. *ITE Journal, 60*(3), 27–31.

Scarponcini, P. (2002). Generalized Model for Linear Referencing in Transportation. *GeoInformatica, 6*(1), 35–55. http://doi.org/10.1023/A:1013716130838.

Timpf, S. (2002). Ontologies of Wayfinding: a Traveler's Perspective. *Networks and Spatial Economics, 2*(1), 9–33. http://doi.org/10.1023/A:1014563113112.

Timpf, S., Volta, G. S., Pollock, D. W., & Egenhofer, M. J. (1992). A conceptual model of wayfinding using multiple levels of abstraction (pp. 348–367). Springer Berlin Heidelberg. http://doi.org/10.1007/3-540-55966-3_21.

Vonderohe, A., Chou, C., Sun, F., Adams, T., Civil, D. O., & Engineering, E. (1997). A generic data model for linear referencing systems. *Research Results Digest 218. National Cooperative Highway Research Programme. Transportation Research Board*. Retrieved from http://citeseerx.ist.psu.edu/viewdoc/summary?doi=10.1.1.117.11.

Wang, Y., & Meng, H. (2009). Hierarchical Ontology on Multi-scale Road Model for Cartographical Application. In *2009 International Conference on Environmental Science and Information Application Technology* (Vol. 3, pp. 330–333). IEEE. http://doi.org/10.1109/ESIAT.2009.247.

Zhou, C., Lu, F., & Wan, Q. (2000). A Conceptual Model for a Feature-Based Virtual Network. *GeoInformatica*, *4*(3), 271–286. http://doi.org/10.1023/A:1009853309757.

Efficient Computation of Bypass Areas

Jörg Roth

Abstract Route planning in road networks is a basic operation in the area of location-based services. Very often, the knowledge of the optimal route is not the only important information for a driver. Complex services could also present points of interest (e.g. hotels or gas stations) nearby the optimal route as stop-over. Here, 'nearby' means: the bypass route from a start to target that passes that point does not exceed certain costs. In this paper, we present an efficient approach to compute all bypasses that are within a given cost limit. We may additionally request only locally optimal bypasses, e.g., that reach an intermediate point without driving U-turns. The set of all bypasses called *bypass area* can be used for further queries, in e.g. geo databases to find nearby points of interest for a certain application or service. Our approach is fully implemented and evaluated and computes the respective bypass areas very runtime-efficient, whereas it re-uses similar structures as for optimal route planning.

Keywords Route planning · A* · Alternative routes · Bypasses

1 Introduction

Route planning on road networks is a well understood problem. Network services or on-board applications are able to compute a route from position to position that is optimal according to given demands. Resulting routes minimize 'costs' such as time, distance, fuel consumptions or road charges.

For a certain navigation task, the optimal route may not be the only useful information. A driver could refuse the optimal route due to several reasons. From former knowledge, the driver may know the given route does not meet her or his

---

J. Roth (✉)
Department of Computer Science,
Nuremberg Institute of Technology, Nuremberg, Germany
e-mail: Joerg.Roth@th-nuernberg.de

demands on an optimal route, e.g., there is a certain probability for a traffic jam. Or, the driver wants to drive through another region to view a certain sight.

On the other hand, the route planning process should directly incorporate a certain intermediate point. e.g., we want to drive to a target, but on the route we want to stop at an arbitrary gas station. Driving to downtown we may ask: how to drive, if we want to pass a post box, bakery and supermarket? Important: the actual intermediate positions are variable (*any* post box or *any* bakery will do and also any ordering of stops). Thus, we cannot simply split a route and execute individual point-to-point planning tasks to known points. Thus, we have to compute *all* routes from a start to a target and check, whether they touch the respective points of interest.

The following approach to solve this problem may be part of a larger navigation service or route planning application. We present an approach that efficiently computes all respective bypasses. The approach can be configured to reflect the user's or application's demands. In particular we express the quality of a bypass with the help of a cost limit.

2 Related Work

Route planning algorithms usually are based on Dijkstra's shortest path approach (Dijkstra 1959) or the A* algorithm (Hart et al. 1968). A* takes into account additional knowledge about road networks, modeled as future path cost estimation. With this, the computation is significantly faster while the optimality of results is kept. For large road networks, we usually combine the benefits of A* with additional techniques to reduce the runtime. They can be distinguished in approaches that do not change the optimal result (i.e. only speed up the computation) and those that may lead to sub-optimal results. An approach of the first type is to provide precise future path cost estimations, e.g., *ALT (A* search, landmarks, and triangle inequality,* Goldberg and Harrelson 2005). An approach of the second type is to consider only fast roads (e.g. highways), in the middle of the route, i.e. when we exceed a certain distance from start or target (Geisberger et al. 2008).

Bypasses or alternative routes are *non*-optimal routes. Usually, there is a *single* optimal route for a certain cost measure (if we ignore the unlike case of multiple identical minimal costs). In contrast, the number of routes with non-optimal costs is unlimited. In order to limit the set of reasonable bypasses, we thus have to introduce additional conditions on bypasses.

An approach is to order all possible routes by their increasing costs and take the first k routes. The so-called *k-shortest path routing* has a long tradition (Hoffman and Pavley 1959; Bellman and Kalaba 1960); several subsequent approaches speed up the execution (Yen 1971; Eppstein 1994). The problem turned out to be harder, if we want to avoid routes with loops—but this is usually a condition for a reasonable bypass.

Abraham et al. (2010) pointed out a problem of the k-shortest path approach: reasonable alternatives are probably not among the first thousands shortest paths. They suggest an additional condition for alternative routes: *local optimality*. This means, every subroute up to a certain length must be an optimal route. This reflects usual driving behavior: even though a driver might not use an optimal total route according to a given cost function, short parts of a route usually *are* optimal. Luxen and Schieferdecker (2012) combined the idea of locally optimal alternatives with an optimized road network that considers faster roads in the middle of a route. Paraskevopoulos and Zaroliagis (2013) introduced additional filtering methods to remove unwanted alternatives.

The following approach is heavily based on former research on route planning algorithms in den *HomeRun* environment (Roth 2013). The primary goal of former research was to provide algorithms *beyond* the typical task of optimal route planning. All of the following approaches are based on A* but reused the underlying structures in a novel manner:

- For a set of n starts and m targets: find all $n \cdot m$ optimal routes for any combination of start and target. This function is called *multirouting*. We presented an approach to provide multirouting in a single step that is far more efficient than executing $n \cdot m$ single route planning (Roth 2016).
- Supervising a real driven route up to a certain point: what is the area of *possible* targets, if we assume locally optimal driving? This is somehow the opposite question of route planning as we want to predict a locally optimal route from a partly supervised route. This function is called *target prediction*—we provide an efficient solution for it in (Roth 2014).
- Given a set of measured positions: what is the most probable route that approximates all positions—this is an extended version of the so-called *map matching problem* known from navigation systems. But in contrast to existing approaches, we consider all possible routes and take the one with the highest probability (Roth 2016b).

Based on the experience with unusual navigation questions, we wanted to solve the bypass problem. Even though strongly related to alternative path computation as described above, we can state certain differences. First, we strongly base on A* and thus can incorporate useful speed-up techniques, foremost the estimation based on ALT. Second, we do not limit the set of bypasses (e.g. only the shortest k) but return all. This allows an application or service to formulate any condition on its own and is not limited to any concept of alternative routes modeled by the algorithm. Even though 'all' instead of k routes seems to cause a high overhead, we present a structure that represents these routes, meanwhile it enables an efficient computation.

3 Computing Bypass Areas

3.1 Problem Statement

The concept of bypasses requires the knowledge of the optimal route that is implicitly defined by a *start*, *target* and *cost function* (e.g. driving time). We can formulate the bypass problem as follows:

> What are all bypasses that do not exceed the minimal costs from start to target by a given factor?

From these bypasses we want to know their paths, but also all passed positions (i.e. possible intermediate targets). Inevitable for any approach is a road graph that contains all crossings and connections between crossings. The latter specify road characteristics to compute costs, e.g. road type and speed limit.

More formally: for each crossing q_i we know its *directly* connected crossings q_j and the driving costs $c(q_i, q_j)$ to get there. We call a connection between crossings q_i, q_j a *link*. Link costs $c(q_i, q_j)$ can be any positive number. In reality $c(q_i, q_j) \neq c(q_j, q_i)$ for many links due to, e.g., different speed limits or one-way roads.

The optimal route is a sequence of crossings (*start*, q_{i1}, q_{i2}, ..., *target*) that minimizes the sum of link costs. We call the minimal costs between two (not necessarily connected) crossings c^*; in particular, the costs for our optimal route are $c^*(start, target)$.

3.2 Computing Optimal Paths

In order to present our bypass approach, we first describe A* that computes optimal single routes from a crossing to a crossing. In order to execute route planning in a target-oriented manner, A* requires a future path cost estimation function $h(q_i, target) \leq c^*(q_i, target)$ that provides a lower bound of costs for the route termination. We request h to be *monotone*, i.e. for two connected crossings q_i, q_j we get

$$\left|h(q_i, target) - h(q_j, target)\right| \leq c(q_i, q_j). \tag{1}$$

Important: also for bad estimations h, A* always produces optimal results. But the more h reaches the actual costs, the better A* performs, because it only deals with fewer unwanted crossings. Algorithm 1 presents the original A* approach as known from the literature.

Efficient Computation of Bypass Areas 197

Algorithm 1. Standard A*

A_star(*start*, *target*)
$g[start] \leftarrow 0$; $f[start] \leftarrow 0$; $state[start] \leftarrow open$;
openList.add(*start*);   // add *start* to open
for all $q_i \neq start$ {   // initialize fields
   $g[q_i] \leftarrow -1$; $f[q_i] \leftarrow 0$; $state[q_i] \leftarrow not\_visited$;
}
do {
   $q_i \leftarrow openList.poll()$;   // get q_i with state *open* and minimal $f[q_i]$
   if $q_i = target$ return *success*;   // route to target found: finish
   $state[q_i] \leftarrow closed$;
   for all neighbors q_j of q_i {   // expand crossing
      if $state[q_j] \neq closed$ {
         $g_{new} \leftarrow g[q_i] + c(q_i, q_j)$;
         $f_{new} \leftarrow g_{new} + h(q_j, target)$;
         if $state[q_j] = not\_visited$ or $f_{new} < f[q_j]$ {
            $g[q_j] \leftarrow g_{new}$; $f[q_j] \leftarrow f_{new}$;   // the route *start*→...→q_i→q_j is less costly
            $backLink[q_j] \leftarrow q_i$;   // than the formerly stored route
            $state[q_j] \leftarrow open$;   // *start*→...→$q_j$ (if any)
            openList.add(q_j);
         }
      }
   }
} while not openList.isEmpty();
return *failure*;   // no route at all from *start* to *target*

Some remarks:

- We assign one of three states to each crossing: *not_visited*, *open* (*g* not finally computed), *closed* (optimal route from *start* discovered). The section '*expand crossing*' in the loop means: take an *open* crossing and check all its neighbors.
- $g[q_i]$ contains the currently *assumed* costs from *start* to crossing q_i. If q_i is *closed*, $g[q_i]$ contains the minimal costs from *start* to q_i. If q_i is *open*, $g[q_i]$ may be still larger than the minimal costs.
- $f[q_i]$ contains the currently *estimated* total costs from *start* to *target*, if a route goes through q_i.
- To efficiently get the open crossing with minimal f, we additionally need an *openList* that internally keeps the list sorted whenever an *open* crossing is added.
- For *closed* crossings q_i, the link ($backLink[q_i]$, q_i) is the last link of the optimal route from *start* to q_i.
- Once we polled *target* from the *openList*, the optimal route is discovered. We then can easily collect the optimal route from *start*, following the *backLink* entries.

Even though the primary computation result is the *backLink* array, we also consider the *g* array as important output. For each crossing with state *closed*, the *g* array provides the minimal costs to get there from the *start*, also, if the crossing is not part of the optimal route. We take advantage of this property in our approach (see next section).

A last important property of f that need in our approach: if we check the values of $f[q_i]$ in the do-loop, $f[q_i]$ cannot get smaller for a new iteration. In other words: if we expand a node q_i, none if its non-closed neighbors q_j can receive a smaller f-value. This is because

$$f[q_i] \leq f[q_j] \Leftrightarrow g[q_i] + h(q_i, target) \leq g[q_j] + h(q_j, target)$$
$$\Leftrightarrow h(q_i, target) \leq c(q_i, q_j) + h(q_j, target) \Leftrightarrow (q_i, target) - h(q_j, target) \leq c(q_i, q_j) \quad (2)$$

which is true because h is monotone (formula (1)).

3.3 Bypass Areas—Basic Considerations

Once we are able to compute an optimal route between two crossings, we can think about alternative routes—our *bypasses*. Let $opt = c*(start, target)$ denote the costs for the optimal route from *start* to *target* and $v \geq 1$ the *extension factor*. We are looking for all routes from *start* to *target* with costs of not more than $v \cdot opt$. As we are looking for the entire *bypass area*, we are actually looking for *all crossings* that are part of all such routes.

Another view to the problem (Fig. 1a): an *intermediate crossing I* of the network is part of the required bypass area, if

$$c^*(start, I) + c^*(I, target) \leq v \cdot opt. \quad (3)$$

Based on this consideration, we are able to provide a definition of a bypass area B:

$$B(start, target, v) = \{I | c^*(start, I) + c^*(I, target) \leq v \cdot c^*(start, target)\}. \quad (4)$$

Even though easy to formulate, a real implementation is not obvious. A naïve computation of B would iterate through all possible crossings of the road network with e.g. some million crossings. As the check, whether I belongs to B requires to compute c^* (and thus to execute A*) two times, this approach would require an unacceptable long computation. We thus present a solution that discovers all I more efficiently in the next section.

A further issue: even though the respective routes $start \rightarrow I \rightarrow target$ do not exceed the given cost limit, they sometimes do not meet the drivers expectation of an appropriate alternative route (Fig. 1b). The problem: the route can be split in two optimal subroutes, however the total route may lose optimality in the range of I. This contradicts the *local optimality* as described in Sect. 2. The example in Fig. 1b is a worst case of a *not* locally optimal route, as the subroutes to and from I contain reverse links and the driver has to U-turn at I.

Efficient Computation of Bypass Areas

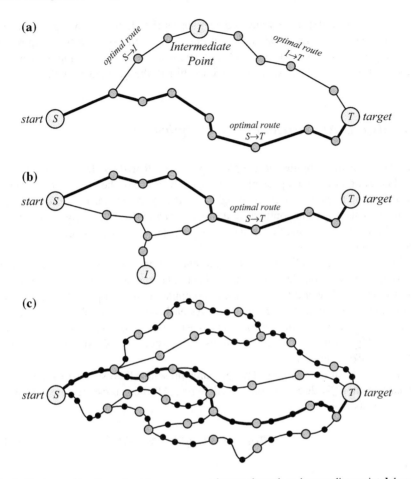

Fig. 1 Basic considerations about bypass area—**a** bypass through an intermediate point; **b** bypass without local optimality; **c** all bypasses with geometric representation

We thus introduce a second type of bypass area B_L that only contains intermediate crossings of routes with local optimality. It depends on the application scenario and user's expectation, whether local optimality is required (B_L) or not (B). To control the degree of local optimality we define a parameter ℓ with the meaning: all subroutes starting with a distance up to ℓ meters from I (on the route $start \to I$) and terminate up to ℓ meters to I (on the route $I \to target$) go through I. We can thus define B_L as follows:

$$B_L(start, target, v, \ell) = \{I | c^*(start, I) + c^*(I, target) \leq v \cdot c^*(start, target) \quad (5)$$
$$\text{and the route is locally optimal according to } \ell\}.$$

We provide an efficient approach to compute this set in the next section.

Figure 1c illustrates a last consideration: no matter whether we compute B or B_L, both sets only contain *crossings*. This is not sufficient if we want to compute an *area*. As an example: if we want to detect appropriate hotels for a trip, we have to compute a geometric area that covers all line string points of corresponding routes, not only the crossings. Until now, all operations are based on the topological road network. For further operation, we have to shift to the geometric model.

3.4 Bypass Areas Without Local Optimality

We first want to compute the simple bypass area *B without* local optimality. To avoid iterating through all possible *I*, we take advantage of the way A* computes an optimal path: from the *start* crossing it iteratively considers new crossings and simultaneously maintains optimal routes from *start* to these crossings, even though they will not be part of the final route. A* terminates when it considers *target*, as the optimal route was found.

We modify the condition to terminate and we are able to create a field of all crossings *I* that *may* be element of *B*. As this field computes paths from the *start*, we call it *start field*. We additionally have to compute the second part of the route from *I* to *target*, represented by a *target field*. As each field is only able to consider its contribution to the condition in formula (4), each field contains more crossings than required. However, the set of crossings to consider is by far smaller than *all* crossings of the network.

Figure 2 illustrates the idea. The gray area presents all crossings that are *open* or *closed* and their *f*-values do not exceed $v \cdot opt$. All these crossings are reasonable candidates as from

$$c^*(start, q_i) + c^*(q_i, target) \leq v \cdot opt \qquad (6)$$

and

$$f[q_i] = g[q_i] + h(q_i, target) = c^*(start, q_i) + h(q_i, target) \leq c^*(start, q_i) + c^*(q_i, target) \qquad (7)$$

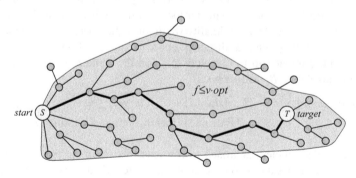

Fig. 2 Idea of start field generation

(which is true, because $h(q_i, target) \leq c^*(q_i, target)$) follows

$$f[q_i] \leq v \cdot opt. \tag{8}$$

As the f-value of the next crossing in the open list cannot get smaller (see formula (2)), we can stop when we poll a crossing with an f-value larger than $v \cdot opt$. As we do not know opt from the beginning, we first expand crossings as described in Algorithm 1 until we polled $target$. Then, we visit more crossings as far as we get the first crossing with $f > v \cdot opt$. As a consequence, the field goes beyond the $target$. This is reasonable, as those crossings are also candidates for I. Algorithm 2 sums up these considerations and shows how to compute the start field. To clearly distinguish the respective structures, we now apply index s to all arrays of the start field (e.g. g_s).

In a second round, we have to generate a target field with indexes t (Algorithm 3). The approach is similar to Algorithm 2, but in order to apply the appropriate driving direction, we have to incorporate these changes:

- The first *open* crossing is *target*.
- Whenever we expand a crossing q_i, we check the distance *from* the neighbor q_j, i.e. $c(q_j, q_i)$, not *to* the neighbor.
- The estimation is computed from *start* to the respective crossing.
- A sequence of *backLink* entries points to the *target* not to the *start*.

Algorithm 2.   Start field generation

```
start_field(start, target, v)
g_s[start]←0; f_s[start]←0; state_s[start]←open;
openList.add(start);                // add start to open
for all q_i≠start {                 // initialize fields
    g_s[q_i]← -1; f_s[q_i]←0; state_s[q_i]←not_visited;
}
opt←undefined;                      // costs for optimal route unknown for now
do {
    q_i←openList.poll();            // get q_i with state open and minimal f_s[q_i]
    if opt is defined and f_s[q_i]> v · opt  // this cannot be an intermediate crossing
        return success;             // as the bypass would be too costly: finish
    if q_i=target opt←g_s[q_i];     // route to target found: set opt
    state_s[q_i]←closed;
    for all neighbors q_j of q_i {  // expand crossing
        ...                         // see Algorithm 1
    }
} while not openList.isEmpty();
return failure;                     // no route at all from start to target
```

In addition, we can directly use the opt value of the start field. The target field generation can benefit from a much better estimation h compared to the start field (see (*) in Algorithm 3): we set $h = g_s[q_j]$ whenever $g_s[q_j] \geq 0$. This does not change the result, but significantly reduces the number of visited crossings for the target field, thus improves the runtime. Using the g_s for h does not only provide an estimation, but returns the real costs, thus is the best considerable estimation.

The start and target field generation produce $g_s[q_i] = c^*(start, q_i)$ for all q_i with $state_s[q_i] = closed$ and $g_t[q_i] = c^*(q_i, target)$ for all q_i with $state_t[q_i] = closed$. As a consequence, we are now able to provide an efficient formula for B:

$$B(start, target, v) = \{q_i | state_s[q_i] = state_t[q_i] = closed \text{ and } g_s[q_i] + g_t[q_i] \leq v \cdot opt\} \quad (9)$$

This approach is by far more efficient than an approach that computes two optimal routes for *every* crossing in the network. The runtime of such a naïve approach would be million times longer than the single route planning. In contrast, our approach only causes a runtime of approx. 2 times more (see evaluation). Again note that in this time the information of *any bypass* in the road network with the required cost limit is discovered.

Algorithm 3. Target field generation

---

target_field(*start, target, v*)
$g_t[target] \leftarrow 0; f_t[target] \leftarrow 0; state_t[target] \leftarrow open;$
openList.add(target);   // add *target* to open
for all $q_i \neq target$ {   // initialize fields
    $g_t[q_i] \leftarrow -1; f_t[q_i] \leftarrow 0; state_t[q_i] \leftarrow not\_visited;$
}
do {   // Note: *opt* is known from start field
    $q_i \leftarrow openList.poll();$ // get q_i with state *open* and minimal $f[q_i]$
    if $f_t[q_i] > v \cdot opt$ // this cannot be an intermediate crossing
        return *success*;   // as the bypass would be too costly: finish
    $state_t[q_i] \leftarrow closed;$
    for all neighbors q_j of q_i {   // Note: driving direction is $q_j \rightarrow q_i$!
        if $state_t[q_j] \neq closed$ {   // expand crossing
            $g_{new} \leftarrow g_t[q_i] + c(q_j, q_i);$ // Note: costs from $q_j \rightarrow q_i$!
            if $g_s[q_j] \geq 0$ // (*) ideal estimation from *start* to q_j available
                $h \leftarrow g_s[q_j]$ // i.e. the *real* costs, taken from start field, g_s
            else
                $h \leftarrow h(start, q_j)$ // not available: compute h yourself
            $f_{new} \leftarrow g_{new} + h;$ // Note: h estimates *start* $\rightarrow q_j$!
            if $state_t[q_j] = not\_visited$ or $f_{new} < f_t[q_j]$ {
                $g_t[q_j] \leftarrow g_{new}; f_t[q_j] \leftarrow f_{new};$ // the route $q_j \rightarrow q_i \rightarrow ... \rightarrow target$ is less costly
                $backLink_t[q_j] \leftarrow q_i;$ // than the formerly stored route
                $state_t[q_j] \leftarrow open;$ // $q_j \rightarrow ... \rightarrow target$ (if any)
                openList.add(q_j);
            }
        }
    }
} while not openList.isEmpty();
return *failure*;   // no route at all from *start* to *target*

3.5 Bypass Areas with Local Optimality

In a second step, we may reduce B to B_L that only considers locally optimal bypasses. Again note that this operation is application-dependent. In particular, we need to consider the user's expectation of a bypass.

Obviously $B_L \subseteq B$, thus we can iterate through all $I \in B$ and check, if the respective bypass is locally optimal. As the size of B can by large (e.g. some hundred thousand crossings), we must avoid multiple executions of A* to check the local optimality.

Some considerations: From a bypass through I, the parts $start \to I$ and $I \to target$ are already optimal, thus also *locally* optimal. The crucial part of the route is thus around I. If we require all subroutes up to length ℓ to be optimal, we can formulate this condition as follows:

- for a crossing q_i on the route $start \to I$ with a distance to I not more than ℓ, I is on the optimal route from $q_i \to target$ and
- for a crossing q_i on the route $I \to target$ with a distance to I not more than ℓ, I is on the optimal route from $start \to q_i$.

The distance ℓ can be expressed in any unit, e.g. meters for driving distances, but also in the same unit as c^*, e.g. the driving time. It depends on the application to choose an appropriate measure and value.

To avoid the costly check with the help of A* for every I we make use of an approach as illustrated in Fig. 3. We see the start and target field and a crossing $I \in B$.

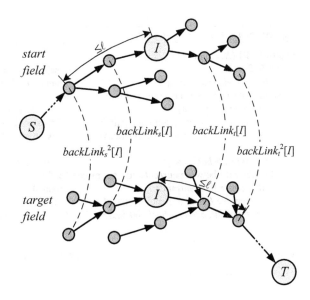

Fig. 3 Checking local optimality

Let $backLink^n[q_i]$ denote the n-times application of $backLink$, e.g., $backLink^2[q_i] = backLink[backLink[q_i]]$. We further define $backLink^0[q_i] = q_i$.

Obviously, $backLink_s^n[I] \in B$ for all n for which $backLink_s^n$ is defined. This is because

$$c^*\left(start, backLink_s^n[I]\right) + c^*\left(backLink_s^n[I], target\right) \leq c^*(start, I) + c^*(I, target). \tag{10}$$

As a consequence, $backLink_s^n[I]$ is also a *closed* crossing in the target field and its $backLink_t$ entry is defined.

On the subroute $start \rightarrow I$ we are interested in all crossings $backLink_s^n[I]$ that do not exceed distance ℓ to I. One of these crossings may violate the local optimality condition, i.e. the optimal route to $target$ may not go over I. We thus get local optimality, if

$$backLink_t\left[backLink_s^n[I]\right] = backLink_s^{n-1}[I], \tag{11}$$

that can easily be checked for all n for which the corresponding entries are defined. In a second round we also have to check subroute $I \rightarrow target$ with corresponding crossings $backLink_t^n[I]$:

$$backLink_s\left[backLink_t^n[I]\right] = backLink_t^{n-1}[I]. \tag{12}$$

In the example in Fig. 3, the corresponding $backLink$ entries are all equal, i.e. they all fulfil formulas (11) and (12). This proves the local optimality of the bypass over I. Algorithm 4 shows the final approach to compute B_L.

Algorithm 4. Computation of B_L

---

compute_BL(B, ℓ, $backLink_s$, $backLink_t$)
for all $I \in B$ {   // main loop: check all I
    $last = I$; $current = backLink_s[I]$;   // inner loop 1: $I \rightarrow start$
    while distance between *current* and I not more than ℓ {
        if $backlink_t[current] \neq last$ // route does not go over I
            remove I from B; continue main loop;   // I does not belong to B_L
        $last = current$; $current = backLink_s[current]$;   // next in *start* direction
    }
    $last = I$; $current = backLink_t[I]$;   // inner loop 2: $I \rightarrow target$
    while distance between *current* and I not more than ℓ {
        if $backlink_s[current] \neq last$ // route does not go over I
            remove I from B; continue main loop;   // I does not belong to B_L
        $last = current$; $current = backLink_t[current]$;   // next in *target* direction
    }
}
return B;   // this set now is B_L

Some remarks about the runtime behavior: even though we have to loop over all $I \in B$, the two inner loops are not time-critical. This is because they only iterate up to a distance of ℓ. For an average link length, this is a constant, thus we have a complexity of $O(|B|)$ to compute B_L.

3.6 Computing Geometric Areas, Samples

Once we computed our sets of bypass crossings B or B_L respectively, we now have to build structures that represent the bypass areas. We start with the definition of a route that goes through a bypass crossing q_i:

$$route(q_i) = (start, backLink_s^{n-1}[q_i], \ldots, backLink_s^1[q_i], q_i, \\ backLink_t^1[q_i], \ldots, backLink_t^{m-1}[q_i], target) \quad (12)$$

whereas $start = backLink_s^n[q_i]$ and $target = backLink_t^m[q_i]$. Note that these routes are ordered sequences of crossings. We further define

$$routes(B^*) = \{route(q_i) | q_i \in B^*\} \quad (13)$$

for B^* that either is B or B_L. It represents all distinct routes through our bypass crossings. To switch to a geometric view, we further define $linestring(q_i, q_j)$ that represents the line string geometry of link (q_i, q_j). The $area$ is the union of distinct line strings

$$area(B^*) = \{linestring(q_i, q_j) | (q_i, q_j) \text{ is link in } routes(B^*)\} \quad (14)$$

again for B^* that either is B or B_L.

Figure 4 presents the output of $area(B_L)$, i.e. the geometry of $routes(B_L)$, whereas B_L is computed for a given $start$, $target$, v and ℓ. As B_L only keeps intermediate points with local optimal driving, $routes(B_L)$ only contains reasonable *alternative routes*. Thus, we have an effective means to compute all alternative routes to a target that do not exceed a certain cost limit (10 % in Fig. 4). This is a useful service for all types of route planning. A navigation environment is able to immediately present the optimal routes as well as *all* reasonable alternatives on a map. The driver may decide to use a suboptimal route, e.g. to avoid a certain region or road. The navigation application could also provide a slider that allows a user to change v with immediate feedback of alternatives on the map.

We can also use the $area$ function to query geo databases for interesting objects near this area. Figure 5 shows an example. It illustrates a search for gas stations near the route to the target, again with additional bypass costs that do not exceed 10 %. We use $area(B)$ here, as we accept to drive to and from the gas station *without* local optimality. Note that a real application would take into account the

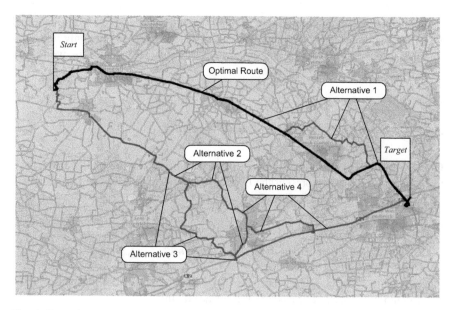

Fig. 4 Example computation of alternative routes (opt. route length = 35 km; $v = 1.1$; $\ell = 5000$ m)

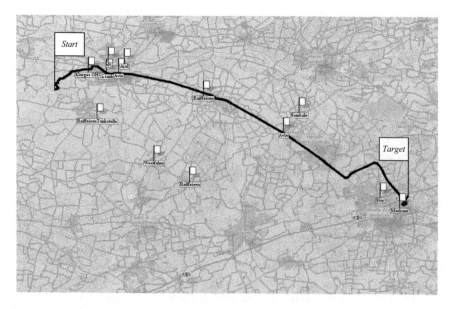

Fig. 5 Search for gas stations with limited bypass costs ($v = 1.1$; no local optimality)

current tank content to query for gas stations. If we are able to travel up to 80 % of the route with our tank content, we are only interested in gas stations that are 20–30 % from the target. This fine-tuning of queries, however, depends on the respective application.

We can additionally convert the set of line strings to a polygonal area using a *concave hull* operation. A concave hull is a closed polygon that encloses a set of given positions. In contrast to the *convex* hull, there is no unique concave hull—usually a parameter α specifies how much the polygon geometrically adapts to the given coordinates. We use an approach based on the Delaunay Triangulation (Duckham et al. 2008). Figure 6 shows concave hulls with different values for v for our example.

Note that a concave hull also covers intermediate positions that do not belong to our bypass areas, as concave hulls do not create holes. On the other hand, all intermediate points that hold the cost limit have to be inside a hull. As a result, the concave hull region only provides an impression of reachable points, e.g. to be painted on a map. It also can be used to query a superset of results with a much simpler geometry than all line strings. This is reasonable as most spatial indexing mechanisms are based on a simplified geometry (e.g., Roth 2009) and an exact geometric check has to be performed in a second step.

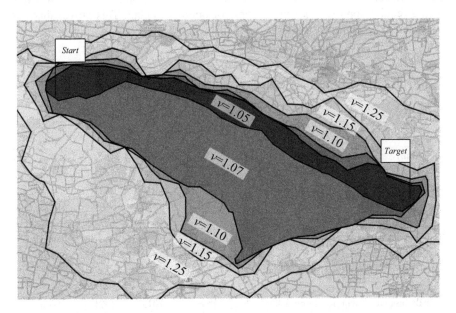

Fig. 6 Concave hulls of all bypass routes (parameter v; no local optimality)

4 Evaluation

We evaluated our implementation of the bypass functions in the HomeRun environment. The goal was to answer the following questions:

- For typical road networks and routes: what are the runtime costs to compute bypass areas?
- Does the approach to speed up the target field generation pays off?
- How costly is the detection of bypasses that are locally optimal?
- Finally: what are typical values for v and ℓ and what are their impact on the number of bypass crossings and alternative routes?

The following tests are based on the road network of Germany, imported from Open Street Map to our HomeRun navigation structure. The network contains 12.8 million crossings and 33.7 million links. For execution time measurements we used a PC with i7-4790 CPU, 3.6 GHz. We conducted 5 tests with different route lengths and values for v and ℓ. For each of these tests we selected 2000 random routes. Table 1 shows the results.

We first compared the time to compute the optimal route with the time to compute all bypasses (rows 4 and 5). The bypass computation takes 1.9–2.3 more computation time. This is a great result, as the bypass approach has to compute two fields, whereas we only need a single field to compute the optimal route—only this would require approx. twice the runtime of optimal routing. This proves that our approach to compute all bypasses is efficient.

Table 1 Bypass execution statistics (average values for 2000 executions per test)

	Test1	Test2	Test3	Test4	Test5
1: Distance start-target (km)	1–10	10–25	25–100	100–250	250–500
2: v	1.2	1.12	1.08	1.05	1.03
3: ℓ (km)	0.5	1.2	5	10	20
4: Execution time for opt. route (ms)	40	42	61	119	249
5: Bypasses execution time (ms)	77	83	134	282	587
6: Time to check local optimality (ms)	0.172	0.423	3.143	8.96	13.41
7: Visited crossings start field	4059	12 651	69 303	222 997	533 906
8: Visited crossings target field	1722	4127	21 013	66 633	145 651
9: Total bypass crossings	1425	3096	14 190	38 425	69 610
10: Ratio start field/bypass crossings	2.8	4.1	4.9	5.8	7.7
11: Ratio target field/bypass crossings	1.2	1.3	1.5	1.7	2.1
12: Bypass cross. with local optimality	306	516	1013	1471	1940
13: Alternative routes	18	17	16	12	6

Second, we checked, whether the speed up of the target field generation (marked by (*) in Algorithm 3) really pays off. Comparing rows 7 and 8 (also 10 and 11) we see the numbers of visited crossings of the target field are significantly lower than the numbers in the start field.

Third: we proved that the approach to detect locally optimal bypasses (Algorithm 4) is very efficient (row 6). The time compared to the total runtime can nearly be ignored.

Finally, we got an impression of bypasses (row 9), bypasses with local optimality (row 12) and alternative routes (row 13) in typical scenarios on real road network. The number of real alternatives is considerably low in our tests (less than 20 alternatives). This small list of routes could, e.g., be painted on a single map with different colors.

To sum up: the evaluation shows the effectiveness of the overall approach. It is possible with a considerable small execution time to compute the set of bypasses and, if required, to keep those with local optimality.

5 Conclusions

Our approach is an efficient solution for the bypass-problem: *what are the positions than can be reached with limited additional driving costs*? In contrast to similar questions (k-shortest paths, alternative routes), we are not limited to certain bypasses, but directly compute all within the cost limit. This enables applications to perform their own queries based on the bypass areas, such as: 'On the way to my target; which gas stations can be reached within 10 % additional driving costs?' And also: 'What are all alternative routes from start to target within 5 % more costs than optimal?'.

This approach is heavily based on A* and its structures. In particular, the approach benefits from the estimation function that does not affect the correct result but significantly speeds up the computation. The overall runtime is comparable to optimal path planning, thus a respective service could directly call the computation of bypass areas, whereas the optimal route can be considered as special type of bypass.

We fully implemented and evaluated our approach that is part of the donavio navigation environment in the HomeRun project.

References

Abraham I., Delling D., Goldberg A. V., Werneck R. F. (2010): Alternative Routes in Road Networks, in Proc. of the 9th Intern. Symposium on Experimental Algorithms (SEA '10).

Bellman R. and Kalaba R. (1960): On kth best policies, Journal of the Society for Industrial and Applied Mathematics, 582–588.

Dijkstra E. W. (1959): A note on two problems in connexion with graphs, Numerische Mathematik. 1, 1959, 269–271.

Duckham M., Kulik L., Worboys M. and Galton A. (2008): Efficient generation of simple polygons for characterizing the shape of a set of points in the plane, Journal Pattern Recognition, Vol. 41, Issue 10, Oct. 2008, 3224–3236.

Eppstein D. (1994): Finding the k shortest paths, in Foundations of Computer Science, Proc of the 35th Annual Symposium on. IEEE, 154–165.

Geisberger R., Sanders P., Schultes D. and Delling D. (2008): Contraction Hierarchies: Faster and Simpler Hierarchical Routing in Road Networks, WEA 2008, LNCS 5038, 319–333.

Goldberg A. V., Harrelson C. (2005): Computing the Shortest Path: A* Search Meets Graph Theory, in Proc. 16th ACMSIAM, Symposium on Discrete Algorithms, pp. 156–165.

Hart P. E., Nilsson N. J. and Raphael B. (1968): A Formal Basis for the Heuristic Determination of Minimum Cost Paths, IEEE Transactions on Systems Science and Cybernetics SSC4 (2), 1968, 100–107.

Hoffman W. and Pavley R. (1959): A method for the solution of the n th best path problem, Journal of the ACM (JACM), vol. 6, no. 4, 506–514.

Luxen D., Schieferdecker D. (2012): Candidate Sets for Alternative Routes in Road Networks, in Proc. of the 11th Intern. Symposium on Experimental Algorithms (SEA '12).

Paraskevopoulos A., Zaroliagis C. (2013): Improved Alternative Route Planning, 13th Workshop on Algorithmic Approaches for Transportation Modelling, Optimization, and Systems (ATMOS'13). 108–122.

Roth J. (2009): The Extended Split Index to Efficiently Store and Retrieve Spatial Data With Standard Databases, IADIS International Conference Applied Computing 2009, Rome, Nov. 19–21, 2009, Vol. I, 85-92.

Roth J. (2013): Combining Symbolic and Spatial Exploratory Search – the Homerun Explorer, Innovative Internet Computing Systems (I2CS), Hagen, June 19–21. 2013, VDI, Vol. 10, No. 826, 94–108.

Roth J. (2014): Predicting Route Targets Based on Optimality Considerations, International Conference on Innovations for Community Services (I4CS), Reims (France) June 4–6, 2014, IEEE xplore, 61-68.

Roth J. (2016): Efficient Many-to-Many Path Planning and the Traveling Salesman Problem on Road Networks, KES Journal: Innovation in Knowledge-Based and Intelligent Engineering Systems, 20 (2016), IOS Press, 135–148.

Roth J. (2016b): The Offline Map Matching Problem and its Efficient Solution, International Conference on Innovations for Community Services (I4CS), Vienna, under review.

Yen J. Y. (1971): Finding the k shortest loopless paths in a network, Management Science, vol. 17, no. 11, 712–716.

A Heuristic for Multi-modal Route Planning

Dominik Bucher, David Jonietz and Martin Raubal

Abstract Current popular multi-modal routing systems often do not move beyond combining regularly scheduled public transportation with walking, cycling or car driving. Seldom included are other travel options such as carpooling, carsharing, or bikesharing, as well as the possibility to compute personalized results tailored to the specific needs and preferences of the individual user. Partially, this is due to the fact that the inclusion of various modes of transportation and user requirements quickly leads to complex, semantically enriched graph structures, which to a certain degree impede downstream procedures such as dynamic graph updates or route queries. In this paper, we aim to reduce the computational effort and specification complexity of personalized multi-modal routing by use of a preceding heuristic, which, based on information stored in a user profile, derives a set of feasible candidate travel options, which can then be evaluated by a traditional routing algorithm. We demonstrate the applicability of the proposed system with two practical examples.

Keywords Multi-modal routing · Routing algorithms · Personalization

1 Introduction

Whereas the computation of an optimal driving path through a static road network is a well-researched topic, there has been far less work on the still challenging tasks of integrating personalization options and multiple modes of transportation (Sanders and Schultes 2007; Bast et al. 2015). Particular problems result from the fact that

---

D. Bucher (✉) · D. Jonietz (✉) · M. Raubal (✉)
Institute of Cartography and Geoinformation, ETH Zurich, Stefano-Franscini-Platz 5, 8093 Zurich, Switzerland
e-mail: dobucher@ethz.ch

D. Jonietz
e-mail: jonietzd@ethz.ch

M. Raubal
e-mail: mraubal@ethz.ch

different means of transport operate on specific sub-networks, e.g. a walking path versus a road or train network, which have to be interconnected via distinct transfer nodes which are partly time-dependent, for instance in the case of public transportation or car-pooling (Delling et al. 2014; Raubal et al. 2007). Apart from an increasing complexity of the subsequent graph, which can lead to higher computational effort required for routing queries, a typical result are drastic inequalities with regards to node degrees, which in turn reduce the effectivity of several algorithms developed to speed up graph network computations (Bast 2009). In addition, the dynamic nature of modalities such as car-pooling or public transport imply further growth of the graph and the attributes of its elements as well as call for more complex updating and routing algorithms (cf. Bast et al. 2015 for a list of algorithms).

Personalization of a routing service, i.e. tailoring the results to the specific requirements of a particular user, is especially relevant in the context of multi-modal routing (e.g. Arentze 2013; Liu et al. 2014; Funke and Storandt 2015), but further complicates graph operations. The recommended route is typically individualized by means of a set of user-specific numerical values, which serve as parameters of link cost functions. In our view, this method has the following drawbacks:

- Semantically richer user profiles lead to more complex (and computationally intensive) edge cost functions.
- While user preferences can be easily and intuitively expressed by means of weight coefficients, they are not well-suited for representing hard restrictions (e.g., by means of setting the weight coefficients to ∞).
- Although due to hard restrictions, some travel options might not be feasible for a specific user, they are normally still checked by the routing algorithm.
- The system cannot easily be adapted to new requirements, e.g. additional user characteristics.

With regards to these issues, in this paper we propose an alternative approach for personalized multi-modal route computation, namely to use a rule-based heuristic which precedes the actual routing algorithm, and which, based on a user profile and the situational context, identifies a preliminary set of feasible candidate travel options. In a second step, rather than analyzing the entire graph for the optimal multi-modal route, a routing algorithm focuses merely on the set of pre-computed alternatives, thereby computing detailed routes and assessing their total cost. This procedure, we argue, can dramatically reduce the computational load of the route calculation process, allows for the incorporation of semantically rich user profiles with preferences and restrictions, and can be easily extended. Finally, we demonstrate the applicability of our method with a practical scenario in an urban area, and discuss the resulting routes.

This paper is structured as follows: In Sect. 2, we present relevant prior work related to multi-modal routing and the personalization of route recommendation systems. Our method is described in Sect. 3, starting with a formal description of the rule-based heuristic and moving on towards the general system architecture and details of the implementation. In Sect. 4, we show a practical example before the results of this study are presented in Sect. 5. The final Sect. 6 provides a discussion of the results and concludes this paper.

2 Related Work

This section presents prior work of relevance in the areas of route planning and personalization, in particular with regards to techniques which address the challenges of multi-modal route planning discussed previously.

2.1 Techniques for Multi-modal Route Planning

Routing (i.e., finding the shortest path between nodes) on graphs is a well-researched area (cf. Bast et al. 2015). For that purpose, the famous algorithm of Dijkstra (1959) can be used on a broad range of graphs (directed and undirected, as well as cyclic and acyclic), but is only suitable for graphs with positive edge weights. In case of negative edge weights, the algorithm of Bellman-Ford (1956) is widely used as an alternative.

An application of such algorithms to transportation route planning, however, often requires specific means to speed up processing times, which is due to domain-specific aspects such as large, dynamic road networks and complex calculation criteria. For this reason, various algorithms have been developed. The A* algorithm, for instance, uses a simple heuristic to direct the search towards the target, e.g. by incorporating the Euclidean distance to the destination as an additional factor (Hart et al. 1968). Although this technique has proven to be very useful for routing through road networks, it still requires large parts of the graph to be traversed. Faster speedup techniques include for example Contraction Hierarchies (Geisberger et al. 2008) and related algorithms, which, however, require extensive preprocessing of the graph, during which shortcuts in the graph are inserted whenever a connection between two nodes is suitable. A graph which has been reduced in this manner allows quick routing, and can be unfolded after a route has been identified. Contraction Hierarchies work best for balanced graphs with evenly distributed nodes and vertices (such as a street network). When modeling a typically rather unbalanced public transport network, however, Contraction Hierarchies have to be applied with care (Bast 2009). A similar, but even faster method is proposed with the Transit Node algorithm, which, however, requires more intensive preprocessing and auxiliary space (Bast et al. 2007).

In general, the choice of a suitable routing algorithm requires a trade-off between query processing time, preprocessing time, and graph size, ranging from no preprocessing at all (using Dijkstra or similar), to exhaustive pre-calculation of the optimal route between all vertex pairs, thereby reducing route queries to a simple table lookup.

Naturally, these techniques can be used for multi-modal routing. However, since the underlying graph can change its structure dynamically, query times can increase dramatically, depending on the deployed unfolding technique (Bast 2009). While graph unfolding is successfully applied to a variety of modal combinations (Bast et al. 2015; Dibbelt et al. 2015; Sanders and Schultes 2007), there are also other approaches: For example, RAPTOR (Round-Based Public Transport Routing) does not use a graph at all, but instead relies on data structures that hold individual public transport trips with stops, departure times, and transfer possibilities (Delling et al. 2014). In each iteration, newly reached stops are computed without a change of the transport vehicle. In the next round, the means of transport is changed in order to reach new stops. This procedure continues until a route has been identified. To the best of our knowledge, the (hypothetical) expansion of RAPTOR to multi-modal routing comes closest to our approach, however, without any means of personalization or acknowledgement of the different requirements posed by the various means of transport.

Another noteworthy and comparable approach combines spatial analysis and network routing to reach the desired result (Eiter et al. 2012): In a first step, possible intermediate stops are evaluated based on a spatial join, a standard analysis method to be found in many geographic information systems (GIS). These are then connected, using "traditional" routing algorithms. In contrast to our work, Eiter et al. (2012) do not use this technique for the routing itself, but rather to find places that adhere to some user preferences or choices. Finally, they only consider a single intermediate stop.

2.2 Personalization and User Profiling

The concepts of personalization and user profiling are very closely related. Thus, a user profile, according to the European Telecommunications Standards Institute (ETSI 2005, p. 17), "the total set of user-related information, preferences, rules and settings, which affects the way in which a user experiences terminals, devices and services", represents a prerequisite for successfully personalizing systems or applications (Foerster et al. 2012; Aoidh et al. 2009). Personalization, in turn, describes the tailoring and adjustment of such services to the specific needs and preferences of a particular user, and might serve to filter irrelevant content, alter data visualization methods, make suitable recommendations or adapt the functionality or visual appearance of the program itself (Aoidh et al. 2009; Aissi and Gouider 2012; Raubal 2009).

User profiling and personalization are not only of relevance for routing applications, but for a wide range of Information and Communications Technology (ICT) applications, including Location Based Services (LBS) (Foerster et al. 2012; Raubal 2011). In a routing context, apart from calculating the shortest path, as has traditionally been a key functionality of GPS-enabled navigation devices, alternative approaches have developed algorithms to compute the fastest, the least complicated in terms of wayfinding, the most attractive, the safest, or the easiest route (Golledge 1995; Huang et al. 2014; Krisp and Kehler 2015). A more substantial personalization, however, is achieved when individual preferences and constraints are incorporated into the routing algorithm, which are usually represented as parameters of link costs, stored as a vector of individual weight coefficients (e.g. Arentze 2013; Liu et al. 2014; Funke and Storandt 2015; Nuzzolo et al. 2015; Yang et al. 2015; Campigotto et al. 2016).

The user profile, therefore, occupies a central role in the process of computing personalized routes. It is usually represented as a set of numerical values, which serve as parameters for the route or edge cost calculation. Arentze (2013), for instance, acknowledges individual preferences with regards to travel-time (mode-dependent), financial travel costs (situation-dependent, e.g. fuel costs are generally valued less than ticket costs), preferred traffic mode and service quality (e.g. seat availability and avoiding transfers). In their work on a personalized routing system for tourists, Liu et al. (2014) represent preferences with regards to the degree of interest in specific points of interest (POI), maximum distance, fee, minimum road conditions and traffic conditions. The user profile derived by Nuzzolo et al. (2015) relates to differences between actual and desired arrival times, waiting times, sum of access and egress times, on-board time spent, on-board crowding degree, preference for transit service, number of transits. Campigotto et al. (2016) distinguish between features of routes related to the distances covered and time spent with regards to each individual mode of transportation, financial cost, and others, including the number of transportation means changes, the amount of uphill cycling, as well as the number of public transport stops.

In order to create a user profile, information related to aspects such as personal needs, preferences, interests or habits must be collected. This can occur either by explicitly requesting a user's feedback, or implicitly by inferring such unobservable information from observable behavior (Aoidh et al. 2009) or a combination of both (e.g. Allemann and Raubal 2015). In the context of routing applications, users are often explicitly asked for their general preferences prior to the route recommendation, however, in some cases their actual choice behavior is also monitored and taken into account (e.g. Nuzzolo et al. 2015; Campigotto et al. 2016). There are also studies which use GPS-trajectories from previous travels to extract user preferences and restrictions, however mostly focusing on unimodal transportation (e.g. Letchner et al. 2006; Dai et al. 2015; Yang et al. 2015; Jonietz 2016; Broach et al. 2012; Tsui and Shalaby 2006).

3 The Pre-processing Heuristic

In this section, we propose a heuristic as part of a larger multi-modal personalized routing system. The heuristic allows to pre-calculate a set of plausible routes (where each route is defined by a list of route segments, consisting of start and end point, and a transport modality), which can then be evaluated in more detail by a conventional route planner, in order to check for actual feasibility and user acceptance. In contrast to conventional route planning, we argue, our heuristic is easily adaptable to user preferences (e.g. with regards to modes of transport) or routing constraints (e.g., critical threshold values, such as maximal walking distances, or excluded modalities), and, other than algorithms based on extensive graph pre-processing, does not require algorithm-dependent updates of the transport graph.

In fact, the proposed travel options could already be useful for presenting a user the different possibilities to reach the destination on a less detailed level, for instance by taking a particular train line in combination with a tram, or going by bus to a car-sharing station. Nevertheless, on this general level, our heuristic does not consider the actual road network or public transport timetables. Thus, in a second step, the individual segments of a route should be analyzed and validated by use of a conventional route planner, for instance based on an existing solution such as the Google Maps Direction API[1] or ESRI routing services,[2] and run independently of the heuristic itself.

Exemplary use cases for application of our heuristic are the following:

- A user is interested in evaluating different options to get from Zurich to Munich, without any particular constraints.
- A user has a bike at home, and is interested in getting to work in an energy-efficient manner (i.e., emitting the least amount of CO_2).
- A backpacker tourist would like to use car-pooling or any other cheap means of transport.

Naturally, such use cases could also be addressed via traditional routing systems, however, such queries would necessarily involve multiple web services and platforms, and require the user to manually assess the feasibility of the derived route options. This is partially due to the fast increasing complexity of multi-modal route requests, which, by use of our heuristic, can be decreased by the possibility to formulate pre-set queries, build a dynamic transport graph, and prune large parts of this graph before evaluating it in detail.

3.1 Formalizing Route Calculations

Formally, a route request R can be defined as a tuple consisting of an origin O, represented by a coordinate pair (O_{lat}, O_{lon}) and a set of labels O_L, which denote selected characteristics of the location, such as the number of bikes at the location or the train lines that stop at the location, and a destination D of the same structure as O. More complex requests (which we cover in this paper) also contain a set of

---

[1] https://developers.google.com/maps/documentation/directions.
[2] https://developers.arcgis.com/features/directions.

user preferences P, a set of user constraints A and a description of the context C, i.e., the current state of the world, such as traffic conditions or the weather. Both preferences and constraints can have effect on a variety of route characteristics, such as mode choice, number of inter-modal transfers, time or financial cost, walking distances, or energy consumption.

The expected result of a route request is a list of routes L_R, ordered by some criteria, for instance a combination of various route characteristics, such as the number of mode transfers and energy consumption. A single route contains all intermediate trip segments S_i as well as summary values V_i, such as distance covered with the different means of transport, overall and detailed energy consumption, or the incurring monetary cost of the route. Summarizing, the function of calculating routing results can be denoted as follows:

$$\{(S_{0,i}, V_0), (S_{1,j}, V_1), \ldots\} = f_{routing}(O, D, P, A, C)$$

The aim of our heuristic is not to produce an exact description of the route (i.e., a detailed path on the transport network), but instead a list of trip segments denoted by a start and end point, and an associated mode of transport. For instance, the returned result could tell a user to take the bike from home to a particular train station, then a train line to another station, a tram to a stop close to the destination, and walk the final part of the journey. At this stage, however, it is important to note that the results are approximate, and not yet verified by a route planner. Thus, in order to ensure that the proposed trip is actually feasible, a traditional route planner needs to be deployed, which takes the exact transport network into account.

3.2 Rule Base for Different Modes of Transport

The proposed heuristic evaluates the feasibility of different trip segments and used modes of transport based on their individual preconditions in combination with user preferences and constraints. Thus, for example, driving is only possible if a car is available at the current location of the user; a train ride between two stations depends on the existence of at least one connecting line.

Since all modes of transport are marked by an individual set of preconditions and outcomes, we formulate individual rules of the general form:

$$O[condition] \rightarrow M[condition] \rightarrow D[condition] : [outcomes]$$

where both the origin O and the destination D locations have to fulfill some conditions in order for the mode M to be potentially available for a user. Examples for location conditions would be the availability of a car or the labelling as a train station. The mode condition, in contrast, concerns the action of traveling per se. Walking between locations, for instance, is only possible if the total distance is below the maximum acceptable walking distance for a particular user. If a user moves between two locations, the resulting outcomes mostly affect herself (e.g. she will have walked a certain distance if the mode is "WALK"), but can also influence the context (time will inevitably pass, and the weather might change). Naturally, a further outcome is that the user changed its position to the destination location.

To state an example, consider the following rule, simplified for better reading (for a comprehensive list of all rules, we refer to Sect. 4):

$$A[\emptyset] \rightarrow$$
$$WALK[(user[distWalked] + dist(A, B) < user[maxDist])$$
$$AND\ (NOT\ context[rainyWeather])$$
$$AND\ (context[currentTime]\ IN\ user[acceptableTimeIntvlWalk])]$$
$$\rightarrow B[\emptyset]:$$
$$user[distWalked] \mathrel{+}= dist(A, B)],\ context[time] \mathrel{+}= time(A, B)]$$

The rule describes the locational and mode preconditions as well as outcomes for walking between an origin and a destination. A lack of locational preconditions denotes that in principle, it is always possible to walk between locations. With regards to the mode conditions of walk, however, one can see that there are three preconditions which need to be fulfilled for the journey to be possible: Firstly, the maximal distance covered by walking must not exceed a certain threshold, namely the maximum acceptable walking distance for a particular user. It would, of course, also be thinkable to incorporate other aspects here, such as the steepness of terrain, which, however, is omitted at the present stage of our work. The second and third preconditions are related to the context, thus walking is only possible if the weather is not rainy and the current time is within an acceptable time interval for walking set by the user (e.g., no walking at night time due to fear of crime). With regards to the outcome of walking from A to B, the walked distance needs to be added to the total walked distance of a particular user. Further, the passed time has to be noted. As it has been stated earlier, for these calculations, the heuristic does not consider the exact street network distance between *A* and *B*, but rather refers to the Euclidean distance. This approximation eliminates the need to have an underlying transport graph, and makes the system very flexible with regards to inserting and updating locations. A similar approach has been used in transport planning, e.g., by Raubal et al. (2007).

3.3 Procedure

Based on rules such as the one discussed in the previous section, an ever-increasing graph of modal change nodes can be computed via several functions. The first function, called *checkReachability*, computes the modes of transport that can potentially be used to reach a set of destination locations from a set of origin locations, if any are available. This is done by iterating through the set of origin locations, and, for each mode of transport, determining whether all origin conditions are fulfilled. The destination locations are analyzed in a similar manner. Afterwards, the mode conditions are tested for valid origin and destination pairs. The rule for WALK, for instance, allows all locations as valid origins and destinations, and then checks for every combination whether the destination is reachable with regards to the walking distance of the origin, the weather conditions, and the time of day.

Given an origin (or destination) location, and a mode of transport, the second function, called *expand*, is used to produce a new set of locations. For example, applying the function to a location which is connected to the public transport network, and setting public transport as mode, would yield all other locations connected by the same lines (respecting one-way properties of individual lines), i.e., those which are reachable without a transfer.

Based on these two functions, the following procedure can produce potential routes:

1. An initial check tests whether the destination can be reached directly (i.e., without any transfers) from the origin location. This is done by calling the *checkReachability* function with origin O and destination D. If potential routes are detected, for each mode M that allows a direct transition between origin and destination, a triple (O, D, M) is added to the set of solutions. In most cases, this will include taking the car (if a user does not explicitly discourage this solution, or does not own a car), as most locations can be reached by driving.
2. The function *expand* is applied to the origin O, and, based on all modes M of transport allowed by the user, calculates a set of reachable locations $\{L_1, L_2, ...\}$. This implies a user to travel to L_i by using the respective mode. For this reason, the quadruple (O, L_i, M, S) is stored with the location L_i, which will, in a later step, allow the algorithm to backtrack and build an optimal route. The state S contains more information on the route to reach L_i, such as the number of mode transfers, or the distance walked. This information is necessary since a location L_i can usually be reached by more than one mode.
3. For each member of this new set of starting nodes $\{L_i\}$, we check if the destination D can be reached, using *checkReachability*. If true, the triples (L_i, D, M) are added to the set of preliminary results. These need to be backtracked in a later step in order to calculate the optimal route.
4. Starting from the destination location, *expand* is used to identify a set of new reachable locations L_i and their corresponding modes. This step is similar to 2, but takes place in reversed order. The found quadruples (L_i, D, M, S) are again added to the respective location L_i, to allow backtracking later on.
5. *checkReachability* is applied to this new set of nodes in order to determine possible starting points from within the previously defined set of reached locations when having started from the origin. Similar to 3, found triples (L_i, L_j, M) are added to the set of preliminary solutions. The algorithm then continues at point 2 until a maximum number of transfers are reached, or another stopping criteria has been met (e.g., a valid route has been found, or a route which is substantially better than simply using the car).

6. The preliminary set of solutions is used to calculate a final set of solutions, which includes all trip segments. For this, a third function *unfold* is needed. Here, the stored quadruples are used to backtrack and find a way to the origin, respective to the destination. At the current stage, the tuple with the least number of transfers is chosen, since this is generally assumed to be a useful proxy for route quality (users are often interested in transferring as little as possible, because transfers always lead to slack times). However, more elaborate backtracking functions are imaginable, which take into account the energy efficiency or the travelled distance.

The presented approach dynamically constructs a mode transfer graph which is personalized with regards to user preferences and constraints. Technically, steps 4 and 5 are not required, but instead one could continue expanding the graph from one side. However, comparable to bidirectional Djikstra search strategies (cf. Bast 2009), this leads to a smaller set of locations needed to be checked, which in turn reduces computational complexity. Also, due to the fact that travel routes usually follow a hierarchical scheme (for instance one needs to walk to a higher-order hub of a means of mass-transit, use this mode, and finally cover the "last-mile" by walking again), bidirectional search simplifies the incorporation of strategies which favor such hierarchical route calculation. For instance, long segments associated with "WALK" could only be allowed at origin and destination.

The constructed graph is drastically reduced in comparison to the full transportation network graph, which is due to the *expand* function, which only considers locations that can be reached via the allowed modes of transport, and by considering user preferences, the context, and the state stored in the quadruple (O, D, M, S). For example, if a user has reached her maximum walking distance, further nodes that would involve walking are not considered anymore. The same is true if the weather conditions, as stored in the context, are rainy.

4 Implementation

In this study, the proposed method was implemented as a preprocessing step for a routing system aimed at supporting sustainable ways of traveling. This is embedded in the larger context of *GoEco!*, a project which aims at developing a mobile app designed to provide feedback to users with regards to energy-efficient ways to alter their mobility behavior (Cellina et al. 2016; Bucher et al. 2016). This includes the computation of multi-modal, personalized routes which favor sustainable mobility options such as walking, biking or public transport.

The described concepts were implemented in Python, using a PostGIS database for storing and querying spatial data. The database contains all transfer locations, with respective labels denoting properties of the location. Querying this database

A Heuristic for Multi-modal Route Planning

Table 1 A selection of rules implemented in our prototype system

Mode	Rule	Description
Walk	$d_w = user[distWalked] + dist(A, B)$ $A[\emptyset] \to WALK[(d_w < user[maxDist])$ and $(NOT\ context[rainyWeather]\ OR\ (d_w < user[maxDistRain]))$ and $(context[currentTime]\ IN\ user[acceptableTimeIntvlWalk])] \to B[\emptyset]$: $user[distWalked] += dist(A, B)$, $context[time += time(A, B)]$	Every node provides walking, however, a user can only walk up to a maximal distance (which gets decreased if it is raining), and if the current time is within an accepted time interval for walking. As a result of walking, the total walked distance is updated as well as the context
Bike	$A[user[bikeLocation] = A] \to BIKE[(NOT\ context[rainyWeather])\ AND\ (context[currentTime]\ IN\ user[acceptableTimeIntvlWalk])] \to B[bikeParking = true]$: $user[bikeLocation] = B$, $user[distBiked] += dist(A, B)$, $context[time += time(A, B)]$	A user can only take the bike, if her bike currently is at the location. Further, the destination needs to have a bike parking spot available. Concerning contextual variables similar to walking
Car	$A[user[carLocation] = A] \to CAR[\emptyset] \to B[\#parkingSpots > 0]$: $user[carLocation] = B$, $context[time += time(A, B)]$	Taking the car is only possible from the location where the user currently has parked her car to locations with a parking spot available. As a result, the car is at location B
Train	$A[connectsLineX = true] \to TRAIN[\emptyset] \to B[connectsLineX = true]$: $[\emptyset]$, $context[time += time(A, B)]$	Similar to bus
Tram	$A[connectsLineX = true] \to TRAM[\emptyset] \to B[connectsLineX = true]$: $[\emptyset]$, $context[time += time(A, B)]$	Similar to bus
Carshare	$A[carSharing = true, \#cars > 0] \to CARSHARE[\emptyset] \to B[\#parkingSpots > 0]$: $A[\#cars -= 1]$, $context[time += time(A, B)]$	Carsharing is possible from carsharing locations, where enough cars are available. The destination needs to have free parking spots
Carpool	$A[intersects(A, C) = true] \to CARPOOL[\emptyset] \to A[intersects(B, D) = true]$: $[\emptyset]$	Carpooling is possible from locations that intersect with a spatio-temporal corridor of a carpooler
Bike-Share	$A[bikeSharing = true, \#bikes > 0] \to BIKESHARE[context[weather]\ !=\ "rain"] \to B[bikeParking = true]$: $A[\#bikes -= 1]$, $user[distBiked] += dist(A, B)$, $context[time += time(A, B)]$	Bikesharing is possible from bikesharing locations, where enough bikes are available

for appropriate locations, our heuristic dynamically builds up the transport graph. Table 1 shows the different modes of transport supported by our prototype system, and the according rules currently implemented. For practical reasons, this list does not include the entirety of the updated variables (i.e., the outcomes of the actions). Thus, for example, the distances traveled with the respective mode are always updated (which also makes it easy to introduce rules depending on those distances, e.g., a maximal car distance), as well as the number of transfers taken.

The modes car and carpool represent special cases, as the simple application of the above heuristic would quickly lead to a very large set of nodes due to the fact that by car, all locations are reachable. While this makes sense for trips of the pattern WALK → CAR → WALK, it seems an edge case to assume that a user would for instance drive the major part of the total distance to the destination, only to switch to carpooling for the last part of the trip. To reduce the search space, we constrain the maximal distance traveled by car in the *expand* function to half the total distance between the car location and destination, even if the user specified a larger maximal car distance (note that the patterns WALK → CAR → WALK, or even WALK → CAR → TRAIN → WALK, as commonly found in Park and Ride systems, are not excluded from the solutions here, as they are still identified during *checkReachability*). In general, carpooling, meaning shared journeys in one privately owned car, is implemented as follows: assuming that a driver is willing to drive short detours to pick up other people, and drop them off somewhere along her original route, we model a detour corridor by buffering the original route. Any location within this buffer is a possible pick-up spot, whereas any location downstream becomes a valid drop-off location.

In our implementation, upon calculation of a set of viable routes, a weighting and ranking by approximate energy consumption is applied (using a distance based energy consumption model[3]). Of course, for other applications, a weighting by distance walked, number of transfers taken, or a combination thereof could be more suitable. Finally, from these sorted route alternatives, a predefined number of the top distinct ones are selected and presented to the user, where distinct means they at least differ in the mode of one segment. This distinction is necessary, because it is often possible to change the transport vehicle at different locations, even when using the same sequence of transport means (e.g., changing from tram number X to tram number Y might be possible at different stops, because they share a part of their routes).

In our approach, the search space is therefore reduced as follows: First, the heuristic only has to consider a graph consisting of transfer nodes, which is mostly made up from public transport stops (a reduction of two to three orders of magnitude). Second, for validation of individual route segments, only the respective sub-graphs have to be queried (e.g., the street network, or the public transport network). Because the linking of these sub-graphs causes many speedup techniques to fail, it is beneficial to query them separately.

---

[3] We use the energy consumption values for different modes provided by http://www.mobitool.ch.

A Heuristic for Multi-modal Route Planning

5 Case Studies

In this section, we describe an exemplary application of the proposed heuristic to two case studies set within the city of Zurich. To illustrate the potential of our method for personalization, for each scenario, two users with differing preferences and constraints are defined:

1. Users A and B would both like to travel from Seebach (a suburb in the northeast of Zurich) to a place close to the city center at daytime. The weather is fine at the time the users leave but heavy rain is forecasted for a later point in time (15 min after leaving). User A has a car, but no bike available and would like to walk no more than 1 km in total. User B has no car and no bike but is willing to walk the longer distance of 3 km. For both, the maximum walking distance decreases by a factor 5 during rain. If multiple alternatives are available, they would like to choose the ones with the smallest amount of CO_2 emissions at the expense of longer walking or biking distances.
2. Users C and D would both like to travel from Altstetten, located on the periphery of Zurich, to the city center during late evening hours, and fine weather conditions. User C only has a bike (no car), but is afraid of walking or biking at nighttime (so C tries to keep the respective distances small). User D also has a bike, and has no problem with walking or biking at night. User D likes to do sports, tries to save energy, and does not mind biking for longer distances.

We stored all public transport stations together with the served lines (in the vicinity of Zurich, taken from the GTFS[4] feed from the Swiss federal railways[5]), all carsharing stations (obtained from http://www.mobility.ch) and bikesharing stations (obtained from the city of Zurich) in a PostGIS database, and ran all queries on an Intel Core i7-4910MQ CPU (2.9 GHz), on a laptop with 16 GB of RAM. Query responses of our Python implementation were in a range of approximately 100 ms– 3 s, depending on the number of routes found, and the number of modal changes involved. After the heuristic had been completed, the routes were refined using a routing system comparable to Google Maps Directions. Individual route segments (start and end location, as well as modality) were fed into the system, and the returned routes drawn on an OpenStreetMap base map, using the Leaflet library[6].

For case study one, we additionally assume that user A has parked her car approximately 400 m down the road. The heuristic is able to pick up arbitrary car and bike locations, similar to carsharing, or bikesharing stops. Figure 1 shows the routing result for user A. While the left part shows the raw output of our preprocessing heuristic, we refined the route by querying a conventional route planner for better visualization, as described previously. The route planner computes a path for

---

[4]General Transit Feed Specification, see https://developers.google.com/transit/gtfs.
[5]http://www.fahrplanfelder.ch, retrieved via http://gtfs.geops.ch.
[6]http://www.openstreetmap.org, http://leafletjs.com.

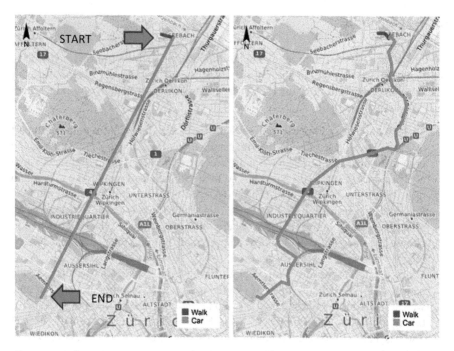

Fig. 1 User A of case study one. The *left* figure shows the raw output of the heuristic, the *right side* the exact route. The *red* WALK segment in the beginning shows the distance walked to the car

each segment, using the mode of transport proposed by the heuristic. The output is shown in the right part of the figure.

It can be seen that the heuristic identifies only the car as a viable travel option, primarily because of the user's reluctance to walk for longer distances. Even though public transport stops can be found approximately 450 m from the start location, the maximum walking distance of only 1,000 m, which is further reduced due to the forecasted rainy weather conditions (max. walking distance during rain 200 m), conflicts with the fact that the closest public transport stop to the destination is around 300 m away, which ultimately leads to the sole preference of the car.

Figure 2 shows the result for user B (max. walking distance of 3,000 m, which during heavy rain is reduced to 600 m). Whereas the left figure shows the results under consideration of the forecasted rainy weather, this is neglected on the opposite map. It is clearly visible that the upcoming rain limits the options of user B, primarily because it leads to a reduced number of modes considered, namely solely the ones which do not require a long final walk to the destination.

It is noteworthy, however, that the right figure does not depict the routes with the smallest walking distances. This is due to the fact that for the sake of energy efficiency, user B is willing to walk or bike longer distances (as long as it is not raining). In comparison to user A, user B has a completely different set of options, resulting from the lack of a car, and the willingness to walk further. A final note

A Heuristic for Multi-modal Route Planning 225

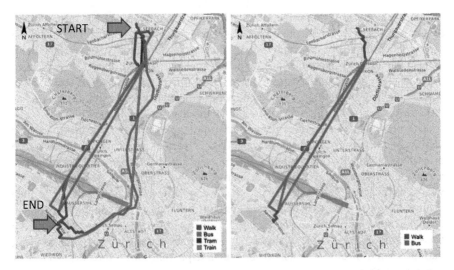

Fig. 2 User B of case study one. The *left* figure shows possible routes without consideration of weather restrictions. The right figure assumes good weather in the beginning, but high chances of rain at a later point

concerns the different route alternatives: In the left figure, it is easy to identify different ways of reaching the destination. User B could for example cover most of the route with the train, or use a combination of train, tram, or bus to reach a variety of stops close to the destination. Of course, it is not guaranteed that all of these routes are still feasible once time schedules are considered. However, such output could already assist a user in planning a route.

Figure 3 compares the results of users C and D. User C's options are very limited: Since C would like to minimize walking during the night (the maximal walking distance during the night was set to 200 m), the only viable travel option computed by the heuristic includes renting a car from a nearby carsharing station.

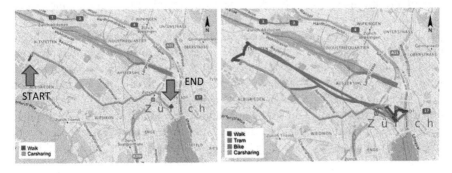

Fig. 3 User C (*left side*) and user D (*right side*) of case study two. While C is restricted from walking too far during the night by fear, D has many more options

Using this car, user C can drive to the city, park it, and walk the last meters to the destination. User D, on the other hand, has plenty of options: Apart from using the same carsharing station, D could directly ride the bike to the city center (blue line), or take a variety of combinations of public transport (green lines). Note that the order of the returned alternatives for user D is *"bike, public transport (route 1–3), carsharing"*, as D would like to use an energy efficient route (where bike is rated as the most energy efficient, and carsharing as the least efficient alternative).

The presented case studies demonstrate the applicability of the heuristic to a variety of personalized use cases, resulting in various multi-modal routes to be proposed. Depending on the preferences of users and the available transport modes, a completely different set of route alternatives are suggested. While the routes which have actually been computed by a conventional route planner resemble solutions known from various routing systems available to the public, the heuristic is able to drastically reduce the search space prior to detailed computations on the transport graph, and is able to suggest less conventional alternatives, such as using various shared means of transport, or routes depending on the weather and other context variables.

6 Conclusion

In this paper we proposed the use of a preset rule-based heuristic for the computation of personalized multi-modal routes through complex, dynamic transportation networks. We discussed how the user profile and the situation-specific context can be deployed in order to reduce the search space of the route planner to a subset of pre-defined feasible origin-destination pairs and the according modes of transport. The functionality and practical value of our method was demonstrated on the example of different users traveling in the city of Zurich.

The use of a preceding heuristic has several advantages, including a reduction of computational load needed for the route planning process. By providing the route planner with a set of pre-defined start- and end-nodes, as well as the mode-specific part of the transportation network to be used for this trip segment, the search options necessary to be checked are minimized. Further, this strategy allows for parallelizing the route planning process. Thus, while the heuristic is still being executed, the actual route planner can already evaluate the resulting route options. Also, individual trip segments might be part of several different trip options, but only be evaluated once (for example, a journey by bike from home to a train station needs to be evaluated only once, even though the overall journey might continue differently). Additionally, the procedure allows for more detailed and dynamic personalization due to the fact that different user profiles merely require the set of rules in the heuristic to be updated, whereas altering of the graph itself is not needed.

There are, however, also some shortcomings of our approach. For example, with more complex personalization, the rules can get complicated and lead to an increased computational overhead. This can partly be addressed by implementing a "transfer graph" (i.e., a network connecting different transfer points, such as public transport stops, car- or bikesharing facilities, etc.) as part of the heuristic which could improve the efficiency, however, at the cost of reverting to a more rigid structure. Further, the scalability of such a system was only briefly discussed in this paper. The heuristic, at its present stage, can quickly generate a high number of possibly reachable states. However, by restricting modes to maximal distances, this can be avoided to some degree (e.g., if both origin and destination are far away from an intermediate stop, it does not make sense to consider using a tram at this stop, except to reach a "hub" for faster transportation).

Finally, our work can possibly be extended by expanding the heuristic to include other points of interest (POIs) or general user needs (cf. Eiter et al. 2012; Raubal et al. 2004; Bucher et al. 2015). For example, a user might be interested in a journey that includes a stop at a pharmacy, or knowing possibilities for meeting friends during a journey. Using concepts from time geography, we can compute sets of intermediate reachable POIs, and integrate them into the rule-based approach.

Acknowledgments This research was supported by the Swiss National Science Foundation (SNF) within NRP 71 "Managing energy consumption" and by the Commission for Technology and Innovation (CTI) within the Swiss Competence Center for Energy Research (SCCER) Mobility.

References

Aissi S, Gouider MS (2012) Personalization in geographic information systems: A survey. International Journal of Computer Science Issues (IJCSI) 9(4): 291–299.
Allemann D, Raubal M (2015) Usage differences between bikes and E-bikes. In Bacao F, Santos M, Painho M (eds) Geographic information science as an enabler of smarter cities and communities, Springer, Heidelberg, 201–217.
Aoidh EM, McArdle G, Petit M, Ray C, Bertolotto M, Claramunt C, Wilson D (2009) Personalization in adaptive and interactive GIS. Annals of GIS 15(1): 23–33.
Arentze TA (2013) Adaptive personalized travel information systems: a Bayesian method to learn users' personal preferences in multimodal transport networks. IEEE Transactions on Intelligent Transportation Systems 14(4): 1957–1966.
Bast H (2009) Car or public transport—two worlds. Springer, Berlin, Heidelberg, 355–367.
Bast H, Delling D, Goldberg A, Müller-Hannemann M, Pajor T, Sanders P, Wagner D, Werneck RF (2015) Route planning in transportation networks. Technical Report, Microsoft Research.
Bast H, Funke S, Sanders P, Schultes D (2007) Fast routing in road networks with transit nodes. Science, 316(5824): 566–566.
Bellman R (1956) On a routing problem (No. RAND-P-1000). Rand Corp, Santa Monica Ca.
Broach J, Dill J, Gliebe J (2012) Where do cyclists ride? A bicycle route choice model developed with revealed preference GPS data. Transportation Research Part A: Policy and Practice 46 (10): 1730–1740.
Bucher D, Cellina F, Mangili F, Raubal M, Rudel R, Rizzoli A.E, Elabed O (2016). Exploiting Fitness Apps for Sustainable Mobility – Challenges Deploying the GoEco! App. ICT4S 2016.

Bucher D, Weiser P, Scheider S, Raubal M (2015). Matching complementary spatio-temporal needs of people. In Online proceedings of the 12th international symposium on location-based services, Augsburg, Germany.

Campigotto P, Rudloff C, Leodolter M, Bauer D (2016) Personalized and situation-aware multimodal route recommendations: the FAVOUR algorithm. IEEE Transactions on Intelligent Transportation Systems, arXiv:1602.09076.

Cellina F, Rudel R, De Luca V, Rizzoli A E, Botta M, Raubal M, Bucher D, Weiser P (2016) Eco-feedback and gamification elements for sustainability: the GoEco! living lab experiment. Poster presented at the 6th European Transport Research Conference (TRA 2016). Warsaw. Poland.

Dai J, Yang B, Guo C, Ding Z (2015) Personalized route recommendation using big trajectory data. In: IEEE 31st International Conference on Data Engineering, Seoul, 543–554.

Delling D, Pajor T, Werneck RF (2014) Round-based public transit routing. Transportation Science 49(3): 591–604.

Dibbelt J, Pajor T, Wagner D (2015) User-constrained multimodal route planning. Journal of Experimental Algorithmics (JEA) 19: 3–2.

Dijkstra EW (1959) A note on two problems in connexion with graphs. Numerische Mathematik 1 (1): 269–271.

Eiter T, Krennwallner T, Prandtstetter M, Rudloff C, Schneider P, Straub M (2012) Semantically enriched multi-modal routing. International Journal of Intelligent Transportation Systems Research 14(1): 20–35.

ETSI (2005) Human factors (HF): user profile management. ETSI, Nice.

Foerster T, Stoter J, van Oosterom P (2012) On-demand base maps on the web generalized according to user profiles. International Journal of Geographical Information Science 26(1): 99–121.

Funke S, Storandt S (2015) Personalized route planning in road networks. In: Proceedings of the 23rd SIGSPATIAL International Conference on Advances in Geographic Information Systems (GIS '15). ACM, New York.

Geisberger R, Sanders P, Schultes D, Delling D (2008) Contraction hierarchies: Faster and simpler hierarchical routing in road networks. In: Experimental Algorithms, Springer, Heidelberg, 319–333.

Golledge R (1995) Path selection and route preference in human navigation: A progress report. In: Frank A, Kuhn W (eds) Spatial Information Theory-A Theoretical Basis for GIS, Vol. 988, Springer, Berlin, Heidelberg, New York, 207–222.

Hart PE, Nilsson NJ, Raphael B (1968) A formal basis for the heuristic determination of minimum cost paths. IEEE Transactions on Systems Science and Cybernetics 4(2): 100–107.

Huang H, Klettner S, Schmidt M, Gartner G, Leitinger S, Wagner A, Steinmann R (2014) AffectRoute – considering people's affective responses to environments for enhancing route-planning services. IJGIS 28(12): 2456–2473.

Jonietz D (2016) Personalizing Walkability: A Concept for Pedestrian Needs Profiling Based on Movement Trajectories. In: Sarjakoski T, Santos MY, Sarjakoski LT (eds) Geospatial data in a changing world. Springer, Heidelberg, 279–297.

Krisp J, Keler A (2015) Car navigation – computing routes that avoid complicated crossings. IJGIS 29(11):1988–2000.

Letchner J, Krumm J, Horvitz E (2006) Trip router with individualized preferences (TRIP): incorporating personalization into route planning. In: Porter B (ed) Proceedings of the 18th conference on Innovative applications of artificial intelligence - Volume 2 (IAAI'06), AAAI Press, 1795–1800.

Liu L, Xu J, Liao SS, Chen H (2014) A real-time personalized route recommendation system for self-drive tourists based on vehicle to vehicle communication. Expert Systems with Applications 41(7): 3409–3417.

Nuzzolo A, Crisalli U, Comi A, Rosati L (2015) Individual behavioural models for personal transit pre-trip planners. Transportation Research Procedia 5: 30–43.

Raubal M (2009) Cognitive engineering for geographic information science. Geography Compass 3(3): 1087–1104.
Raubal M (2011) Cogito ergo mobilis sum: The impact of Location-Based Services on our mobile lives. In: Nyerges T, Couclelis H, McMaster R (eds) The SAGE Handbook of GIS and Society, Sage Publications, Los Angeles, London, 159–173.
Raubal M, Miller H, Bridwell S (2004) User-Centred Time Geography For Location-Based Services. Geografiska Annaler B, 86(4), 245–265.
Raubal M, Winter S, Teßmann S, Gaisbauer C (2007) Time geography for ad-hoc shared-ride trip planning in mobile geosensor networks. ISPRS Journal of Photogrammetry and Remote Sensing 62(5): 366–381.
Sanders P, Schultes D (2007) Engineering fast route planning algorithms. In: Experimental Algorithms, Springer, Berlin, Heidelberg, 23–36.
Tsui A, Shalaby A (2006) Enhanced system for link and mode identification for personal travel surveys based on global positioning systems. Transportation Research Record: Journal of the Transportation Research Board 1972: 38–45.
Yang B, Guo C, Ma Y, Jensen CS (2015) Toward personalized, context-aware routing. The VLDB Journal 24(2): 297–318.

Part III
Spatial-Temporal Data Processing and Analysis

Development of a Road Deficiency GIS Using Data from Automated Multi-sensor Systems

Alexander Mraz and Abdenour Nazef

Abstract Traditional survey methods have long been used in the field of highway engineering to measure the cross-slope, longitudinal grade, rut depth, and ride quality of existing roadways. However, these methods are slow, tedious, labor intensive, and almost always require partial or full lane closure resulting in traffic delays, increase in costs, and inconvenience to the traveling public. Advances in inertial sensor and inertial navigation technologies have allowed their implementation as state-of-the-art mobile data collection systems. The Florida Department of Transportation (FDOT) operates two mobile data collection systems referred to as Multi-Purpose Survey Vehicles (MPSVs). They collect pavement data including but not limited to cross-slope, longitudinal grade, and wheel-paths' rut depth at typical highway speeds. The MPSVs are equipped with a position and orientation system (POS) coupled with an inertial profiler unit. The core of the POS consists of a tightly-coupled Inertial Measurement Unit (IMU) and a Differential Global Positioning System (DGPS). This paper presents a methodology for the development of an Automated Roadway Deficiency Information System using Geographical Information System (GIS) software to map areas prone to hydroplaning. The functionality of the developed information system was tested on a pilot project using MPSV collected data. Highway agencies can successfully implement this methodology to complement and enhance their existing safety and pavement management programs.

Keywords Transportation · Multi-sensor data collection system · GIS · Hydroplaning · Safety

---

A. Mraz (✉)
Palacky University Olomouc, Olomouc, Czech Republic
e-mail: alexander.mraz@upol.cz

A. Nazef
Florida Department of Transportation, Gainesville, FL, USA

© Springer International Publishing AG 2017
G. Gartner and H. Huang (eds.), *Progress in Location-Based Services 2016*, Lecture Notes in Geoinformation and Cartography, DOI 10.1007/978-3-319-47289-8_12

1 Introduction

Traditional pavement survey methods have long been used in the field of highway engineering to measure and evaluate the cross-slope, longitudinal grade, rut depth, and ride quality of existing roadways. Analysis of the data helps engineers to identify problem-prone areas or hot spots with deficient cross slope and inadequate surface drainage. However, these traditional methods are costly, tedious, and labor intensive. Moreover, survey crews often operate in a travel lane, creating potentially hazardous conditions. A high-speed automated data collection system equipped with multiple sensors subsystems offers an efficient and cost-effective alternative with the added safety of the field survey personnel and motorists.

The Florida Department of Transportation (FDOT) conducts annual pavement condition surveys to assess the condition and performance of its State Highway System and to predict future rehabilitation needs. The main pavement condition parameters used for the assessment include ride quality, rut depth, and cracking. Pavement friction characteristics are also evaluated to identify spot hazard conditions.

In 2003, the FDOT acquired its first automated multi-sensor system, referred to as Multi-purpose Survey Vehicle (MPSV) to measure cross-slope (CS), longitudinal grade (LG), smoothness, and rut depth at typical highway speeds (Gunaratne et al. 2003). Currently, the FDOT operates two upgraded MPSVs that incorporate position and orientation systems (POS) and two additional outside-angled laser sensors as an enhancement to the original 3-sensor inertial profiling system. Furthermore, the Florida Department of Transportation 2012 Plans Preparation Manual (PPM) recommends using a vehicle-mounted system for CS verification in the pre-design phase of projects and to identify roadway limits where the CS is out of tolerance (FDOT 2016).

The FDOT develops, maintains, and makes available a plethora of GIS tools, base data layers, maps, and services throughout its offices. However, there is no specific tool or Geographic Information System (GIS) data layer to assess areas prone to hydroplaning due to improper roadway geometry, or excessive rutting, or both. This paper presents a methodology and computer application for detecting and mapping accident prone areas by fusing multi-sensor system rich data and the functionality of GIS.

2 Related Work

The implementation of sensor and navigation technologies into the pavement evaluation area allows highway engineers to capitalize on a large amount of information to detect problem prone areas. In 2003, the Iowa DOT investigated the use of LIDAR technology to extract cross-slope (CS) and longitudinal grade (LG) data for its roadway network inventory. The study results indicated that longitudinal grade could be estimated to within 1 %, but cross-slope could not practically be estimated using LIDAR (Souleyrette et al. 2003). Another 2007 study in Italy compared two different automated methods of measuring roadway CS. The first method computed the CS from Inertial Navigation System (INS) data using a simplified algorithm. The second method used a single-axis laser scanner synchronized with an INS/Differential Global Positioning System (INS/DGPS) (Bolzon et al. 2007).

Transportation agencies collect roughness and rut depth measurements using inertial profiling systems as a standard practice. These systems obtain longitudinal profiles by measuring vertical elevations as a function of the longitudinal distance along the traveled path (Gillespie and Sayers 1987). Rut depth which is defined as a depression worn into a road by the wheels can be calculated using elevation information captured by the measured longitudinal or transverse profiles. Currently, there is no consensus on the number of sensors required for rut depth measurement. Most transportation agencies have implemented 3- and 5-sensor systems with some organizations adopting the systems with more than seven sensors (TRB 2004). Recently, several researchers have taken a position that 3-sensor systems do not measure rutting with sufficient accuracy (FHWA 2001). Consequently, some vendors and highway agencies have implemented laser-based systems that project a laser line across the pavement with a lateral resolution of 1,024 elevation points across the pavement width to determine the deepest rutting in each wheel path regardless of vehicle position.

The longitudinal profiles are typically used to calculate profile-based roughness index like the International Roughness Index (IRI). Since its introduction in 1986, the IRI has become a universal roughness index for evaluating and managing road systems (Sayers et al. 1986). The IRI summarizes the roughness qualities that impact vehicle response and is most appropriate when a roughness measure is desired that relates to overall vehicle operating cost, overall ride quality, and dynamic wheel loads (Sayers and Karamihas 1998).

Another road surface characteristic that affects safety is skid resistance which determines the friction between the road surface and a vehicle tire. Andriejauskasa et al. (2014) give a description of skid resistance methods and devices operating at different operating speeds, wheel loads, water film thicknesses, and using different test tire types. FDOT typically assesses skid resistance and mean profile depth using a high-speed lock-wheel testing device. These measurements allow for calculating the International Friction Index as described in ASTM E 1960 (Jackson 2008).

Many transportation agencies specify the location of road attribute values by referencing them along the roadway using a linear referencing system (LRS). It is defined as a way to identify a particular location with respect to a known point (NCHRP 1974). The Distance Measuring Instrument (DMI) is a vehicle-mounted electronic instrument that precisely measures distances along traveled path, and can be calibrated to match an agency's LRS. The LRS is a support system for storage and maintenance of information on events that occur along a roadway network. In this context, it consists of an underlying transportation network that supplies the geographic backbone for the location of events (Curtin et al. 2007). Several GIS software packages currently offer tools to assist in the generalization of spatial features and events for the purpose of an LRS. ArcGIS geodatabase supports a new model for dynamic segmentation route system using "Polyline with Measure" feature class (Goodman 2001). The dynamic segmentation methods divide network links into segments that are homogenous for the specified set of link attributes. It is dynamic because the linear feature is split into a new set of segments whenever its attributes change (Jelokhani-Niaraki et al. 2009). Some LRSs use absolute distance to describe location which is based on the distance measured from the beginning of the linear elements. Other LRSs use relative distance where the event location is described as an offset distance from a pre-defined reference point along the linear element. Relative measurements can be achieved using mileposts or reference posts (Scarponcini 2002).

Many agencies create transportation GIS to query and analyze spatially-enabled data, display trends and patterns in the data, and create maps and reports. A 2003 study by Tsai et al. (2013) evaluated the use of LiDAR technology presented a GIS-based application to identify road sections with sub-standard cross-slope using a 1 % cross-slope threshold. The Georgia DOT has integrated pavement condition survey data with GIS technology to visualize statewide pavement condition and to perform spatial data analysis (Tsai and Gratton 2004). The Illinois DOT has developed the ILLIPIMS system containing unique mapping and graphing capabilities allowing the user to display, analyze, and update data for the entire Illinois Interstate Highway Network (Bham et al. 2001).

3 Objective

Since 2003, FDOT has been collecting CS, LG, rut depth, and ride data using the MPSV. The collected data have been evaluated on a project basis, and the results have typically been provided to the Districts in the form of tables and graphs. Integrating MPSV data into a GIS-based transportation information system for the entire roadway network would greatly improve this process.

The primary objective of this paper is to present a GIS application for mapping areas prone to hydroplaning. The secondary objective is to present a GIS application design that provides a seamless integration with existing FDOT pavement evaluation databases. The paper also presents the results of a pilot project used to test the functionality of this application using MPSV collected data.

4 Applied Methodology

4.1 Evaluation of Roadway Geometry

Poor pavement surface drainage may lead to hydroplaning. Highway agencies interested in minimizing this problem have minimal control over drivers, vehicles, and environmental factors. However, they have more control identifying causal or contributing factors to roadway departure or hydroplaning by analyzing roadway geometry and surface drainage characteristics.

The cross-slope influences water drainage across the pavement and thus affects the operational characteristics of a vehicle moving along the road. Other factors that can also affect surface drainage are longitudinal grade and rutting. The wheel path CS may change over time due to surface deformation caused by traffic loads, environmental effects, and pavement settlement. Also, the in situ CS magnitude can differ significantly from the design values. Therefore, knowledge of existing CS and LG is essential for the evaluation of roadway geometry of existing and newly built or resurfaced roads.

Mobile data collection systems like MPSVs collect CS and LG data simultaneously. These data provide critical information about a roadway's ability to drain surface runoff and, therefore, minimize the potential for hydroplaning. The drainage path length (*DP*) relates cross-slope (*CS*), longitudinal grade (*LG*), and pavement drainage width (*W*) using the following equation (Guven and Melville 1999):

$$DP = W \sqrt{\left[1 + \left(\frac{LG}{CS}\right)^2\right]} \qquad (1)$$

where:

DP drainage path length (ft or m)
W   pavement drainage width (ft or m)
CS cross-slope (ft/ft or m/m)
LG longitudinal grade (ft/ft or m/m)

Drainage path length, which is computed at discrete station locations, is the relative distance runoff travels on the sloped pavement surface before it leaves the pavement. A longer DP, sometimes combined with roadway depressions like wheel ruts, creates areas susceptible to hydroplaning especially on high-speed roads. DP increases with steeper LG and decreases with increasing CS. It is also significantly influenced by the pavement drainage width (Glennon 2007). Figure 1 illustrates an example of pavement with good surface drainage characterized by short DP (*DP₁*) due to adequate cross-slope while the pavement area with poor surface drainage is represented by a relatively long DP (*DP₂*). Since the CS runs in the same direction in both lanes, all surface runoff is carried by *lane 2* which amplifies the drainage problem in *lane 2* resulting in a relatively long *DP₂* before draining off the pavement. The longitudinal grade prolongs the DP when it has a substantially larger magnitude than the CS. It also impacts heavy vehicle operation by reducing their stopping and passing distances. Therefore, the longitudinal grade as shown in Eq. 1 is a significant factor in roadway geometry evaluation.

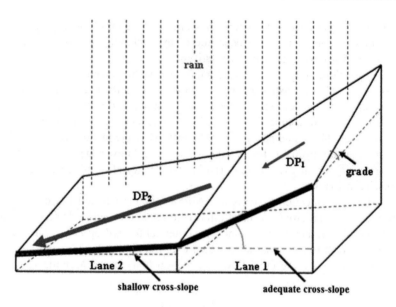

Fig. 1 Example of long drainage path (DP$_2$) due to shallow cross-slope

4.2 Evaluation of Rut Depth

Wheelpath depressions, caused by poor mix design, compaction, or loading, interrupt the normal flow of surface runoff to the edge of the pavement. Ruts filled with water can contribute or cause hydroplaning. Virtually all inertial profiler systems are designed to measure the relative elevation of the host vehicle reference plane to the pavement surface, using a 3-point or 5-point laser sensor system as shown in Figs. 2 and 3, respectively. Assuming the pavement surface does not experience rutting in the middle of the lane (Fig. 2), the rut depth in each wheel path can be calculated using 3-point inertial profiler system as presented in Eqs. 2 and 3.

$$R_L = h_2 - h_1 \qquad (2)$$

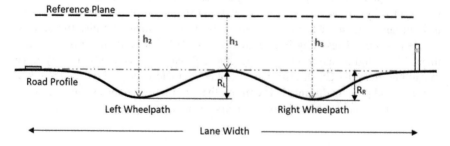

Fig. 2 Sketch of a transverse profile from a 3-point laser sensor configuration

Development of a Road Deficiency GIS ... 239

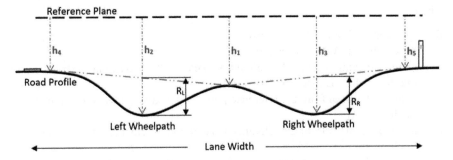

Fig. 3 Sketch of transverse profile from a 5-point laser sensor configuration

$$R_R = h_3 - h_1 \qquad (3)$$

where:

R_L rut depth in left wheelpath, LWP (inch or mm)
R_R rut depth in right wheelpath, RWP (inch or mm)
h_1 relative elevation from the reference plane to the pavement surface close to the roadway centerline (inch or mm)
h_2 relative elevation from the reference plane to the pavement surface in the LWP (inch or mm)
h_3 relative elevation from the reference plane to the pavement surface in the RWP (inch or mm)

The rut depth calculations presented in Eqs. 2 and 3 cannot be used if pavement surfaces experience rutting in the center of a lane. In such situation, only average rut depth (R_{AVE}) can be calculated as shown in Eq. 4:

$$R_{AVE} = \frac{h_2 + h_3}{2} - h_1 \qquad (4)$$

An inertial profiler system with at least five laser sensors is required when rut depth is calculated for each wheel path under any surface conditions. The left wheelpath (LWP), right wheelpath (RWP), and average rut depth can then be calculated as:

$$R_L = h_2 - \frac{h_4 + h_1}{2} \qquad (5)$$

$$R_R = h_3 - \frac{h_5 + h_1}{2} \qquad (6)$$

$$R_{AVE} = \frac{1}{2}(h_2 + h_3 - h_1) - \frac{1}{4}(h_5 + h_4) \qquad (7)$$

where:

R_L rut depth in LWP (inch or mm)
R_R rut depth in RWP (inch or mm)
h_1 relative elevation from the reference plane to the pavement surface in the middle of the roadway (inch or mm)
h_2 relative elevation from the reference plane to the pavement surface in the LWP (inch or mm)
h_3 relative elevation from the reference plane to the pavement surface in the RWP (inch or mm)
h_4 relative elevation from the reference plane to the crown of pavement surface (inch or mm)
h_5 relative elevation from the reference plane to the edge of pavement surface (inch or mm)

The rut depth calculated from the data collected with an inertial profiler system does not directly account for the ability of the road rutting to carry or hold the water. Therefore, the measured rut depth must be adjusted for the cross-slope (Glennon 2007). The maximum water depth (WD) which determines the capacity of water a roadway rutting can hold is calculated as:

$$WD = d - 1/2 W \tan(\alpha) \qquad (8)$$

where:

d rut depth calculated using Eqs. 2–7 (inch or mm)
W wheel path rut width, Fig. 4 (inch or mm)
α cross-slope measured with inertial navigation system (deg)

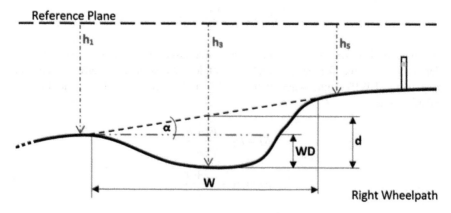

Fig. 4 Measuring RWP rutting and maximum water depth

5 Implementation

5.1 Data Collection

The MPSV, shown in Fig. 5, is an automated multi-sensor system which consists of a host vehicle equipped with several sensor subsystems. It includes an inertial profiler for measuring rut depth and roughness, a POS for measuring roadway geometry and vehicle's position, a right-of-way imaging system, and a pavement imaging system for capturing images of pavement deficiencies such as cracking.

The core components of the POS are the Inertial Measurement Unit (IMU), multi-GNSS receivers, and Distance Measuring Indicator (DMI). The multi-GNSS receivers provide improved position updates and allow processing signals from multiple systems, including Global Position System (GPS) and GLONASS (APPLANIX 2015). GNSS have excellent low-frequency performance resulting in relatively constant position error over longer periods of time. However, their performance from one second to another may change by up to several meters. In contrast, the IMU has excellent high-frequency performance, but poor low-frequency performance. Thus, inertial navigation solely based on the IMU suffers from instrument's drift error which accumulates over time. Combining the two systems in a tightly-coupled manner where the GPS receiver provides raw pseudorange data and the IMU supplies acceleration and angular rate measurements results in excellent low and high-frequency performance and allows the overall solution to use GNSS data even when fewer than four satellites are available (Neu 2014). The POS also

Fig. 5 MPSV with a 5-point laser sensor setup

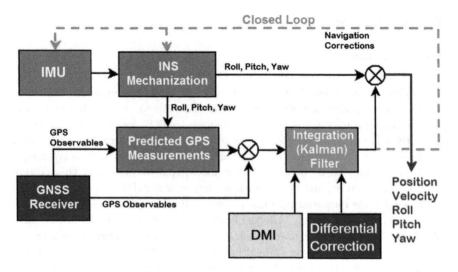

Fig. 6 Simplified tightly-coupled POS scheme

incorporates real-time differential correction to further improve positional accuracy. To constrain the IMU's drift error during GPS signal outages and determine when the vehicle is at rest, the POS uses additional information provided by the DMI. A simplified scheme of the tightly-coupled POS is shown in Fig. 6. When the GPS receivers are unable to provide position information due to partial or total signal outage, the IMU continues to provide the position and orientation information unaided. The indicated position error is relatively small over a short period but within minutes, the position begins to drift significantly. The commonly used Kalman integration filter estimates the errors in the IMU. The algorithm is recursive and works as a two-step process.

The IMU is mounted on the vehicle frame and undergoes the same movements as the vehicle. Modern cars use a suspension system that connects the vehicle frame to its wheels and allows relative motion between the two vehicle components. Therefore, vehicle bounce introduced by the suspension system is also sensed by the IMU and adds an error to the measured CS and LG. To minimize the error, the IMU measured CS and LG are compensated for the relative elevation information provided by the inertial profiler system (Fig. 7). Measured CS can be compensated using the equations shown below. Similar equations can be derived for LG.

$$CS = CS_{IMU} - CS_{IP} \qquad (9)$$

$$CS_{IP} = \tan^{-1}((h_5 - h_4)/L) \qquad (10)$$

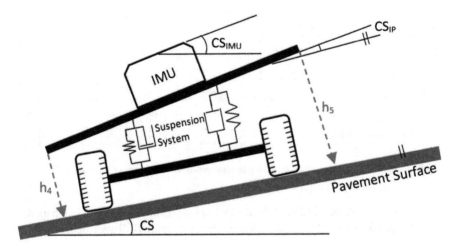

Fig. 7 Correction of IMU measured CS for suspension movement using laser sensors

where:

CS	roadway CS (deg)
CS_{IMU}	CS sensed by the IMU (deg)
CS_{IP}	CS correction calculated from inertial profiler's side laser sensors (deg)
h_4	relative elevation from the reference plane to one side of the pavement surface (inch or mm)
h_5	relative elevation from the reference plane to the other side of the pavement surface (inch or mm)
L	lane width (inch or mm)

5.2 Data Processing

The MPSV collects a large amount of sensor data which, if processed manually, would require a considerable amount of time. The authors used FDOT developed Automated Cross-slope Analysis Program (ACAP) to import and simultaneously process POS, GNSS, and profile data. The program outputs CS, LG, DP, rutting, and ride measurements in tabular and in graphical form at any user-defined distance interval (Mraz and Nazef 2008). Each measurement record is provided with a geographic coordinates provided by the POS, as illustrated in Fig. 8. The ACAP was

COUNTY:	99	DATE COLLECTED:	02/24/2015
ROUTE:	SR9	TIME COLLECTED:	13:41:16
DIRECTION:	East(+)	OPERATOR:	HL
LANE:	R2	DRIVER:	HL
UNITS:	ENGLISH	WAVELENGTH LONG:	none
SECTION:	009	RPT VERSION:	2.3.7.18
SUB SEC:	000	MDR VERSION:	WP3.7.6.0
RUNOPTIONS:		SENSOR CAL DATE:	02/24/2015
SAMPLE DISTANCE:	0.73 in (1)	ACCEL CAL DATE:	02/24/2015

Miles Ref Post	Deg Slope	Deg Grade	Feet Radius Curv	Deg Rad Curv Dir	Deg IMU Slope	Latitude	Longitude	Q	Sat	HDOP	Height
15.7991	1.320	-0.510			1.280	+29.493387167	-082.973296667	2	07	1.3	144.2585
15.7991	1.329	-0.490			1.280	+29.493387167	-082.973296667	2	07	1.3	144.2585
15.7996	1.210	-0.480	2332.32	207.45	1.280	+29.493396667	-082.973317333	2	07	1.3	144.2913
15.8001	1.160	-0.490	2783.12	207.40	1.270	+29.493399230	-082.973322865	2	07	1.3	144.3179
15.8005	1.270	-0.530	2306.21	207.34	1.250	+29.493402833	-082.973331000	2	07	1.3	144.3241

Fig. 8 Sample MPSV cross-slope and longitudinal grade raw data output from ACAP

updated to calculate maximum water depth from rut data, as described in Chap. 4.2. Also, the application uploads the collected data into an SQL Server database.

5.3 Development of Automated Roadway Deficiency Information System

5.3.1 Methodology

The Geographic Information System (GIS) is an information system that deals with digital geographic data (Longley et al. 2005). The GIS can pre-process the data for the further query, perform spatial data analysis, display trends and patterns in the data, and create maps and reports. The spatial nature of roadway data also promotes the use of GIS (Kiema and Mwangi 2009). Therefore, GIS has a potential for modeling processes in pavement evaluation, including the detection of areas prone to hydroplaning.

The authors developed the Automated Roadway Deficiency Information System (ARDIS) in Esri ArcMap 10.3.1 by importing a shapefile of the Florida State Highway System and database tables holding MPSV collected data. Development of the ARDIS involved the collection of spatial data with MPSV, the design and creation of attribute tables for Microsoft SQL Server database, followed by GIS implementation, data analysis, and reporting for determining areas prone to hydroplaning due to poor roadway geometry, or excessive rutting, or both using the GIS software.

5.3.2 Data Sources

The spatial data included FDOT GIS Basemap Route and MPSV collected data that were processed using the ACAP. The FDOT GIS Basemap Route feature class provides spatial information on Florida roads and attributes compatible with the Roadway Characteristics Inventory database which contains information about roadway location, classification, physical characteristics, traffic information, and traffic control device inventory. The basemap dataset includes lines representing each roadway within the inventory. It is a representation of the roadway system which is comprised of individual features (intersections, bridges, auxiliary lanes, medians, etc.) positioned along the roadway segments (Jacobs 2015). The data collected with MPSVs included information about linear location, road geometry, rutting, and geographical coordinates for each tested segment and they were processed to 52.8 ft (16.1 m) interval. The ACAP program calculated drainage path length and maximum water depth for measured road rutting using the MPSV data. The collected data were linearly referenced to the basemap dataset using "ROADWAY," "BEGINPOST," and "ENDPOST" fields. The location is determined with regards to the particular roadway at a distance from a documented feature along the roadway (Jacobs 2015).

5.3.3 Database Design

The database was designed to contain all attribute information about roadway geometry and pavement condition including rutting and roughness. It was designed to give pavement engineers all necessary information to identify pavement segments prone to hydroplaning. After identifying the potential users and their information needs, the logical model in the form of tables and relations was developed and is presented in Fig. 9. The final step involved populating the database with MPSV collected data using the ACAP program.

5.3.4 GIS Implementation

The authors used the Esri ArcMap software to implement the spatial data into GIS. This choice was made to take advantage of the data management, display functionality, and analysis tools provided by the Esri product. Also, FDOT creates and maintains its organization-wide GIS using the Esri software and employees have the most experience working with it. The ARDIS was set up in Esri ArcMap 10.3.1 by connecting to the Microsoft SQL Server database and adding *cs_dp* and *rut_wd* tables to the map as layers. The *cs_dp_SELECT* and *rut_wd_SELECT* layers were created to show pavement sections with drainage path length equal to or greater than 55 ft (16.7 m) and maximum water depth equal to or greater than 0.1 in (2.5 mm), respectively. The *pavement_ride_cond* layer was also created following four IRI categories: (1) excellent for IRI $\leq$ 70 inch/mile, (2) good for 70 inch/mile < IRI $\leq$ 100 inch/mile, (3) fair for 100 inch/mile < IRI $\leq$ 150 inch/mile, and (4) poor for IRI > 151 inch/mile. Finally, the *cs_dp_SELECT*

Fig. 9 Database logical model

layer was intersected with the *rut_wd_SELECT* layer to capture features where both drainage path length and maximum water depth were deficient. The results were saved in *hp_analysis* and *pavement_ride_cond* layers. The ARDIS components including layers and their attributes are presented in Fig. 10.

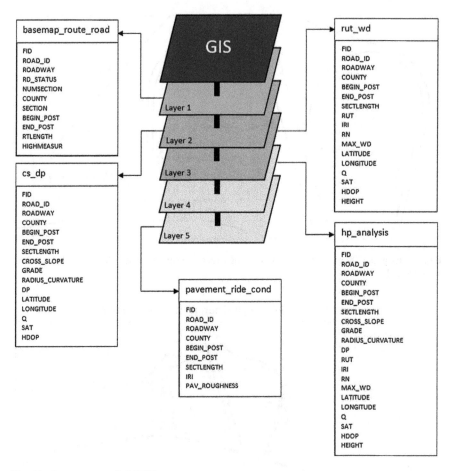

Fig. 10 Components of ARDIS

6 Results

The ARDIS database contains both the general view of the entire road network and specific information about sections with roadway geometry or rut deficiencies. It also provides information about road sections prone to hydroplaning. Figure 11 shows a typical query on a road with a road segment prone to hydroplaning selected on the map and the respective attributes displayed in the attribute table. Similarly, roadway segments prone to hydroplaning and road ride condition can be displayed, as shown in Figs. 12 and 13, respectively. Such a display of results can help pavement engineers to localize quickly areas prone to hydroplaning and plan for short-term and long-term mitigating solutions.

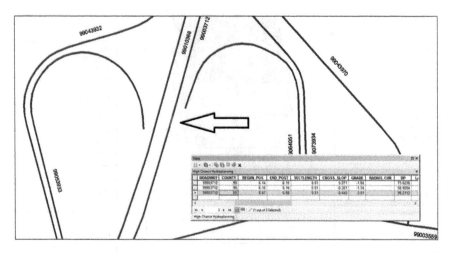

Fig. 11 Example showing the selection of road section prone to hydroplaning in the map and attribute table

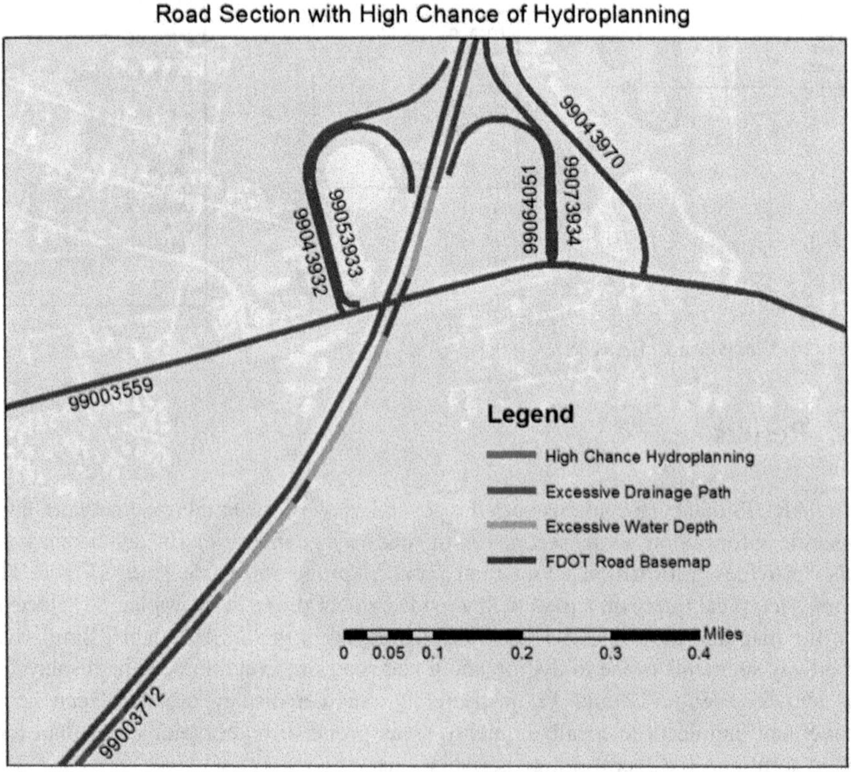

Fig. 12 Results of identifying road sections prone to hydroplaning obtained for a sample road section

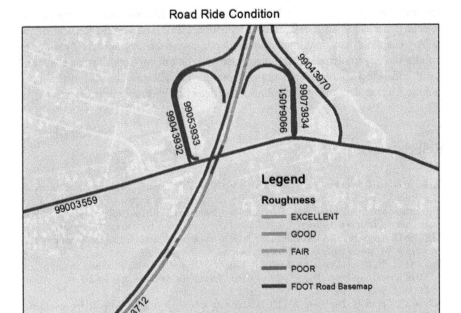

Fig. 13 Display of road ride condition for a sample road section

7 Conclusion and Discussion

The ARDIS coupled with a state-of-the-art data collection system provides pavement engineers and highway practitioners a tool for identifying areas with deficiencies in roadway geometry and excessive road rutting that contributes to vehicle departure. Information can further be used to formulate short-term and long-term mitigating solutions to reduce the likelihood of roadway departures. Collecting data using high-speed data collection systems also improves safety while reduces traffic delays and operational costs because their use does not require partial or full lane closures. The ARDIS can also be used as an enhancement to existing pavement evaluation programs as it provides the ability to quickly identify and output the results for existing and newly resurfaced pavements. Moreover, analytical and display functions provided by the GIS software gives transportation agencies an overall view of the road network for future fund allocation.

The ARDIS has been under development and it was tested only on a small number of projects. The plan is to import all more projects already collected with MPSVs. The authors have also recognized several areas of improvements to be implemented in the next software update. The authors realized that some drainage path length peaks detected as "deficient" locations represent transitions from a tangent section into and out of super-elevated areas. Therefore, the radius of curvature data collected with MPSVs will be an additional input parameter in detecting drainage path peaks within transition areas. The current version of the ACAP calculates the maximum water depth using the average rut depth. The authors plan to implement a routine which will output rut depth in the LWP and RWP and select the one with the deeper rut. The authors are also planning to include friction data to further enhance the detection of problem-prone areas.

References

Andriejauskasa T, Vorobjovasa V, Mielonasb V (2014) Evaluation of skid resistance characteristics and measurement methods. In Environmental Engineering. Proceedings of the International Conference on Environmental Engineering. ICEE (Vol. 9, p. 1). Vilnius Gediminas Technical University, Department of Construction Economics & Property.

APPLANIX (2015) POS-LV Specifications, http://www.applanix.com/media/downloads/products/specs/poslv_specifications12032012.pdf, Accessed 5 August 2016.

Bham G H, Nasir G, Darter M. I. (2001) Illinois's Experience with Pavement Analysis and Management Systems. Paper presented to Transportation Research Board 80th Annual Meeting, Washington, D.C.

Bolzon G, Caroti G, Piemonte A (2007) Accuracy Check of Road's Cross Slope Evaluation Using MMS Vehicle. The 5th International Symposium on Mobile Mapping Technology, Padua, Italy.

Curtin K M, Nicoara G, Arifin R R (2007) A comprehensive process for linear referencing. URISA Journal, 19(2), 41–50.

FDOT (2016) FDOT Plans Preparation Manual – Design Criteria and Process, Florida Department of Transportation (Volume 1), Roadway Design Office, Tallahassee, Florida.

FHWA (2001) Adequacy of Rut Bar Data Collection, FHWA, Office of Policy, Federal Highway Administration, Washington, D.C., TechBrief RD-01–027.

Gillespie, T.D., Sayers, M.W (1987) Methodology for Road Roughness Profiling and Rut Depth Measurement, World Bank Technical Paper No. 46, The World Bank, Washington, D.C.

Glennon J. D. (2007) Hydroplaning – The Trouble with Highway Cross Slope, http://www.johncglennon.com/papers.cfm?PaperID=8, Accessed 1 June 2016.

Goodman, J E (2001) Maps in the fast lane—linear referencing and dynamic segmentation. http://www.directionsmag.com/article.php?article_id=126, Accessed 2 August 2016.

Gunaratne M, Mraz A, Sokolic I (2003) Study of the Feasibility of Video Logging with Pavement Condition Evaluation, Florida Department of Transportation, Report No. BC-965, Florida.

Guven O, Melville J (1999) Pavement Cross Slope Design – A Technical Review, Auburn University, Highway Research Center, Auburn, Alabama.

Jackson N M (2008) Harmonization of Texture and Skid Resistance Measurements. Florida Department of Transportation Research Report, Fl.DOT/SMO/08-BDH-23, University of North Florida, College of Computing, Engineering and Construction, Florida, USA.

Jacobs B (2015) Crash GIS Data Creation, https://fdotewp1.dot.state.fl.us/TrafficSafetyWebPortal/post/Post_694_WhitePaper_CrashGISDataCreation_v7_1_FINAL_2015-05-18.pdf, Accessed 2 August 2016.

Jelokhani-Niaraki M R, Alesheikh A A, Alimohammadi A, Sadeghi-Niaraki A (2009) Designing road maintenance data model using dynamic segmentation technique. In International Conference on Computational Science and Its Applications (pp. 442–452). Springer Berlin Heidelberg.

Kiema J B K, Mwangi J M (2009) A Prototype GIS-Based Road Pavement System, Journal of Civil Engineering Research and Practice, Vol. 6, No. 1.

Longley P A, Goodchild M F, Maguire D J, Rhind D W (2005) Geographic Information Systems and Science. John Wiley and Sons, West Sussex, England.

Mraz A, Nazef A (2008) Innovative Techniques with Multi-Purpose Survey Vehicle for Automated Analysis of Cross-Slope Data, Transportation Research Record 2068, Transportation Research Board, Washington, D.C.

NCHRP (1974) Highway Location Reference Method, Synthesis of Highway Practice 21, National Academy of Sciences, Washington D. C.

Neu J (2014) A Tightly-Coupled INS/GPS Integration Using a MEMS IMU – Thesis, Air Force Institute of Technology, Department of the Air Force, Air University, Ohio.

Sayers M W, Gillespie T D, Queiroz C A V (1986) The International Road Roughness Experiment: Establishing Correlation and a Calibration Standard for Measurements, World Bank Technical Paper No. 45, The World Bank, Washington D. C.

Sayers M W, Karamihas S M (1998) The Little Book of Profiling: Basic Information about Measuring and Interpreting Road Profiles. University of Michigan Transportation Research Institute, Ann Arbor, MI, USA.

Scarponcini, P (2002) Generalized Model for Linear Referencing in Transportation. GeoInformatica, 6(1), 35–55.

Souleyrette R, Hallmark S, Pattnaik S, O'Brien M, Veneziano D (2003) Grade and Cross Slope Estimation from LIDAR-based Surface Models. U.S. Department of Transportation, Iowa Department of Transportation. MTC-2001-02.

TRB (2004) NCHRP Synthesis 334: Automated Pavement Distress Collection Techniques. A Synthesis of Highway Practice, Transportation Research Board, Washington, D.C.

Tsai Y, Chengbo A, Wang Z, Pitts E (2013) Mobile Cross-Slope Measurement Method Using Lidar Technology, Transportation Research Record Journal of the Transportation Research Board 2 (2367), pp. 53–59.

Tsai Y, Gratton B (2004) Successful Implementation of a GIS-Based Pavement Management System. Applications of Advanced Technologies in Transportation Engineering, pp. 513–518.

Identifying Origin/Destination Hotspots in Floating Car Data for Visual Analysis of Traveling Behavior

Mathias Jahnke, Linfang Ding, Katre Karja and Shirui Wang

Abstract In this paper, we present the results of developing a geo-visual analytics application to support urban services. The goal is to allow non-GIS users to explore the taxi traveler's hot spots in Shanghai extracted from one week taxi floating car data (FCD). To achieve this, we proposed a workflow based on the visualization pipeline. Firstly, we preprocess the data to extract the origins (o) and destinations (d) from the FCD and apply data mining methods to detect taxi traveler's hot spots, to which semantics are further tagged using point of interest (POI) data extracted from OpenStreetMap (OSM) project. The detected hot spots are selected to show in the application for the user to conduct further visual analysis. Furthermore, we implement a web-based interactive visual explorative system, in which the graphic user interface contains multiple views (spatial, temporal and thematic) and interactive components are built up using the current web technologies. Finally, a possible use case of the application is introduced. Our results show that the developed geo-visual analytics application enables studying traveler's activity patterns. The visual analysis can be conducted with this tool for several aspects. The visual queries help to detect when and where hot spots occur and to compare the temporal distributions for nearby hot spots.

Keywords Visual analytics · Floating car data · Web visualization · Decision support system · Smart city

---

M. Jahnke (✉) · L. Ding · K. Karja · S. Wang
Chair of Cartography, Technical University of Munich, Munich, Germany
e-mail: mathias.jahnke@tum.de

L. Ding
Group of Applied Geoinformatics, University of Augsburg, Augsburg, Germany

© Springer International Publishing AG 2017
G. Gartner and H. Huang (eds.), *Progress in Location-Based Services 2016*, Lecture Notes in Geoinformation and Cartography,
DOI 10.1007/978-3-319-47289-8_13

1 Introduction

With the development of telecommunication technology researchers gained new methods for studying a variety of movement phenomena. In urban areas the spatial researchers have studied people's activities and movement dynamics collected with the use of mobile technologies over a decade (Mooses 2013; Hu et al. 2009; Messelodi et al. 2009; Ahas et al. 2008). The citizen's movement has drawn particular attention for urban service providers (Ahas and Mark 2005) in the context of smart city for offering enhanced services e.g. taxi fleet management or traffic control. The visualization of origin-destination (o/d) flows are investigated by Wood et al. (2010) while the visualization of travel pattern of the London bicycle hire system users are presented by Wood et al. (2011).

Nevertheless, the trait of the analysis of movement data involves knowledge of spatial data processing, spatial data mining and GIS skills. With the inherent spatial, temporal and thematic components of movement data, the complexity of information and visualization modes increases. When it comes to presenting results for the decision makers, there is a need for a presentation mode that supports gaining insights into different aspects of the data.

Nowadays information technology offers various methods and technologies, like data processing, data mining and web visualization methods, for analyzing human movement patterns from movement data, for instance, floating car data (FCD). In this work, floating car data, in particular those collected from taxis, is of main interest to study taxi behavioral patterns within the city to improve the quality of services offered in a city. For instance, we can derive from FCD the probability of finding a taxi around as well as the average time of taxis to reach the users location as a specific application of location-based services for tourists and citizens planning their trips. Origin/destination (o/d) spots can be extracted from floating car data to analyze the spatial and temporal distribution of taxi traveler' origin/destination hot spots which can be further used for a variety of research and applications, for instance to infer different functional areas (Yuan et al. 2012; Mazimpaka and Timpf 2015; Ding et al. 2015). In some areas similar patterns appear routinely every day, in others they appear more randomly. Typical example areas to study these different patterns are shopping malls or conference centers. Before representing the data in a feasible and suitable way to the user, data mining techniques like spatial and temporal clustering have to be applied to feature multifold analysis on different aspects of the data. To make these analysis more advantageous, the raw FCD could be enriched and supplemented with volunteered geographic information (VGI) of the studying areas to investigate diverse human behavioral pattern in urban environments.

Clustering belongs to the exploratory data analysis (EDA) tools and is one among many data mining techniques for discovering patterns in data. It has been frequently used for detecting dense regions in data space (Suh 2012). The dense regions of the clustering results reflect the spatial hotspots of the investigated events or phenomena.

Based on the mined patterns from the clusters, visual and interactive visualization of those hotspots can help urban service providers or decision makers to conduct surveys on several scales of the city and to improve their services as well as to make it possible to the non-expert user to gather information for their daily travel.

Usability is one of the main issues when developing a user interface. In particular web-based visualizations need a thoughtful approach to attract the user and to deliver the desired information. Usability describes how well a user is supported by the system or visualization to gather the desired information (Nielsen 1993, 2004). Different methods like questionnaires, user observations, thinking aloud etc. are able to estimate a usability measure (Heidmann et al. 2003; Lobben et al. 2005; Swienty et al. 2008; Burghardt and Wirth, 2011; Sarodnick and Brau 2011) which leads into usability engineering lifecycle (Mayhew 1999).

The outcomes of temporal and spatial changing of human behavioral patterns can be used for planning purposes (Ahas and Mark 2005) if they are visualized in a suitable manner, especially in the context of a smart city with the ideas of offering better services for the citizens. The nowadays information and communication technology offers tools and methods for developing analytical support services to enhance the public transport services in cities. Similarly, the daily travel behavior of citizens can benefit from this kind of visual analytics (Andrienko et al. 2007; Andrienko and Andrienko 2012) of taxi-taker's travelling patterns.

Within this contribution, we aim to mainly answer the question when and where high intensity hotspots occur and nearby which types of venues they take place. The user interface to the analytical results of the submission is implemented as a web-based visualization service using up-to-date web technologies.

2 Information Visualization, Geovisual Analytics and Geovisualization

Since the human visual system is effective at recognizing patterns, visualization of data is considered as a powerful method in knowledge extraction process (Swienty et al. 2008; Miller and Han 2009). Information visualization integrates interaction and visual representation and is considered to be useful for developing insights and understanding complex datasets (Fekete et al. 2008).

Geovisualization refers to the visual exploration, analysis, synthesis, and presentation of geospatial data for which it provides theory, methods, and tools (MacEachren and Kraak 2001). Geovisualization integrates approaches from different fields which are geovisual analytics, cartography, exploratory data analysis, scientific visualization and other related fields (Nöllenburg 2007). According to MacEachrens (1995) conceptual model there exist different functions of geovisualization. They depend on the different purposes of visualization, which can be on one scale the construction of new knowledge by supporting exploration over data or on another scale is sharing the already generated knowledge with public via

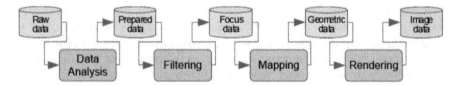

Fig. 1 The visualization pipeline (Dos Santos and Brodile 2004)

presenting the results (MacEachren and Kraak 2001). During the process of knowledge construction, the search is being done for detecting patterns and associations in data. According to this model, during the stage of exploration, the specialist has a need for high level and more sophisticated interaction tools with the visualization application, while for presentation purpose low level interactivity can be offered.

Visual analytics is the science of analytical reasoning facilitated by interactive visual interfaces (Cook and Thomas 2005). According to (Keim et al. 2008) visual analytics provides technology that combines both the strengths of human and electronic data processing. So that visualization is used as a medium of a semi-automated analytical process together with human-computer interaction techniques. Analogously, geovisual analytics, having the visual, map-based interfaces with computational methods, supports human reasoning and decision making process (Cook and Thomas 2005; Andrienko et al. 2007).

In virtual mediums many interactivity modes are being used. Depending on the purpose of the visualization, different interaction techniques can be built for the visualization applications. Some of the components of the interaction design of geospatial visualization application involve navigation over the information space and the design how to access and manipulate data (Cartwright et al. 2001). Methods for enabling interaction in web-based visualization are e.g. linked views, focusing, brushing etc. (Roth 2012).

Generally speaking, the visualization community adopts the visualization pipeline to describe the process of generating visual representation of data (Tominski 2006; Fry 2008). (Dos Santos and Brodlie 2004) proposed four main steps, which are data analysis, filtering, mapping and rendering (Fig. 1) to gain insights into data and to construct knowledge.

3 Methods

Within this contribution, detection of street level hot spots is of main interests. The origins and destinations (o/d) of the trips are not uniformly distributed in space and time. The web based map visualization of the floating car data reveals areas with high as well as low concentration of pick-up and drop-off events of taxi passengers. In space and time, events take place with different variations in intensity. For detecting hot-spots of taxi traveler's origin (pick-up) and destination (drop-off) events, the focus is in detecting space-time subsets in which the concentration of events is higher than in other subsets.

To reach the stage of visualizing the origin/destination hotspots from the floating car data, we need to conduct different analysis steps. Within a preprocessing step the data was cleaned up, the origins and destinations were derived and stored in a database system. Based on this pre-processing step the data was temporally grouped and spatially clustered to identify the origin and destination hotspots. The cluster results lead to the web based visualization. Within the following sections these steps are described in more detail.

3.1 Floating Car Data (FCD) and Preprocessing

In this contribution we used floating car data of taxis collected in Shanghai, China as our test data. Data of one week (17.05.2010–25.05.2010) is used for analyzing and visualizing origin and destination hotspots of the travelers. The data set consists of point locations of 1690 taxis, which are sampled using 10 s interval over the mentioned time period. Each data record represents one GPS point with different attributes: timestamp, company-id, car-id, longitude, latitude, speed, altitude, car-status, GPS-effectiveness and orientation.

The attributes from which we can extract pick-up or drop-off events are of main interest and they are the spatial attributes longitude and latitude, the temporal attribute timestamp as well as the semantic attributes car-status and car-id. The car-status indicates whether a taxi cab is occupied or not. The changes of "car-status" from occupied to non-occupied indicate drop-off events and vice versa pick-up events (Table 1).

Furthermore, the location of 3150 POI's distributed over the area of Shanghai will be used for labelling, visualization and knowledge extraction and are integrated into the web based visualization service.

The first exploration to FCD revealed that sometimes the car-status is changing rapidly, repeatedly and very close to each other in time and space or there are isolated instances of values 1 or 0. These unexpected changes or isolated values are due to data acquisition errors and have to be cleaned up. In our case cleaning the data means smoothing out these isoloated and rapidly changing car-status. At the end we got a cleaned dataset where the car-status stayed constant for at least for

Table 1 Relevant attributes for extracting pick-up and drop-off events

Attribute name	Description
lon	Longitude
lat	Latitude
timestamp	Time and date information
car_status	1 if occupied; 0 if non-occupied
car_id	The id of a car

40 s. The cleaned data set is a prerequisite for extracting origin/destination (o/d) points.

The pick-up or drop-off events is indicated by the change in the car-status, in our case if the car-status changes from 0 to 1 it is probably a pick-up event if the status changes from 1 to 0 it is probably a drop-off event. These changes were extracted for the whole dataset of the one-week period.

After derivation of the o/d points, those points will be aggregated into temporal intervals of different granularities. We used 1 h up to 4 h intervals. A pre-clustering revealed a 3-h interval for aggregating as sufficient to get the most suitable cluster results. Within the 3-h interval o/d events are more distributed over the city. If the interval is too short only a few o/d concentrations occurred on the other hand if the interval is too long the o/d's are too condensed for a suitable clustering, visualization and knowledge extraction.

3.2 Clustering

Cluster analysis belong to the knowledge discovery from databases (KDD) methods, and is helpful in finding patterns in a large collection of data by grouping data items based on their similarities in data space (non-geographic space). To find similarities within geographic space Tobler (1979) already mentioned that "everything is related to everything else, but near things are more related than distant things". Bringing both together the determination of clusters is being done based on relationships between data items by finding items that are closer to each other in geographic as well as in non-geographic space.

The major categories of clustering techniques include partitioning methods, hierarchical methods, density based methods, and grid-based methods (Miller and Han 2009). For the sake of investigating movement data, clustering methods can be applied to spatial, temporal, or spatial-temporal analysis. The k-means, agglomerative hierarchical clustering and density–based spatial clustering of applications with noise (DBSCAN) have been used for detecting hot-spots of taxi trips origins (Chang et al. 2008).

Detecting accumulations of origin and destination events was the aim of the clustering. The derived clusters are our hotspots of o/d events. In a city, only areas with high densities were of interest while areas with low densities were not taken into further consideration. Another issue we expect within the data set was a varying number of clusters in space and time and that there was no previous knowledge concerning the number of clusters. Therefore, the DBSCAN clustering algorithm (Ester et al. 1996) seemed to be suitable for our purpose.

The clustering algorithm should be able to take into account the following assumptions

- Detecting high intensity gathering of o/d events all over the city
- Clusters can have varying shapes

- For each interval, there can be a different distribution of data points
- The number of cluster can vary and is unknown for each time interval (3 h).

The DBSCAN algorithm was considered suitable for the clustering task of detecting o/d hot-spots in the data set. Two parameters are required to form a dense region, they are namely epsilon (*Eps* or ε) and the number of minimum points (*minPts*) per cluster. Therefore, the DBSCAN algorithm detects only clusters in which the density of points is higher than that specified by their parameters. The density of a data point is defined by a number of points that lie within a radius *Eps* of that point. The data points are classified into core, border, or noise points. A point is specified as core point, if a data point contains at least a number of within a radius *Eps*. A point is a border point, if a data point contains less than *minPts*, but it also contains at least one core point within the radius *Eps*. Finally, a noise point if a data point is neither a core points nor a border point as shown in Fig. 2 (Ester et al. 1996).

For each 3-h interval spatial clusters of o/d hotspots were detected using the before mentioned DBSCAN algorithm based on their spatial location. The parameters for DBSCAN were chosen considering the previously listed assumptions for cluster detection. According to the assumptions a relatively small radius *Eps* should be chosen. To detect hotspots, we have chosen 300, 200, 100 and 25 m for *Eps* as well as 10 points and 6 points for the *minPts* in a pre-calculation of clusters. Finally, the visualization relevant data was extracted with the parameters of *Eps* = 25 m and *minPts* = 6. These values were suitable for detecting high intensity locations in a given test area. Higher *Eps* value should be chosen to detect clusters with varying densities on block level.

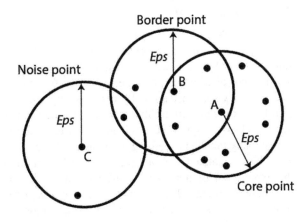

Fig. 2 DBSCAN clustering algorithm: core, border and noise points (Ester et al. 1996)

3.3 Adding Semantic Information to the Hotspots

After detecting the o/d hotspots the next issue is to add semantic information to the hotspots. Adding semantics to places is a challenging task. Zhang et al. (2012) used a grid based approach to attach the semantics of nearby buildings to every cell while Andrienko et al. (2013) added semantics to places based of GPS, GSM and Twitter data. Another example of context enrichment is presented by Krueger et al. (2014) who linked FCD and Foursquare data for automatic event extraction. The POI data describing the different venues were derived from the OSM data layers' *buildings* and *landuse*. In the area of Shanghai these layers contain about 11500 buildings and approximately 2700 records with different land use. Most of the points of the interests (POI) are located within the inner districts of Shanghai while a smaller number of objects are located outside. Approximately 7000 not meaningful or unclassified objects were not taken into account. Because of the very sparse OSM data in Shanghai the number of POI's can be increased with the knowledge of local experts. At least similar types of venues were categorized. Therefore, in our paper, we chose the POI's which are fallen into the categories of hotel, commercial, office, exhibition center, restaurant, public, and business to enrich the semantic information. We got approximately 3100 POIs that can be used for labelling the detected cluster.

The POI's were assigned to every origin or destination point within a respective cluster by getting the o/d point and adding a buffer distance to the point. The distance refers to a short walking distance of 100 m with the assumption that pick-up or drop-off areas are very close to the related venues. The POI with the shortest distance of those falling inside the buffer zone were chosen for semantic enrichment. The o/d points which are not nearby a POI were labeled as "undefined" and got their semantic manually annotated.

3.4 Visualization

The last step within the workflow was to visualize the derived information in a suitable way. A web-based visualization was designed to gain insights into the data (o/d hotspot distributions and the additional semantic information). Therefore, the task was to visualize temporally and spatially changing o/d hot-spot occurrences distributed over the city of Shanghai. We distinguish two types of hot-spots, namely the traveler's origin and destination places. Additionally, the hot-spots were semantically labelled according to the nearest place or venue. The time range which should be visualized was one week on a 3-h interval. The main goal is to support the exploration of hot spots in the city and to discover different spatiotemporal distribution patterns. The hot-spots could be discovered in small scale and studied further in large scale views. Thus, the visualization needs to be meaningful when viewed in different zoom levels.

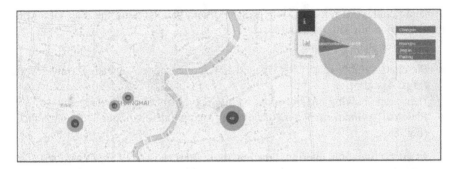

Fig. 3 The user interface with the map view on the *left* and the map tools panel on the *right* showing temporal thematic charts

The detected hot-spots were visualized on a web application that is composed of multiple linked views. The spatial data is queried from the database with ajax requests via a python based web server (python flask web server). The graphical user interface (GUI) was a mashup that was built upon up-to-date web technologies like HTML5, CSS3 and javascript APIs, which are namely leaflet, with its q-cluster plugin (Hallahan 2014), leaflet-heat.js plugin (Agafonkin 2014) and the d3.js based crossfilter.js and dc.js.

The thematic map layers were drawn either as heat maps or symbol maps (pie charts). Both of them were built using the aforementioned leaflet plugins. An area selector was integrated in the visualization to create bounding boxes around areas of interest. The linked map view and charts are changing correspondingly based on actions in the map or the chart view. In addition, the cognitive linking between the views is achieved by using associative colors for the chart and visual map elements.

To sum up, the front-end of the application contains both visualization and interactive components. The visualization approach chosen was to use 2D maps with interactive charts and controllers. The main user-interface components of this application are the map view which shows the spatial distribution of hot-spots and the map tools panel offering temporal and thematic charts, which are linked together to display information synchronously (Fig. 3).

4 Results

4.1 Web Application

The application has three map views with data filtering tools and statistical charts. The charts show concurrently the statistics of temporal distributions and perform as query filters. By selecting different time slots snapshots of different temporal lengths can be chosen for exploring the data. Different time scale choices can be made at daily and hourly levels. In addition to time related information, also semantic context of hot-spots is represented with pie-charts on the map view with the corresponding filtering tool.

The three map views are:

- Pie-chart symbol-map with labelled hot spot locations revealing venue types (Figs. 4 and 5).
- Heat-map showing o/d hot-spots with temporal filtering tools (Fig. 6).
- Pie-chart symbol-map for discovering temporal distribution for hot-spots (Fig. 8).

As interactive tools filtering, brushing techniques and linked views are used. The visualization components are the map views with the thematic map layers on the left and the statistics charts on the right. With using the time slider, a period of time or a moment can be chosen.

It is possible to choose the time period of interest, by selecting all seven days, or studying hourly patterns day by day. It is also possible to choose an hour of interest

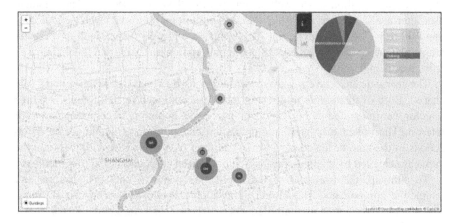

Fig. 4 Filtering by district with the chosen Pudong district

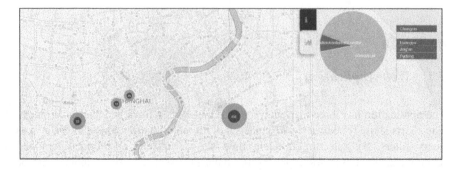

Fig. 5 Filtering by venue type, conference centers and exhibition halls are being chosen

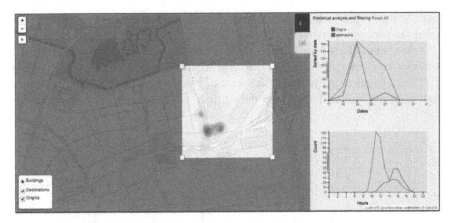

Fig. 6 Area selector concentrating on special places

and go through results for this hour day by day. This investigation is supported by filtering the desired district or different venues by their semantic types. The intensities are displayed as a heat map. When creating a bounding box for the desired area, the statistical charts are showing results corresponding to this area (Fig. 6).

4.2 The Example Study Area

Different hot spot occurs at different areas depending on the daily and weekly rhythm of taxi trips. In some areas, hot spots appear and disappear with a similar pattern every day in one week. In other areas, the high intensity activities take place less regularly. Many types of patterns are triggered by different reasons for travelling, since in each urban zone or place taxi travelers are expected to hold different activities. Hereby, one study area is described to give an insight into temporally changing taxi travelling events during high intensity times. The study area is the Shanghai New International Expo Centre (NIEC) in the Pudong district.

The analyst can answer different types of questions with the use of the application by using filters. It is possible to filter hotspots by district and see what semantical types of hotspots occur. As shown in Fig. 4, within the chart view the proportions of the different venue types for the selected district are shown, while in the map view the hotspots' distribution in space is visualized. In the map view events in nearby areas are clustered and their proportions are shown with pie symbol charts.

Another option is to filter the venues by semantical type, in order to find all hotspots with the selected label, e.g. conference and exhibition centers in Shanghai. In Fig. 5 it is shown that during the study period destination hot spots take place in four districts. A few of them in Jing'an and Huangpu, higher activity occur near the center, in Changnin, and very high activity events are taking place in Pudong.

As we have detected the location of the exhibition hall, near which high intensity activities take place, we can find the same location in the heat map and continue with more detailed analysis of activities taking place near the Shanghai NIEC. As high intensity events took place during four days (Fig. 6), we can brush the

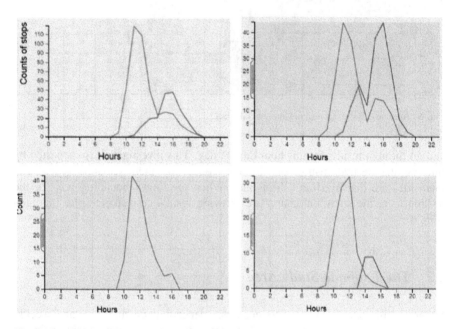

Fig. 7 Selected activities near Shanghai NIEC from 18.05.10–21.05.10

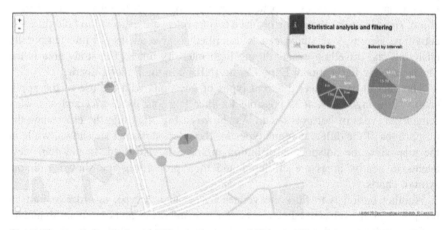

Fig. 8 Temporal distribution in different spots around Shanghai New International Expo Centre for destination hotspots

date-chart showing activities by days and choose each day in order to get an insight of daily activities (Fig. 7).

The alternative map view for discovering temporal distribution in different spots is given with pie-chart symbol map (Fig. 8). This map reveals summaries of different hot spots around the Shanghai NIEC. Their intensities and temporal distribution are described by using pie-chart symbols. It is possible to filter by day or by a 3-h interval. Figure 8 shows where most of the drop off events took place during four days and during which daily interval. In addition, most of the drop off events took place during the morning hour interval from 09:00 till 12:00 (colored in yellow). In this time duration, the spots for drop off events are more distributed. During the one week of time a trade fair tool place at Shanghai NIEC, therefore taxi

Table 2 The summary and description of the example questions that can be answered using the web-based system, the included functionalities, and the implemented visualization techniques

Question	Functionality	Description	Visualization
When did high intensity time for hotspots occurrence take place?	Search for high intensity times	The temporal distribution of hotspot events occurred during a day and a week	Line-charts for daily and weekly distributions
Where did hotspots occur during the high intensity times?	Detect locations of hotspots during a chosen time period	The spatial distribution of hotspot areas for a given time period	Heatmap describing origin and destination hotspots
Where did hotspot of venue type conference hall occur?	Detect locations of hotspots that occurred near a specific type of venue	The spatial distribution of hotspot events	Pie-charts symbol-map describing the locations, sizes and types of hotspots
In which districts did hotspots near the commercial areas occur?	Get a list of districts where hotspots near a chosen venue occurred	The activity districts will be highlighted if the chosen type of event did occur there	A List of district shown on a toolbar area
Near which types of venues do the hotspots take place?	Detect venue types and their total amounts for the hotspot areas and districts	All of the venue types that occurred in a certain district or area are highlighted	Pie-charts symbol-map describing the locations, size and types of hotspot, and a pie-chart in a statistics view showing the summarized statistics for the chosen district
During which time interval did most of the events take place near the conference hall?	Compare temporal patterns for a chosen hotspot	The temporal distribution of events occurred over a day divided by 3-hour intervals. Proportions of intervals can be compared	Pie-chart symbol-map describing the proportions of events for temporal intervals

events can describe well traveller's behaviour during a big event like the trade fair. It shows that in certain days the demand for taxi in this area increases a lot, while on other days it can be very low.

Table 2 shows a summary of different example questions which can be answered with the web application. We also list the functionality needed to answer those questions with their short descriptions as well as the implemented visualization techniques.

5 Conclusion and Outlook

In this work, we proposed an approach of building an interactive geo-visual web application based on current web technologies. The multidimensional data was presented by using a density map in combination with pie-chart symbol maps. Interactive tools support the user in finding the desired point in space and time for further investigations.

The application represents the results of the cluster analysis, which were conducted for determining taxi traveler's o/d hot spots of taxi trips in Shanghai. Furthermore, the FCD data was combined with OSM data of POIs in order to detect the semantics of popular places and possible activity areas. This tool enables gaining further insight with varying level of detail into the temporal pattern of hot spots and their semantics. The application could be used by decision makers as well as by non-expert users like tourists and citizens to gather information for their daily travel.

The application facilitates exploration of the data set by gaining insights in different aspects of multidimensional data sets. Data can be inspected in different level of detail by using contextual information as well as by studying the semantics and time. The choice of map views and filtering tools used for investigation depend on the study problem and question. The user interface has been built considering usability issues, such as the simple map colors and logical color schemes. The charts that show information and enable direct brushing and selection by clicking on the area of interest could be intuitively used. When filtering by a specific venue type, a quick search for the areas with possibly similar activity patterns can be done. But we are aware of a missing user survey, which are planned as the next step after including expert advice's and opinions.

During development and working with the visualization a few drawbacks come across which are namely the sparse POI data for the study area. Only a small number of POI's are mapped within OSM for the Shanghai area and unfortunately the distributions of the POI's are very inhomogeneous. Therefore, a more consistent semantic data layer is needed to get valid results that would describe the study area better. Nevertheless, it is possible to study the semantics for a few districts.

The semantic layer for the application currently offers simplified labelling of the hot-spots, since each point is labelled just with the nearest venue, while there might be more venues in walking distance from the taxi pick up or drop off location.

Because of this, the semantics of the hot-spots remains uncertain. It might be reasonable to apply further analysis for all of the venues in the buffer zone for selecting out the main venue.

The test data set involves one-week taxi FCD data. For gaining further insight and getting a better overview, a larger dataset with a long period of time should be used. If the interest is to distinguish between common and abnormal hot spots. In addition, with this application it is possible to visualize the findings of historical data. The combination of the historical data with real time information would make the analysis more valid and the visualization more suitable to the expert and non-expert users.

To run the visualization interactively in the web, scalability issues have to be taken into account. The filtering is run on the client which could become difficult with larger datasets. To overcome this issues more calculations should be done on the server side.

References

AGAFONKIN, V. 2014. *Leaflet.heat* [Online]. [Accessed 10.10 2015].
AHAS, R., AASA, A., ROOSE, A., MARK, Ü. & SILM, S. 2008. Evaluating passive mobile positioning data for tourism surveys: An Estonian case study. *Tourism Management,* 29, 469–486.
AHAS, R. & MARK, Ü. 2005. Location based services—new challenges for planning and public administration? *Futures,* 37, 547–561.
ANDRIENKO, G., ANDRIENKO, N., FUCHS, G., RAIMOND, A.-M. O., SYMANZIK, J. & ZIEMLICKI, C. Extracting Semantics of Individual Places from Movement Data by Analyzing Temporal Patterns of Visits. First ACM SIGSPATIAL International Workshop on Computational Models of Place, 2013 Orlando, Fl, USA.
ANDRIENKO, G., ANDRIENKO, N. & WROBEL, S. 2007. Visual analytics tools for analysis of movement data. *ACM SIGKDD Explorations Newsletter,* 9, 38–46.
ANDRIENKO, N. & ANDRIENKO, G. 2012. Visual analytics of movement: An overview of methods, tools and procedures. *Information Visualization,* 1473871612457601.
BURGHARDT, D. & WIRTH, K. 2011. Comparison of Evaluation Methods for Field-Based Usability Studies of Mobile Map Applications. *International Cartographic Conference 2011.* Paris.
CARTWRIGHT, W., CRAMPTON, J., GARTNER, G., MILLER, S., MITCHELL, K., SIEKIERSKA, E. & WOOD, J. 2001. Geospatial information visualization user interface issues. *Cartography and Geographic Information Science,* 28, 45–60.
CHANG, H.-W., TAI, Y.-C., CHEN, H.-W. & HSU, J. Y.-J. 2008. iTaxi: Context-Aware Taxi Demand Hotspots Prediction Using Ontology and Data Mining Approaches. *The 13th Conference on Artificila Intelligence and Applications.*
COOK, K. A. & THOMAS, J. J. 2005. Illuminating the path: The research and development agenda for visual analytics. Pacific Northwest National Laboratory (PNNL), Richland, WA (US).
DING, L., YANG, J. & MENG, L. Visual Analytics for Understanding Traffic Flows of Transportation Hub from Movement Data. International Cartographic Conference 2015, 2015 Rio de Janeiro, Brazil.
DOS SANTOS, S. & BRODLIE, K. 2004. Gaining understanding of multivariate and multidimensional data through visualization. *Computers & Graphics,* 28, 311–325.

ESTER, M., KRIEGEL, H.-P., SANDER, J. & XU, X. A density-based algorithm for discovering clusters in large spatial databases with noise. 2nd International Conference on Knowledge Discovery and Data Mining (KDD-96), 1996. 226–331.
FEKETE, J.-D., VAN WIJK, J. J., STASKO, J. T. & NORTH, C. 2008. The value of information visualization. *Information visualization.* Springer.
FRY, B. 2008. *Visualizing Data*, O'Reilly.
HALLAHAN, N. 2014. *q-cluster* [Online]. [Accessed 16.11 2015].
HEIDMANN, F., HERMANN, F. & PEISSNER, M. 2003. Interactive Maps on Mobile, Location-Based Systems: Design Solutions and Usability Testing. *International Cartographic Conference.* Durban, South Africa: ICA.
HU, J., CAO, W., LUO, J. & YU, X. Dynamic modeling of urban population travel behavior based on data fusion of mobile phone positioning data and FCD. Geoinformatics, 2009 17th International Conference on, 2009. IEEE, 1–5.
KEIM, D., ANDRIENKO, G., FEKETE, J.-D., GÖRG, C., KOHLHAMMER, J. & MELANÇON, G. 2008. *Visual analytics: Definition, process, and challenges*, Springer.
KRUEGER, R., THOM, D. & ERTL, T. Visual Analysis of Movement Behavior Using Web Data for Context Enrichment. 2014 IEEE Pacific Visualization Symposium, 2014 Yokohama, Japan. IEEE, 193–200.
LOBBEN, A. K., OLSON, J. M. & HUANG, J. 2005. Using fMRI in Cartographic Research. *Proceedings of the 22nd International Cartographic Conference.* A Coruna, Spain.
MACEACHREN, A. M. 1995. *How Maps Work: Representation, Visualization, and Design*, The Guilford Press.
MACEACHREN, A. M. & KRAAK, M.-J. 2001. Research challenges in geovisualization. *Cartography and Geographic Information Science*, 28, 3–12.
MAYHEW, D. J. 1999. *The Usability Engineering Lifecycle: a practitioner's handbook for user interface design,* San Francisco, USA, Morgan Kaufmann Publishers.
MAZIMPAKA, J. & TIMPF, S. 2015. Exploring the Potential of Combining Taxi GPS and Flickr Data for Discovering Functional Regions. *In:* BACAO, F., SANTOS, M. Y. & PAINHO, M. (eds.) *AGILE 2015.* Springer International Publishing.
MESSELODI, S., MODENA, C. M., ZANIN, M., DE NATALE, F. G., GRANELLI, F., BETTERLE, E. & GUARISE, A. 2009. Intelligent extended floating car data collection. *Expert systems with applications,* 36, 4213–4227.
MILLER, H. J. & HAN, J. 2009. *Geographic data mining and knowledge discovery*, CRC Press.
MOOSES, V. Geographical perspective in city sensing. Proceedings of the 2013 ACM conference on Pervasive and ubiquitous computing adjunct publication, 2013. ACM, 1351–1354.
NIELSEN, A. 2004. User-Centered 3D Geovisualisation. *Geoinformatics 2004: Proc. 12th Int. Conf. on Geoinformatics - Geospatial Information Research: Bridging the Pacific ans Atlantic.* Gävle, Sweden.
NIELSEN, J. 1993. *Usability Engineering*, Academic Press.
NÖLLENBURG, M. 2007. Geographic Visualization. *In:* KERREN, A., EBERT, A. & MEYER, J. (eds.) *Human-Centered Visualization Environments.* Springer.
ROTH, R. E. 2012. Cartographic Interaction Primitives: Framework and Synthesis. *The Cartographic Journal*, 49(4), 376–395.
SARODNICK, F. & BRAU, H. 2011. *Methoden der Usability Evaluation,* Bern, Verlag Hans Huber.
SUH, S. C. 2012. *Practical Applications of Data Mining,* Jones & Bartlett Learning.
SWIENTY, O., JAHNKE, M., KUMKE, H. & REPPERMUND, S. 2008. Effective Visual Scanning of Geographic Information. *In:* SEBILLO, M., VITIELLO, G. & SCHAEFER, G. (eds.) *Visual Information Systems. Web-Based Visual Information Search and Management.* Heidelberg, Berlin: Springer.
TOBLER, W. 1979. Pycnophylactic Interpolation for Geographical Regions. *Journal of the American Statistical Association,* 74, 519–530.
TOMINSKI, C. 2006. *Event based visualization for user centered visual analysis.* Dr.-Ing., Universität Rostock.

WOOD, J., DYKES, J. & SLINGSBY, A. 2010. Visualization of Origins, Destinations and Flows with OD Maps. *The Cartographic Journal,* 47, 117–129.

WOOD, J., SLINGSBY, A. & DYKES, J. 2011. Visualizing the dynamics of London's bicycle hire scheme. *Cartographica - The International Journal for Geographic Information and Geovisualization,* 46, 239–251.

YUAN, J., ZHENG, Y. & XIE, X. Discovering regions of different functions in a city using human mobility and POIs. Proceedings of the 18th ACM SIGKDD international conference on Knowledge discovery and data mining, 2012. ACM, 186–194.

ZHANG, W., LI, S. & PAN, G. Mining the semantics of origin-destination flows using taxi traces. 2012 ACM Conference on Ubiquitous Computing, 2012 Pittsburgh, USA.

Part IV
Innovative LBS Applications

Enhancing Location Recommendation Through Proximity Indicators, Areal Descriptors, and Similarity Clusters

Sebastian Meier

Abstract Location recommendation (LR) or rather location-based recommender systems (LBRS) are an integral part of modern location-based services (LBS). Most LR algorithms only focus on location-specific attributes when calculating recommendations, while completely ignoring the urban structure surrounding the locations. (In this paper we refer to a geographic coordinate (latitude and longitude) as *position*. *Locations* and *places* in contrast refer to physical entities e.g. a restaurant, a bus stop or a lake). This paper demonstrates how the urban structure can be modelled in LR calculations by using data from OpenStreetMap (OSM) and the location data itself. Based on these datasets, we present two approaches to extend the LR process by (1) including the urban structure in direct proximity of the location (Proximity Indicators and Areal Descriptors) and by (2) not only looking for individual locations but location clusters (Similarity Clusters). Thereby we acknowledge the complexity of a location, which can not be perceived as a detached entity. A location is part of a given urban structure and we need to include the parameters of this structure in our algorithms. A prototypical implementation compares locations from four major German cities: Berlin, Hamburg, Munich and Cologne and thereby highlights the applicability of the underlying data structures derived from OSM and the location data itself. We conclude by outlining the potential of the presented approaches in the context of LR as well as their relevancy for urban planning and neighboring disciplines.

Keywords Recommendation · Location-based services · Volunteer geographic information

S. Meier (✉)
Interaction Design Lab, University of Applied Sciences Potsdam, Potsdam, Germany
e-mail: meier@fh-potsdam.de

© Springer International Publishing AG 2017
G. Gartner and H. Huang (eds.), *Progress in Location-Based Services 2016*, Lecture Notes in Geoinformation and Cartography, DOI 10.1007/978-3-319-47289-8_14

1 Introduction

With the emergence and diffusion of smartphones, location-based services found their way into the consumer market. Thanks to powerful processing capabilities, mobile data connections, and embedded global positioning system technology (GPS), users are able to receive information tailored to their current geographic position. There is a variety of location-based services available today, ranging from being able to order food from restaurants nearby (Deliveroo 2016) to dating (Tinder 2016) and socializing (Happn 2016). The majority of apps include techniques for recommending or rather finding locations based on matching criteria. The research in the area of LR in LBS can be divided into four intersecting foci (Fig. 1): Location analysis (1), user analysis (2), the combination of user data and location data (3) and the user experience and interface design of LR systems (4). In this paper we present two approaches for enhancing existing LR systems at the intersection of field (1) and (3). In order to reference and contrast our approaches from existing work, we first elaborate on these four pillars of LR.

2 Related Works

In the early days of LBS, research primarily focused on how LBS work in general. This included the observation of human-computer interaction and computer-assisted interpersonal interaction in location-based social networks (LBSNs) or how volunteered geographic information (VGI) was generated within those networks (location data, ratings, etc.). As systems grew and global players like e.g. Foursquare (2016) or Yelp (2016) emerged, research set a stronger focus on data-centric and algorithmic questions.

Today, most research focuses on one of the four previously highlighted foci (Fig. 1), although they are strongly interconnected and not meant to be exclusive. In the area of **location data and location analysis**, researchers have tried to create ontologies, schemata or topic models to describe, categorize, and thereby organize locations (Chen et al. 2004; Liu et al. 2014; Ye e al. 2011; Yu et al. 2009). This part

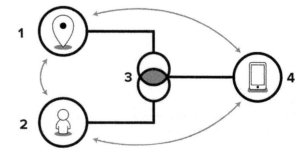

Fig. 1 Location recommendation research *1* location analysis, *2* user analysis, *3* combining user and location data and *4* user experience and interface design

is fundamental to LR and will also be the focus of this publication. A stronger focus, especially in recent work, lies on the **user data**. Early work looked into computing user preferences, which could then be used to find nearby locations that match the preferences (Cheverst et al. 2000; Chincholle et al. 2002; Rao and Minakakis 2003). More recently, user models have become more complex. Methods for analyzing the users' movement (trajectories) and thereby discovering spatial and behavioral patterns are now of growing interest (Pelekis et al. 2008; Zheng 2011; Cao et al. 2015; Chen et al. 2013; Hu et al. 2013). This endeavor is backed by recent developments in machine learning and deep learning methods. At the **intersection of user and location data** the models are being correlated to find locations of interest, organize whole trips or making predictions upon future user behaviors. This also includes so-called collaborative filtering, which combines the data from multiple users within one's network. The first three areas are very technology-centric, but an important part of every LR system is its interface. Without being able to formulate or generate an information request and the ability to display the results in a meaningful way, any LR system is simply an algorithm, but not a service. In regards to LR, the **user experience and interface design** community has focused their attention mostly on usability issues and user studies or applications and use cases for aforementioned algorithms, as well as novel UI principles (Cheverst et al. 2000; Crease and Reichenbacher 2011).

While the primary focus of traditional LR systems is to recommend locations, similar methods have over the last few years also been used to generate recommendations for non-spatial entities, e.g. books, movies, or music (Levandoski et al. 2012). Those approaches are not only looking specifically at individual locations or rather positions but also at areal patterns, e.g. what type of movies do people enjoy who live in Minnesota (Levandoski et al. 2012). What is striking about these hybrid LR systems that also take into account non-spatial data is that they are analyzing locations and their metadata, in order to analyze larger areas and thereby try to describe those areas. When comparing those approaches with traditional LR, we notice that LR algorithms treat locations like entities which are almost detached from their urban environment. The coordinate that determines the position of the location in the real world space is translated into a position in a data space. The data space incorporates locations and user trajectories, while at the same time ignoring the urban surroundings of the locations. Thereby, possible implications derived from the urban structures, which they are embedded in, are lost. This poses the question of **how to reintegrate location data into the urban fabric and account for its structural influence.**

3 Research Question

Through the disentanglement of locations from the surrounding urban structure, locations in LR systems show a lack of descriptive information. Most LR systems include features that help categorize a location e.g. *Food > Restaurant >*

Dining > Italian restaurant, and properties that further describe the location e.g. opening times, accessibility features, menus, ratings, and in rare cases even offer information on the character of the physical space e.g. outdoor/indoor or building type. If we look to architecture and urban planning, it is a prevailing notion that a location is influenced by more than its own properties. Christopher Alexander describes this at many times in his 1977 book "A Pattern language" (Alexander et al. 1977), e.g. in his chapter on the *mosaic of subcultures* (p. 42ff). The environment, the urban fabric, in which a location is embedded, plays a crucial role in shaping the flair or rather atmosphere of a location, e.g. by the amount of nature it is surrounded by Sullivan et al. (2004). One might think of restaurants in a picturesque part of town or a restaurant on the countryside. Following this line of thought, the land use, building types, natural features, the mix of locations, etc. of the surroundings affect the inherent locations. Especially the mix of locations, a feature requiring only a location dataset itself, is one feature that we will further explore. While sometimes a user is looking for a bar in a nightlife area (area with bars, clubs and similar locations), contrasts and outliers might be of interest to another person, like a luxury restaurant in an industrial area. When people go shopping, they might prefer areas where there is more than one shop. Maybe some enjoy areas that show a mix of shops and restaurants or bars. In this case, the individual shop is of increasing interest for the visitor when it is surrounded by the preferred mix of locations. In this context, this paper sets out to suggest (1) a way to **quantify parts of the urban fabric** and (2) to **embed those elements of the urban fabric in LR algorithms** and thereby improve the LR process. The methods as described below are to be integrated within LR research area 1 and 3. The methods are not supposed to be used exclusively, but in combination with existing methods that focus on location specific criteria. Thereby, our methods have the potential to be integrated into multidimensional LR systems as new additional dimensions. In order to help improve recommendations, they can be used as supplementary filters and new search opportunities or as raw input for machine learning algorithms, to help improve recommendations. The development of the approaches is not only technology driven, but has a strong focus on enriching user preferences. Therefore, being able to create **new customized queries** could be an important enhancement, as we will highlight in the implementation section.

Before we are able to embed the urban fabric in LR algorithms, we need to quantify it and find a representation that we can assess and correlate with existing location datasets. We believe that the lack of research and implementations on the urban fabric as described above in part derives from the lack of cross-city or even cross-country data sources. Therefore, the following chapter focuses on assessing open data repositories in order to find sources that allow for a comprehensive implementation of such a quantification process.

4 Quantifying the Urban Fabric

To the authors' knowledge, the urban fabric has not been defined in its full complexity, yet, and due to its intricacy we will not attempt to do so in this section either. Nonetheless, a few characteristics can be derived from research in areas like architecture, sociology or human geography. The dimensions and interrelations forming the urban fabric are highly complex. They range from physical features as broached in the previous section (e.g. built environment or natural features) to social features (e.g. population density, age groups, etc.). While it is possible to theoretically extend the range of parameters almost indefinitely, we decided to focus on data sources that exist today and that would allow us to implement the methods for LR across city borders or even across national borders.

We surveyed open data portals in Germany, the United Kingdom, the United States, the Netherlands and Switzerland[1] as well as global data portals like e.g. FAOSTAT, World Bank and the CIA factbook[2] in order to assess if it is possible to extract georeferenced socio-demographic data, infrastructural data or information on physical features. The major problem we identified is that *international* data providers provide data on a per *country* basis. In the countries listed above, the data provided by the *national* institutions is in most cases only available on a *city*-wide level, data with higher detail (at least *city districts*) is usually only provided by data providers in each *city*. In other words, *the higher the granularity, the more distributed the data is across data providers* (Fig. 2).

As a consequence of these federal data systems the structure and granularity of the data across providing institutions varies strongly. Even within one country (Germany), the data between cities varies.[3] While it would be possible to implement our approach on a city-wide level, the inconsistency will create a limitation when expanding beyond the city level. Especially when we take commercial applications into account that go beyond city borders and even national borders and provide services to people all over the globe, the availability of sufficient and reliable data sources is crucial. To increase the practicability and reproducibility of our

---

[1]Since our goal was not to create a world wide open data survey, but instead learn more about compatibility of various data sources and their internal structures, the authors selected a subset of countries. The selection is influenced by the authors' knowledge of the respective language used by the data provider and thus the ability to process the data. We are aware that the resulting perspective has a Western European bias.

[2]List of International data portals: faostat3.fao.org, cia.gov/library/publications/the-world-factbook, data.worldbank.org. European Data Portals: data.europa.eu/euodp/en/data, europeandataportal.eu. National data portals: data.gov, data.overheid.nl, opendata.swiss, govdata.de, data.gov.uk. German city data portals: daten.berlin.de, opengov-muenchen.de, offenedaten-koeln.de, transparenz.hamburg.de/open-data.

[3]As governmental institutions are moving towards opening their data, we now not only need to advocate for accessible data, but also for consistent exchange formats and data structures. As we will indicate at the end of this paper, being able to create comparisons across city borders holds great potentials for urban planning. But in order to embrace such approaches we need to establish national data formatting guidelines, so that datasets can be combined and compared.

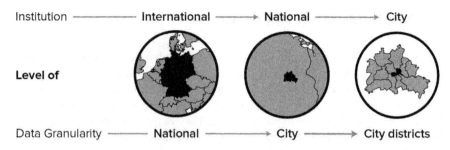

Fig. 2 The higher the granularity of data, the higher the distribution of data

approaches, we went beyond governmental data suppliers and identified OpenStreetMap (OSM) data as a suitable and open data source. The platform, which collects VGI and provides users with it, has established a quite consistent data basis for most of the biggest western European and North American cities. Even beyond that, the network e.g. in Russia, India or China is continuously expanding (Neis 2016).

While OSM does not provide insights into socio-demographic data, it still allows us to collect data[4] on land utilization, road infrastructures, public transportation, parks or natural features like rivers, lakes, forests, and even, if available, building height. For the prototypical implementation we selected a subset of the following elements to enrich location metadata:

- **Nature**: (1) parks and forests as well as (2) rivers and lakes, which are areas within a city that are open to the public and that represent a characteristic quality of a neighborhood (Hammer et al. 1974; Sullivan et al. 2004).
- **Road Networks** are important for two reasons: on the one hand, they represent a way of accessing a location. Thus, roads with a higher priority like e.g. a highway result in better accessibility. On the other hand, bigger roads also represent a negative impact on quality of life, since bigger roads also induce more air and noise pollution. We use the full range of roads and classify them by size into 5 groups, from residential roads to highways.
- **Public Transportation** also relates to the aspect of accessibility of a location. Similar to the road network, we classify the public transportation network into four categories: bus, tram + underground, light rail (S-Bahn), and trains, all taken from the OSM data source. While analyzing the data quality of OSM we observed that underground and tram are sometimes used interchangeable and thus we did not distinguish between these two modes of transport.
- **Land use** is showing the usage of a certain area (park, residential, commercial, industrial ect.), most often with a very low granularity. While the accuracy of land use extraction from OSM varies in level of detail and accuracy, existing

---

[4]Not only the LR algorithms, but also the raw data or rather the code required to aggregate the data from e.g. OSM is available in the GITHUB repository mentioned at the end of the paper.

work by e.g. (Jokar Arsanjani et al. 2013) or (Estima and Painho 2013) showed promising results.

- **Building height**, especially in combination with land use, could proof a viable indicator for population or business density. While subtle differences are hard to calculate, bigger changes between 2-level suburban areas and city center high-rises can be identified from OSM data. While the aforementioned attributes have a good coverage for most western European and North American cities, building height or rather building data is still in the makings, in many countries and cities.

In addition to this external data that we collect for enriching location metadata, we also use the locations themselves to analyze the mix of locations within a certain neighborhood. We therefore include data from the LBSN Foursquare, limited to the German cities Berlin, Hamburg, Cologne, and Munich. Foursquare holds data on more than 223,000 locations in the aforementioned cities. The extracted data contains the geographic position for each location, as well as their categories, allowing us to analyze the location mix in a designated area.

Now, being able to quantify certain aspects of the urban fabric based on OSM and the extracted location categories, we will use this quantitative, spatial data in the following sections in order to enrich a LBSN location dataset, in this case also the previously described Foursquare dataset, and illustrate how this approach can be applied in a novel LR concept.

5 Proximity Indicators and Areal Descriptors

In order to enhance the Foursquare location data with the previously described OSM data, we develop two metrics to connect the OSM dataset and the location dataset. The differentiation between the two approaches is deduced from the data's spatial structure and their descriptive power over a location (see Fig. 3). On the one hand we investigate the distance to and thereby accessibility of entities (e.g. a bus stop) from a certain location (Proximity Indicators). On the other hand, we are interested in the polygonal composition (e.g. water, forests, parks, built environment) of a location's surrounding area (Areal Descriptors).

5.1 Proximity Indicators

The first metric, the Proximity Indicators, calculate the distance to the closest entity of a specific group. The proximity thereby describes access to this entity. For demonstrational purposes we chose the previously described subset: water, parks and forests, public transport locations and roads. The latter two were, as described above, subdivided into several classes. As these entities represent physical

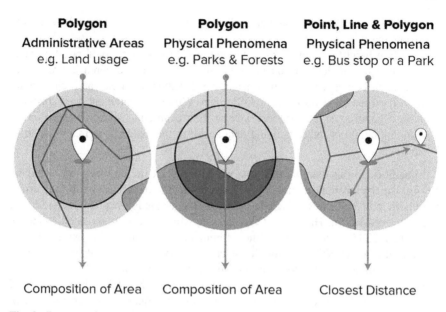

Fig. 3 Structure of the external data derived from OSM

phenomena which are unlikely to change frequently, the distance calculations can be stored with each location's metadata, allowing for an efficient inclusion in future queries. Upon those cached distances, we can dynamically query subsets of locations that match certain characteristics, e.g. locations that are within close reach of a park or a forest.

As we want to allow users to find locations that have desired entities in their direct vicinity, those entities become irrelevant at a certain distance threshold. This threshold of course depends on the user's mode of transportation: a lake in 500-m distance might be relevant when walking, but a lake in 5 km distance might only be of interest when travelling by bike or car. As a preliminary implementation we used a circular distance instead of an actual node-based distance definition (transport network) (Guo and Bhat 2007). It should be noted at this point that by changing the mode of transportation the radius changes. As we are mainly interested in the direct *neighborhood* around a location, we believe walking is the most suitable mode to circumscribe the maximum distance that is still perceived as "close" or "within reach". Walking will therefore be used as the default mode of transport for the remainder of this paper. Literature doesn't provide us with an overarching definition for "walking distances" that would indicate how far people of certain groups are willing to walk and still perceive it as nearby. The only literature close to this question are studies on the access to public transportation (Agrawal and Schlossberg 2008; Burke and Brown 2007; Guerra et al. 2012). These studies proceed on a similar objective, since they are trying to determine the ideal maximum distance between a place of residence and the next public transport stop. The distances mentioned in these papers vary between half a mile and a quarter mile. We chose

Berlin

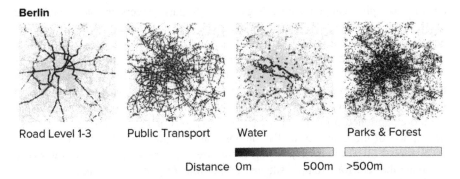

Fig. 4 About 93000 locations (Berlin) from the Foursquare LBSN, and the distance of each location to the next road (level 1–3), public transport stop, stretch of water, or park and forest, illustrated by darkness (from *black* = 0 m, to *grey* = more than 500 ms away)

500 m as a good average, which compares to roughly 6 min walking time in "preferred walking speed" (Browning et al. 2006; Levine and Norenzayan 1999; Mohler et al. 2007).

Nonetheless, this suggested range of 500 ms as "within reach of walking" is only a default value used in the prototypical implementation. Our algorithm allows for this value to be customized on a per-user basis, taking into account e.g. individual abilities, health-related restraints, security precautions like avoiding walks alone at night, or environmental factors like rain or icy conditions that will ultimately have substantial impact on the perception of "closeness". Thus, the walking distance can easily be modified and does not influence the fundamental logic of the Proximity Indicators or Areal Descriptor itself. Beyond providing users with more meaningful results, the distance threshold also allows us to speed up queries by ignoring locations with distances bigger than the threshold.

To illustrate the resulting data based on the 500-m threshold, Fig. 4 maps the distance from each Foursquare location in Berlin to the nearest highway (roads level 1–3), public transport stop, stretch of water, and park or forest. Darker areas show nearby access, while lighter areas represent areas with no close access to these features (further away than the 500-m threshold).

5.2 Areal Descriptors

While the Proximity Indicators describe the distance to a certain entity, Areal Descriptors describe the composition of the "complete" neighborhood, using again the previously defined 500 m threshold. For the calculation of the Areal Descriptors we use polygonal data acquired from the OSM data source. From a semantic and a data granularity point of view, we need to distinguish between two types of Areal Descriptors or rather data sources. On the one hand we have polygons that are derived from physical phenomena (e.g. buildings, parks, rivers or roads) and on the other hand "artificially" generated polygons, e.g. land usage areas created by governmental institutions for statistical reasons, the latter usually have a lower granularity than the physical phenomena (Fig. 3) and thereby introduce higher levels of uncertainty to the interpretation. In the following, these two types are used to calculate the composition of the neighboring area (Areal Descriptors). In order to store the calculated values with our location data, we produce an intersection using the aforementioned 500 m radius with the polygons and create a percentage-wise description of the neighborhood's composition (e.g. 50 % residential, 10 % water, 20 % forest, 20 % industrial). Using these Areal Descriptors, we can now query locations that are for example situated in a dominantly residential area.

The two methods described in the previous section are, from a GIS point of view, not particular novel. The novelty of this section is to demonstrate the potential of OSM data and how to use it to enrich a location dataset. This helps us to create richer query options in LBS for novel location search opportunities, which will be outlined in the implementation section of this article.

6 Similarity Clusters

The concept used to investigate the implications of an area's location mix was inspired by the user experience section carried out in the framework of the HeatTile method by Meier et al. which showed that certain parts of a city carry a strong signature from the embedded locations (Meier et al. 2014). This in turn results in the characterization of certain areas as e.g. a museum district or a nightlife district. The HeatTile method only visualized density maps, it did not identify precise clusters or multi-criteria-clusters, nor did it integrate those into novel interface and search concepts. In the following section we extend the HeatTile approach and show how to easily and efficiently find clusters of similar locations. We use only the location metadata itself to create these Similarity Clusters, whereas the Proximity Indicators and Areal Descriptors draw on external data (OSM) to enrich the location's metadata. Other research by e.g. Liu et al. (2014), Noulas et al. (2011), and Rösler and Liebig (2013) have worked on clusters of LBSN and locations before, but their primary purpose was to determine the characteristic or dominant function of areas within a city (e.g. business district, nightlife district) by looking at locations and their temporal usage patterns. But none of the existing approaches uses the resulting multidimensional cluster data to create additional metadata features for the locations *within* those clusters. In order to extend the existing location metadata of each location we first calculate the number of locations of a certain category within the aforementioned 500-m radius of each location. The distance between locations is accounted as a weight, using a linear scale from 0 m = 1 to 500 m = 0, thereby giving more importance to locations that are nearby (see Fig. 5). The weighted distances are combined in an index, which is stored in the location's metadata. With those pre-calculations in place we can favor locations with similar locations nearby

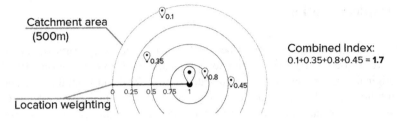

Fig. 5 Weighting of locations of a specific category in the catchment of a location

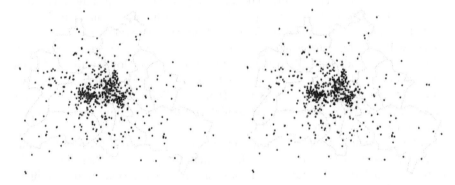

Fig. 6 Both maps show all locations of the category "restaurant" in the city of Berlin. The map on the *left* indicates the density (*red* for high and *black* for low density) of other restaurants in the 500-ms radius, the map on the *right* shows the density of museums in direct vicinity

or even locations that have locations of certain other types nearby (see Fig. 6). By pre-calculating the indices for each category, we create a multidimensional parameter space, which has one index per category per location. As a result, every location has information on how many other locations of a certain category are in the direct neighborhood (500 m).

In order to now calculate the Similarity Clusters, we combine the cached category indices with a clustering algorithm, in this case we used DBSCAN. This allows us to find areal clusters of locations and thereby for example allow users to search for areas of interest rather than exact location matches (see Fig. 6). We can even use multiple indices to identify areas that hold locations of multiple categories, as we will further explore in the next section of the paper.

7 Implementation

The previous two sections described methods for analyzing parts of the urban structure surrounding a location and how to quantify and store this information with the original location metadata. As illustrated above, the result from this operation is a multidimensional dataset that describes various structural features of the urban environment for each location. These additional dimensions can now be used to build queries in order to find locations of interest. To put it into perspective, traditional LBS already allows us to search for a certain category of locations, e.g. "restaurant". We can also include conditions regarding e.g. "opening times" and only look for restaurants that are open right now. With the Proximity Indicators, Areal Descriptors, and Similarity Clusters all stored in individual indices that are added to the location metadata, it is now equally easy to integrate those new dimensions in traditional LBS location search queries. This allows us to e.g. "find a restaurant in a part of town that has a high density of bars". To demonstrate this, we will use the Foursquare dataset (Berlin, Hamburg, Cologne and Munich) to test various querying approaches in order to highlight the opportunity space of implementing these new feature sets.

In the following example we will only focus on the Similarity Clusters as our search parameter. The same process that is illustrated below applies when Areal Descriptors or Proximity Indicators are used as search parameters. In this example we use only three categories: restaurants, bars and museums. From a user perspective the described query approaches should result in a list of locations or rather areas that contain those locations. As in most cases of such search queries we need to define a minimum and maximum number of locations that we want the query to return. We defined that number, based on the Foursquare value, to be roughly 50 locations (± 10 %). To make the queries more realistic, we added a bounding box to the queries, which is in the case of the examples: Berlin. As we are only interested in the best results (in regards to Similarity Clusters: highest density of locations of category X), the results for Berlin are always located in central Berlin. Thus, in order to be able to better visualize the results, the maps in Figs. 8, 9 and 10 only show this central area of Berlin (see Fig. 7, right). The more complex the examples will get (e.g. combining multiple categories), the lower the individual densities will be, since we need to find the highest common denominator. To visualize this phenomenon, we have created a scale, showing the range of densities available for the city of Berlin (see Fig. 7, left).

7.1 Individual Cluster

The least complex example implementation is allowing users to simply define a bounding box, in our example Berlin, and then choose a location category. By doing so, the area with the highest density of locations of the requested category is

Enhancing Location Recommendation Through Proximity ...

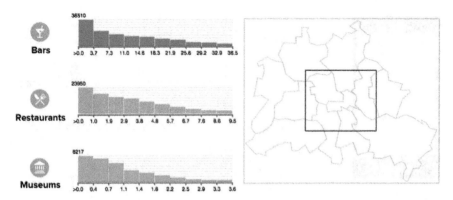

Fig. 7 In Figs. 7–10 *blue* always stands for location category bars, *yellow* for restaurants and *green* for museums. *Left* The *bar charts* show the distribution of locations that have a number of locations of category bar, restaurant, and museum in their neighborhood (500 m). X-axis: density index, y-axis: number of locations that feature a density for this location category (logarithmic scale). The *numbers* show that the highest density of locations can be found among bars and also that bars are the most common thing to find in the neighborhood of any location. *Right* In the following example queries we will try to find the highest densities, those can be found in central Berlin, therefore the maps in Figs. 8–10 only show the central area of Berlin as highlighted in this overview map of Berlin

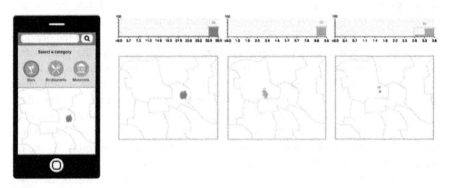

Fig. 8 *Left* An exemplary interface for choosing a location category and the resulting cluster below. *Right* The bar charts visualize the results as described in Fig. 7. The map shows the resulting locations of category bar (*blue*), restaurant (*yellow*) and museum (*green*)

highlighted on the map and the individual locations within this area are displayed. The search algorithm uses a binary search approach on the continuous index value for the requested category. When only searching for one category we only receive locations within very dense clusters (see Fig. 8). Since there are not as many museums as bars in Berlin, the result for museums also includes lower density areas.

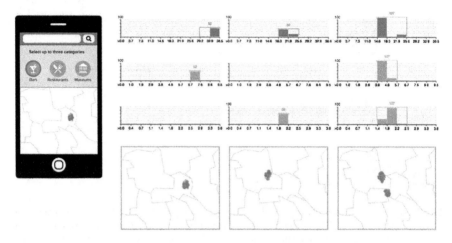

Fig. 9 *Left* An exemplary interface for choosing multiple location categories and the resulting cluster below. *Right* see Fig. 8

7.2 Combined Clusters

Even though existing LBS applications don't offer a density-based search and instead focus on individual locations, the search feature in the previous section is similar to the services that current LBS applications provide. To extend the previous section we are now combining multiple categories. This provides the users with a richer search interface and allows them to find areas that feature multiple categories. This feature could be of particular interest for example for tourist trip-planning applications, a common example in LBS applications. As previously described, when combining multiple categories, we need to find the highest common denominator. Therefore, we increase or decrease the density bounding box across indices using the binary search approach until we find a sufficient result. As the bar charts in Fig. 9 show, the resulting locations are not only in the spectrum of high density, as in the previous example, but also in the spectrum of lower density. We can now also see how the geographic distribution changes (maps in Fig. 9).

When combining three categories, the number of resulting locations is higher than our target 50 (±10 %) results. The reason for the bigger number of results is that we also implemented a maximum for the number of iterations the binary search algorithm is allowed to run, to keep the user's waiting time at a minimum.

7.3 Combined and Favored Cluster

The search methods in the previous section tried to find the highest common denominator, giving the same priority to all categories in question. In this last

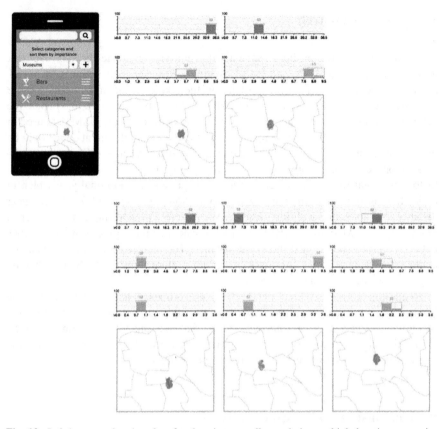

Fig. 10 *Left* An exemplary interface for choosing as well as ordering multiple location categories and the resulting cluster below. *Right* see Fig. 8. The importance ordering of the individual requests is quite visible, as the category with the highest importance also has the locations in areas of highest density

example we will introduce an interface which will allow the user to select multiple categories and order them by importance. The algorithm behind this search approach starts with the category that has the highest importance assigned by the user and starts from the highest density areas and decreases the density until the number of locations in the second most important category reaches a certain threshold. This process is successively applied to all categories included in the search request. As a result, with categories like museums included in the query, that are very sparse across the city, the resulting density in subordinate categories will be less optimal as well (see Fig. 10).

8 Further Research

We plan to further refine the presented work and also evaluate the improvement through our methods when applied to existing LR systems. In regards to refinements, we would first of all like to take on the challenge of implementing more socio-demographic features in defiance of the obstacles mentioned in this paper. We believe socio-demographic data poses a promising addition to the OSM data used in this paper. Especially the age group data, income, and unemployment data would be of interest for us, as they are often referenced by urban planners. Regarding technical improvements, we are working on improving the speed of our algorithm, a factor important to achieve a satisfying user experience. Additionally, we plan to revise the distance measurements, which, right now, only account for direct linear distance, but not actual walking distance. We therefore plan to implement the Open Source Routing Machine (OSRM) in order to calculate more precise walking times between entities and not just the linear distance. The indices described in this paper could provide new input for machine learning algorithms in LRS and thereby improve the recommendation process.

Regarding a user-centered perspective, the proposed approaches need to be implemented in a real-world use case. This will allow us to test traditional location recommendation against an improved model that uses the aforementioned indices. With this test we intend to collect empiric evidence on the performance of these features in regards to more accurate recommendations.

9 Conclusion and Implications

When traditional LR systems query location datasets, the focus lies upon the locations' categories, attributes, and their geographic locations, while ignoring the urban fabric of their direct neighborhood. In this paper we presented an approach for quantifying certain aspects of the urban fabric through OSM data and the location data itself. Furthermore, we described two approaches that use this quantified OSM data to enrich existing location datasets: (1) Areal Descriptors present the composition of the location's neighborhood, e.g. percentage of area covered by residential land use, while Proximity Indicators show the distance to an entity, e.g. the next bus stop. (2) Similarity clusters scan the neighborhood for locations that have been assigned a certain category and thereby identify clusters. The new approaches only use open data sources or data derived from the locations themselves, thereby creating an easy to implement improvement for existing LR systems. The new approaches extend LR and take the urban fabric into account, allowing us to create more comprehensive query possibilities.

A growing amount of data that describes the urban fabric is being published in governmental, corporate, and open data repositories. This trend is fueled by e.g. user-generated data, smaller and cheaper sensors and other technological

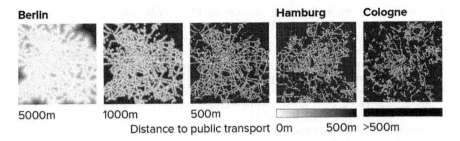

Fig. 11 Access to public transportation (bus, tram, subway, light rail), within a 5000-, 1000- and 500-m radius, in Berlin, Hamburg and Cologne

developments. Urban planners as well as citizens need to develop new tools and approaches in order to be able to profit from the increase of available datasets. Being able to create cross-city comparisons of urban structures could be a valuable opportunity to understand cities and thereby help them learn from each other. Our research could therefore not only help improve LR systems, but also help urban planners in building new tools and processes. To illustrate this with a simple example: we added a gridded set of data points on top of the Berlin, Hamburg and Cologne city model. We then calculated the distance to the nearest public transport access (bus, tram, subway or light rail). In Fig. 11 we visualized undersupplied areas in black, illustrating the cities' public transportation infrastructures. In addition to the 500-m radius, we used 1000 and 5000 m radi for the city of Berlin, in order to highlight how the desired walking distance to the next access point changes the visual analysis outcome. This allows urban planners to compare their public transport system to other systems and thereby generate new insights.

10 Attribution

All the code used to aggregate the date, calculate the indices describe in the paper are published under GNU General Public License V3 on GITHUB: https://github.com/sebastian-meier/Location-Recommendation-OSM.

Most of the datasets are available in the repository as well. Due to copyright issues we can not share the Foursquare data and Munich's FNP data.

References

Agrawal, A. W., & Schlossberg, M. (2008). How far, by which route and why? A spatial analysis of pedestrian preference. *Journal of Urban Design*, *13*(1), 81–98. http://doi.org/10.1080/13574800701804074.

Alexander, C., Ishikawa, S., Silverstein, M., Jacobsen, M., Fiksdahl-King, I., & Angel, S. (1977). A Pattern Language: Towns, Buildings, Construction (1386544400 ed.). New York: Oxford University Press.

Browning, R. C., Baker, E. A., Herron, J. A., & Kram, R. (2006). Effects of obesity and sex on the energetic cost and preferred speed of walking. *Journal of Applied Physiology*, *100*(2), 390–398. http://doi.org/10.1152/japplphysiol.00767.2005.

Burke, M., & Brown, A. L. (2007). Distances People Walk for Transport. *Road & Transport Research: a Journal of Australian and New Zealand Research and Practice*, *16*(3), 16.

Cao, G., Wang, S., Hwang, M., Padmanabhan, A., Zhang, Z., & Soltani, K. (2015). A Scalable Framework for Spatiotemporal Analysis of Location-based Social Media Data. *Computers, Environment and Urban Systems*, *51*, 70–82.

Chen, H., Perich, F., Finin, T., & Joshi, A. (2004). SOUPA: standard ontology for ubiquitous and pervasive applications (pp. 258–267). Presented at the The First Annual International Conference on Mobile and Ubiquitous Systems: Networking and Services, 2004. MOBIQUITOUS 2004, IEEE. http://doi.org/10.1109/MOBIQ.2004.1331732.

Chen, P., Gu, J., Zhu, D., & Shao, F. (2013). A dynamic time warping based algorithm for trajectory matching in LBS. *International Journal of Database Theory and Application*, *6*(3), 39–48.

Cheverst, K., Davies, N., Mitchell, K., Friday, A., & Efstratiou, C. (2000). Developing a context-aware electronic tourist guide: some issues and experiences. *Chi*, 17–24. http://doi.org/10.1145/332040.332047.

Chincholle, D., Goldstein, M., Nyberg, M., & Eriksson, M. (2002). Lost or Found? A Usability Evaluation of a Mobile Navigation and Location-Based Service. Presented at the Mobile HCI'02: Proceedings of the 4th International Symposium on Mobile Human-Computer Interaction, Springer-Verlag.

Crease, P., & Reichenbacher, T. (2011). Adapting Cartographic Representations to Improve the Information Seeking of LBS Users. *Proceedings of The International Cartographic Conference*.

Deliveroo. (2016). Deliveroo. Retrieved May 24, 2016, from https://deliveroo.co.uk/.

Estima, J., & Painho, M. (2013). Exploratory analysis of OpenStreetMap for land use classification (pp. 39–46). Presented at the Second ACM SIGSPATIAL International Workshop, New York, New York, USA: ACM Press. http://doi.org/10.1145/2534732.2534734.

Foursquare. (2016). Berlin | Food, Nightlife, Entertainment. Retrieved May 25, 2016, from https://foursquare.com/.

Guerra, E., Cervero, R., & Tischler, D. (2012). Half-Mile Circle: Does It Best Represent Transit Station Catchments? *Transportation Research Record: Journal of the Transportation Research Board*, *2276*, 101–109. http://doi.org/10.3141/2276-12.

Guo, J. Y., & Bhat, C. R. (2007). Operationalizing the concept of neighborhood: Application to residential location choice analysis. *Journal of Transport Geography*, *15*(1), 31–45. http://doi.org/10.1016/j.jtrangeo.2005.11.001.

Hammer, T. R., Coughlin, R. E., & Horn, E. T., IV. (1974). The Effect of a Large Urban Park on Real Estate Value. *Journal of the American Institute of Planners*, *40*(4), 274–277. http://doi.org/10.1080/01944367408977479.

Happn. (2016). happn. Retrieved May 24, 2016, from https://www.happn.com/en/.

Hu, Y., Janowicz, K., Carral, D., Scheider, S., Kuhn, W., Berg-Cross, G., et al. (2013). A Geo-ontology Design Pattern for Semantic Trajectories (Vol. 8116, pp. 438–456). Presented at the COSIT 2013: Proceedings of the 11th International Conference on Spatial Information Theory - Volume 8116, Cham: Springer-Verlag New York, Inc. http://doi.org/10.1007/978-3-319-01790-7_24.

Jokar Arsanjani, J., Helbich, M., Bakillah, M., Hagenauer, J., & Zipf, A. (2013). Toward mapping land-use patterns from volunteered geographic information. *International Journal of Geographical Information Science*, *27*(12), 2264–2278. http://doi.org/10.1080/13658816.2013.800871.

Levandoski, J. J., Sarwat, M., Eldawy, A., & Mokbel, M. F. (2012). LARS: A Location-Aware Recommender System (pp. 450–461). Presented at the ICDE '12: Proceedings of the 2012 IEEE 28th International Conference on Data Engineering, IEEE Computer Society. http://doi.org/10.1109/ICDE.2012.54.

Levine, R. V., & Norenzayan, A. (1999). The Pace of Life in 31 Countries. *Journal of Cross-Cultural Psychology*, *30*(2), 178–205. http://doi.org/10.1177/0022022199030002003.

Liu, Y., Wei, W., Sun, A., & Miao, C. (2014). Exploiting Geographical Neighborhood Characteristics for Location Recommendation (pp. 739–748). Presented at the 23rd ACM International Conference, New York, New York, USA: ACM Press. http://doi.org/10.1145/2661829.2662002.

Meier, S., Heidmann, F., & Thom, A. (2014). Heattile, a New Method for Heatmap Implementations for Mobile Web-Based Cartographic Applications. In *Thematic Cartography for the Society* (pp. 33–44). Cham: Springer International Publishing. http://doi.org/10.1007/978-3-319-08180-9_4.

Mohler, B. J., Thompson, W. B., Creem-Regehr, S. H., Pick, H. L., Jr, & Warren, W. H., Jr. (2007). Visual flow influences gait transition speed and preferred walking speed. *Experimental Brain Research*, *181*(2), 221–228. http://doi.org/10.1007/s00221-007-0917-0.

Neis, P. (2016). OSMstats - Statistics of the free wiki world map. Retrieved May 1, 2016, from http://osmstats.neis-one.org/.

Noulas, A., Scellato, S., & Mascolo, C. (2011). Exploiting Semantic Annotations for Clustering Geographic Areas and Users in Location-based Social Networks. *The Social Mobile Web*, *11*(2).

Pelekis, N., Frentzos, E., Giatrakos, N., & Theodoridis, Y. (2008). HERMES: aggregative LBS via a trajectory DB engine. *the 2008 ACM SIGMOD international conference* (pp. 1255–1258). New York, New York, USA: ACM. http://doi.org/10.1145/1376616.1376748.

Rao, B., & Minakakis, L. (2003). Evolution of mobile location-based services. *Communications of the ACM*, *46*(12), 61–65. http://doi.org/10.1145/953460.953490.

Rösler, R., & Liebig, T. (2013). Using Data from Location Based Social Networks for Urban Activity Clustering. In *Geographic Information Science at the Heart of Europe* (pp. 55–72). Cham: Springer International Publishing. http://doi.org/10.1007/978-3-319-00615-4_4.

Sullivan, W. C., Kuo, F. E., & DePooter, S. F. (2004). The Fruit of Urban Nature: Vital Neighborhood Spaces. *Environment and Behavior*, *36*(5), 678–700. http://doi.org/10.1177/0193841X04264945.

Tinder. (2016). Tinder - meet interesting people nearby. Retrieved May 24, 2016, from http://tinder.com.

Ye, M., Shou, D., Lee, W.-C., Yin, P., & Janowicz, K. (2011). On the semantic annotation of places in location-based social networks. Presented at the KDD'11: Proceedings of the 17th ACM SIGKDD international conference on Knowledge discovery and data mining, ACM.

Yelp. (2016). Yelp.

Yu, Y., Kim, J., Shin, K., & Jo, G. S. (2009). Recommendation system using location-based ontology on wireless internet: An example of collective intelligence by using "mashup" applications. *Expert Systems with Applications*, *36*(9), 11675–11681. http://doi.org/10.1016/j.eswa.2009.03.017.

Zheng, Y. (2011). Location-Based Social Networks: Users. In *Computing with Spatial Trajectories* (pp. 243–276). New York, NY: Springer New York. http://doi.org/10.1007/978-1-4614-1629-6_8.

Connecting the Dots: Informing Location-Based Services of Space Usage Rules

Pavel Andreevich Samsonov

Abstract Today, a large set of location-based services exists and millions of people use these applications on a daily base. These services are capable of fulfilling and assisting with a variety of tasks, such as fetching points of interest or indicating friends nearby, performing web search requests, which respect the context of the user, or navigating to a destination point. Even though these are very helpful tools, most LBS perform simplistic 'point-to-point' queries. In this paper, we propose to extend the query types of current LBS and show how these also extend their potentials. We explain, why the absence of additional query types is an issue in certain cases, and summarize the work accomplished in this area. In addition, we concentrate on one usage domain which is missing in the majority of current LBS—information about space usage rules, e.g. 'no smoking in the restaurant' or 'no dogs allowed in this park' and illustrate these with various examples.

Keywords Location-based services · Context-aware services · Mobile spatial applications · Space usage rules

1 Introduction and Related Work

LBS have left the research stage and are mainstream now. More than a decade ago, these services have already been predicted to be successful (Schiller and Voisard 2004), and users nowadays increasingly demand them: as such, ABI Research expects advertising LBS application revenues to multiply fourfold in the period from 2014 to 2019 (Gallen 2014). While this growth is mainly expected to happen in North America and Europe, it is not only a western phenomenon. LBS usage rates from BRIC countries—Brazil, Russia, India and China—are growing as well

---

P.A. Samsonov (✉)
Expertise Center for Digital Media, Hasselt University - tUL – iMinds,
Hasselt, Belgium
e-mail: pavel.samsonov@uhasselt.be

© Springer International Publishing AG 2017
G. Gartner and H. Huang (eds.), *Progress in Location-Based Services 2016*, Lecture Notes in Geoinformation and Cartography,
DOI 10.1007/978-3-319-47289-8_15

(Arno 2012). Various companies around the world are also highly interested in location-based services as a promotion tool (Digital News Asia 2015).

Context-aware services in general are clearly mainstream nowadays as well. The most well-known examples of these services are realised in mobile assistance applications developed by three headliner companies: Google Now, Apple Siri and Microsoft Cortana. According to the results of 'Battle of the virtual assistants' survey (Experts Exchange 2015), Siri shows good satisfaction by users (81 %) with Google Now and Cortana just behind (68 % and 57 % respectively). Still, even though users are mostly satisfied with the performance of those services, there is still potential for research in the area of LBS that can help to improve those services to fit for the users' everyday needs. We conducted a study of location-based services and found out that the majority of LBS—not including navigation services—limit their spatial information sources to sets of geopoints. This is also true for the context-aware mobile assistants in particular. These are able to answer user questions such as 'where is the next restaurant?' or 'show me the way to the airport', but cannot process requests that go beyond 'point-to-point' queries or distances. To prove it, we have analysed 54 popular location-based services (see Table 1), which we had extracted from various 'popular LBS review' articles from websites such as Tom's Guide,[1] Slideshare,[2] Connectivity[3] and others. The services were included in the survey, if they have a large user base or got high media coverage. The results showed, that only few LBS used more than point-to-point queries. From the analysed services, the only exceptions to this were those that had a directions service implemented, two geofencing LBS, an advanced weather application, Ingress location-aware game and Line of Sight web application, which shows locations of satellites in the sky above user's location. It is important to note, that in the survey we have only analysed the main functionality of these LBS and did not process their source code. We drew these conclusions from whether the main functionality required more than point-to-point queries—e.g., an app that suggests restaurants nearby and tells you the trigonometric distance to those, is not likely to perform any spatial queries except point-to-point.

Later in the paper, we highlight one usage domain—space usage rules (SURs)—which offers potential for queries that go beyond 'point-to-point' queries. Interestingly, information about these rules is currently not included in the vast majority of LBS.

More and more researchers admit the existence of various problems in modern spatial applications. Even leaders such as *Google Maps* and *Apple Maps* are occasionally criticised by professionals (Crook 2012; Dobson 2012). In another paper (Schöning et al. 2014), the authors highlight the gap between traditional

---

[1]http://www.tomsguide.com/us/best-location-aware-apps,review-2405.html.

[2]http://www.slideshare.net/socialtech/20-hot-locationbased-apps-and-services-you-should-know-about-12841489.

[3]http://www.connectivity.com/blog/2015/07/the-best-location-based-apps-and-services-for-small-business/.

Table 1 Our survey on popular location-based services. *Abbreviations*—**platform:** m = location-aware mobile app, w = web page (or an application for a PC), g = GPS navigation device, i = only Mac/iOS, ML = MirrorLink; **coverage:** ww = supported worldwide or in over than 50 countries, mc = only major cities, p = partially, E = Europe, A = Asia, Af = Africa, Au = Australia, Ir = Ireland, cov(1) = {UK, France, Germany, Spain, China, Japan, Taiwan, Australia}, cov(2) = {USA, UK, Puerto-Rico}; **data source:** c = mainly own company data, v = volunteers, u = users, ext(1) = {OSM, Quattroshapes, OpenAddresses, Geonames}, ext(2) = {Google Maps, Netflix, Yelp, OpenTable, Uber, Lyft, Foursquare, Foursquare, Rotten Tomatoes, Metacritic, Fandango}, ext(3) = {OpenStreetMap, Mapbox, DigitalGlobe, NASA}, ext(4) = {SatNOGS, CelesTask, Mapzen Search, Elevation}, GM = Google Maps; **requests:** ptp = only point-to-point, mptp = more than point-to-point, dir = directions service

App type	Name	Provider	Platform	Coverage	Data source	Requests
Restaurants	OpenTable	OpenTable	w, m	USA	c	ptp
	Zomato	Zomato	m	ww	c	ptp
	Foodspotting	Foodspotting	w, m	ww	c, GM	ptp
Cinemas	Fandango	Fandango	w, m	USA	c	ptp
Gas	GasBuddy	OpenStore LLC	w, m	USA	c	ptp
Discounts	Vouchercloud	Invitation Dig. Ltd	m	USA&E (p)	c	ptp
	Groupon (Whrrl)	Groupon	w, m	USA, E&A (mc)	c	ptp
Events	Eventbrite	Eventbrite	w, m	ww	u	ptp
LBS advertising	Google Ads	Google	w, m	ww	u	–
	Youtube Ads	Google	w, m	ww	u	–
POI search	Google Places	Google	w, m	ww	c	ptp
	Foursquare	Foursquare	w, m	ww	c	ptp
	Yelp	Yelp	m	USA	c	ptp
	Vurb	Vurb	m	ww	ext(2)	ptp, dir
	TripAdvisor	TripAdvisor LLC	w, m	ww	c, GM	ptp
	Pushlocal	Pushlocal	m	USA	c	ptp
	TravelZoo	TravelZoo Ltd.	w, m	cov(1)	c	ptp
	LivingSocial	LivingSocial Limited	w, m	UK, Ir	c	ptp
	Google + Locations	Google	w, m	ww	u, GM	ptp
Social	Facebook NearbyFriends	Facebook	m	USA	u, GM	ptp
	Facebook Places	Facebook	w, m	ww	c, GM	ptp
	Swarm	Foursquare	m	ww	u	ptp
	NearbyFeed	NearbyFeed	w, m	ww	u, GM	ptp
	Twitter Check-in	Twitter	w, m	ww	c	ptp
	Instagram Check-in	Instagram	w, m	ww	c	ptp

(continued)

Table 1 (continued)

App type	Name	Provider	Platform	Coverage	Data source	Requests
	Glympse	Glympse Inc	m, ML	ww	c, GM	ptp, dir
	Strava	Strava, Inc.	w, m, g	ww	ext(3)	ptp
	Yik Yak	Yik Yak	m	ww	c	ptp
Games	Pokemon Go	Niantic Labs	m	ww	c, GM	ptp
	Ingress	Niantic Labs	m	ww	c, GM	mptp
Satellite	Line of Sight	Patricio Onzalez Vivo	w, m	ww	ext(4)	mptp
Weather	Dark Sky	Dark Sky LLC	m	cov(2)	u	mptp
Geofencing	Trigger	Egomotion	Android	ww	u	mptp
	Locate	Pulsate	w, m	ww	u, GM	mptp

cartography and online or mobile map applications. As one of the differences between traditional cartography (including usage of paper maps) and online and mobile map applications, these researchers have identified the lack of SURs.

SURs are activity restrictions, affecting a specific region on the map. The government and local landowners commonly introduce these restrictions to protect health and public safety as well as their own property (Samsonov et al. 2015b, pp. 971–974). People increasingly rely on their mobile context-aware applications instead of legacy paper map brochures in parks, and the ignorance of these services of SURs leads to threats to public safety and occasionally fines for rule violation (PetaPixel 2016). In this paper, we argue for the inclusion of SUR data into LBS. By adding this data, LBS can inform the users of local spatial restrictions and exploit the potential of 'point-to-polygon' and other 2D queries in contrast to simplistic 'point-to-point' queries. In the reminder of the paper, we discuss SUR data sets. To overcome this starting problem we outline several current SURs that are available in digital formats and the lack of consistent and large techniques to mine SURs and show how their inclusion could lead to new classes of LBS.

App type	Name	Provider	Platform	Coverage	Data source	Requests
LBS Supporting Free-form Requests (Mobile Assistants)	Google Now	Google	Mobile	ww	c	ptp, dir
	Siri	Apple	w, m(i)	ww	c	ptp, dir
	Cortana	Microsoft	w, m	ww	c	ptp, dir
	Wolfram|Alpha	Wolfram Alpha LLC	w, m	ww	c	ptp, dir
LBS-enabled search	Google Search	Google	w, m	ww	c	ptp, dir
	Yandex Search	Yandex	w, m	ww	c	–

(continued)

(continued)

App type	Name	Provider	Platform	Coverage	Data source	Requests
	Bing Search	Microsoft	w, m	ww	c	–
	Duckduckgo	Duckduckgo	w	ww	c	–
	Rambler Search	Rambler	w	ww	c	–
	Ask.com	Ask	w	ww	c	ptp
Navigation	Google Maps	Google	w, m	ww	c	ptp, dir
	Apple Maps	Apple	m	ww	c	ptp, dir
	Bing Maps	Microsoft	w, m	ww	c	ptp, dir
	Yandex Maps	Yandex	w, m	ww	c	dir
	OpenRouteService	Dr. D. Luxen	w	E, A, Af&Au	v	dir
	Turn-by-Turn	Mapzen	w	ww	ext(1)	dir
	OSRM	OSRM	w	ww	OSM	dir
	GraphHopper	Peter Karich	w	ww	OSM	dir
	Waze	Waze Mobile Ltd	w, m	ww	v	dir
Shopping	Shopkick	Shopkick	m	ww	c	ptp

2 Existing SUR-Aware Location-Based Services

We believe that the domain of SURs is an interesting field to extend 'point-to-point' queries of most current LBS. SURs are gradually appearing in first LBS. However, SUR datasets in these services are highly sparse and only support a limited number of rules. In this section, we provide several examples of current applications making use of SUR data.

2.1 SURs for Drone Flying Regulations

Drone flying regulations are a very recent SUR amongst location-based services. For example, Sudekum created the *Don't Fly Your Drone Here* application hosted on the Mapbox service. In this app, three governmental rules of restrictions on drone flying in the USA are visualised on the map. Drone flying is forbidden in US national parks, US military bases and additionally within a 5-mile radius around large airports. Even though it is an important step towards SUR-aware applications, there is still a large corpus of work to be accomplished until applications such as context-aware mobile assistants are able to process SUR requests. The application does not take into account other flying restrictions such as weather conditions and allowed flying height, which depend on the size and type of the drone, as well as local restrictions introduced e.g. by park owners. In addition, its coverage is limited to the USA. Other apps also exist, that extend the concept. *Hover* application combines *Don't Fly Drones Here* map with current weather conditions, informing users of whether these satisfy drone flying rules in the US. *RCFlyMaps* is an application for iPhone that visualises drone-related SURs in the United States and Canada. Its rule database is not limited to only airports, national parks and military bases, and has been collected from the information provided by the US Federal Flying Administration and other instances. Furthermore, users can mark their favourite and disliked flying spots. *NoFlyDrones* is an application developed specially for drone hobbyists and professionals in the UK. Their map, also powered by Mapbox, shows in different colours not just where it is allowed to fly a drone, but also danger areas and special restrictions (e.g. flying a drone around airport areas in the UK is only allowed for drones that weight less than 7 kg). In addition to the web interface, it also exists as an app for Android smartphones. However, it does not have data about regulations outside the UK and does not inform users of restrictions introduced by local landowners.

An advanced version of the abovementioned LBS is a web application *RuleMaDrone* (Trippaers 2015). The application's functionality (see Fig. 1) is similar to the ones listed above, but its coverage is expandable to the whole world area and it allows users to suggest new rules in their area. In addition, similarly to *Hover*, the application displays weather conditions and whether the weather is suitable for drone flying. Interestingly, the results of the study accomplished by Trippaers showed that out of 203 surveyed people in Belgium approximately 70 % had little or no knowledge about drone flying restrictions in their country. The application is currently being reworked and in the future is expected to also support other SURs.

Additionally, there are several other maps on the Internet showing drone rules with basic functionality. As such, drone manufacturer company DJI displays a *No Fly Zones* map on their website with no-drone-flying zones around airports. *RC*

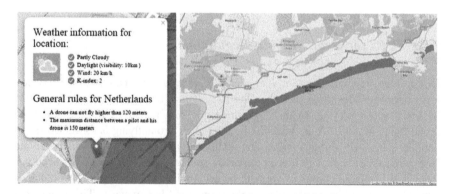

Fig. 1 RuleMaDrone functionality example. The user inspects which flying drone restrictions apply in a certain point (*left*) and areas where drone flying is forbidden (*right*)

Maps application does not explicitly show where drone flying is forbidden, but suggests designated drone flying areas for hobbyists and professionals. *Solar Activity Monitor* app for Android issues a warning when the solar activity in the user's location is too high to fly a drone.

Another SUR-related web application *NoFlyZone* is a tool that allows property owners to restrict flying over their owned buildings or areas. Unfortunately, the map is intended for professional use and is not accessible by hobbyists.

All examples show the need and potential for recent LBS to go beyond simplistic 'point-to-point' queries for a very specific use case.

2.2 Dog Walking Regulations

A more mainstream application that makes use of SURs is *MapMyDogWalk*. It is an application for iOS and Android, the primary function of which is to track dog walking routes and share these with friends. In addition, it also allows users to tag dog-friendly restaurants, leash-free parks, dog waste bags selling points, places where the user can water his or her dog, medical care centres for animals and locations where a trail with dog walking allowed begins as a geopoint marker on the map. Several regions of big cities such as London, Paris, New York and San Francisco are already densely tagged with this dog walking-related SUR data in this application, and this data is available for the application users. Additionally, users can report that a SUR marker is invalid, cross-checking the voluntarily collected information.

Other dog-related applications *Dog Park Finder Plus* designed for iPhone and iPad and *Dog Park Locator* for Android suggest dog-friendly parks nearby the user within the US (*Dog Park Finder Plus* also supports searching for parks in Canada).

2.3 Summary

Drone flying and dog walking are not the only SURs that start to appear in LBS. OpenStreetMap (OSM) already allows 'no smoking', 'no fishing' and 'no dogs' SUR mapping in the form of tags that are attached to spatial features. It is also possible to add certain parameters to the restrictions (days of the week, time and special parameters such as 'dogs should be kept on leash'). Users can also suggest their own tags, potentially expanding covered SUR. However, while the quality of volunteer geographical information (VGI) data overall is satisfactory, especially in populated areas (Haklay 2010; Barron et al. 2014), the SUR tag mapping is still too sparse to be used in actual applications (Samsonov et al. 2015b, pp. 971–974). SUR information in OSM is collected exclusively manually, as website owners are highly concerned about automatic collection of VGI information. Nevertheless, we strongly advocate to include the SURs listed in Table 2 in any spatial data set. We

Table 2 A list of SURs, which we suggest to incorporate into the existing LBS

SUR suggestions		
No smoking	No mobile phone usage	No camping
No drone flying	No photography/filming	No campfires
No dogs	No beach clothes	No open flames
Leash required	No alcohol consumption	No deforestation
No dog poo	No skating/skateboarding	No backpacks
No pets	No entrance/access	No shoes
No eating	Do not use the ladder	Silence (no speaking)
No swimming	No access with pacemakers	No loud sounds
No diving	Do not drink water from it	Do not touch
No firearms	No electronic devices	No pedestrians
No knives	No soccer playing	No running
No children	Credit cards are not accepted	No digging
No heels	No radio transmitters	No littering

discuss different (semi-) automated methods to populate those databases in the next chapter.

3 SUR Mining

To address the problem of sparse SUR data, several ways to collect SUR data have been already suggested above in the paper. We have discussed the existing methods of (semi-) automatic and manual collecting of this data and have analysed whether, from our perspective, they are promising for the future LBS. We do not propose a universal solution to the process of gathering and storing a SUR dataset in this paper; however, we believe that redundancy is the key in building a trustworthy and full SUR database. This can be achieved by combining various approaches listed above (and novel ones) and creating a reverse pipeline, which would motivate the users of location-based services to report errors in the processed rules back to the database owners. Another important aspect of gathering SURs, in our opinion, is analysing and adapting to popular errors that these automatic and semi-automatic approaches produce.

3.1 Volunteer-Based SUR Mapping

The straightforward approach to SUR gathering—volunteers—is already being used by the VGI source OSM. However, apart from the abovementioned issues with gathered data in the current state being very sparse and not supporting many types of

SURs, the complexity of these restrictions that can be included in OSM is narrowed to tag and subtag data (Samsonov et al. 2015b, pp. 971–974). Because of this issue, a large amount of SURs, such as 'no swimming', as well as complex restrictions such as limitations on the weather conditions or complicated if-then-else rules, are not supported by OSM. Another problem with SUR mining in OSM is the fact that this information can only be added in the form of tags for existing areas (parks, reservoirs, etc.). Therefore, even if a 'no drone flying' SUR is added to OSM, there would still be no option to add a rule that restricts drone flying within 2 km of airports without having to manually add this spatial feature. Adding a 2 km radius circular area that does not have other function than showing the no-fly area is confusing for other volunteers. All these limitations restrict the usage of OSM as a SUR data source in mobile applications, but in the future, with the growth of OSM and an expanding of its features, we believe it will turn into a powerful SUR dataset.

3.2 SUR Mining in Location-Enabled Games

Another strategy to accelerate the data collection is gamification (Deterding et al. 2011). The location-aware game *Ingress* and its predecessor *Shadow Cities* had huge success as smartphone games and one could imagine a similar smartphone game that would reward the player for collecting SUR data. To reach a reasonable accuracy from enthusiasts, players would cross-check each other and be punished in-game for submitting wrong or inaccurate data. Additionally, already gathered data can be visualised in game, so the application would inform players of various SURs around them.

3.3 Semi-automatic Text-Based SUR Mining

An alternative work-in-progress method (Samsonov et al. 2015a, pp. 2211–2216) suggests using text to mine SURs. After extracting rules from text sources such as websites (e.g. park behaviour rules) and law books, these rules are transferred into an 'if-this-than-that' interface to be processed by crowdworkers. The interface creates *Lua* script files in a computer-understandable form that expresses SURs as interconnections between two statements, both of which are represented as tag information in OSM. Using this method, various areas in OSM can be tagged using general rules, such as 'no smoking in national parks in the USA', allowing for a reduction of the amount of volunteer work and at the same time attaching the website address as the source of the rule. In the future work section, the authors mention that the crowdworkers' efforts can be cross-checked to achieve higher accuracy, and a checkbox 'this rule cannot be fully tagged using current system' for crowdworkers would indicate cases when the rule should be converted into a Lua script manually by professionals. This approach is an important extension to the image-based method as it allows mining rules that are not represented as a visual sign (such as flying drone regulations) and SURs, photos of which are not uploaded to *Flickr* or other image sharing resources (e.g., due to photographing restrictions). In addition, SUR datasets created with this method can be updated regularly and checked by observing the websites with the rule sources.

3.4 Automatic Image-Based SUR Mining

An automatic method to extract SURs from images was proposed (Samsonov et al. 2015b, pp. 971–974) and later improved (Soll et al. 2016). The method suggests using geotagged photos of locations from image sharing websites such as *Flickr* as a rich, yet unexplored resource. In the first step, the images with corresponding tags (e.g., 'no dogs') are downloaded. Then, for each supported SUR, the images are processed with Viola-Jones object detection framework and filtered via Sparse Coding techniques to find out whether the image contains this specific SUR. As SURs are sometimes represented by text instead of or along with a no-sign (see in Fig. 2a), text in images is also analysed for SURs. In the last step, which is the most challenging, the extracted SURs are assigned to corresponding features in OpenStreetMap. The input to the algorithm is a photo with a known geolocation and a type of SUR, and as an output it generates a polygon which is expected to be the spatial area that is affected by the SUR. In Fig. 2b you can see an example of such assignation of an image to a geographic area: the location of the image is represented by the red marker and the recognized area is highlighted in green. From this example one can see that even for a human it's sometimes unclear, to which exact area this sign applies. To achieve worthy results, ensemble learning is used: several classifiers that are better than random guessing were joined together in one.

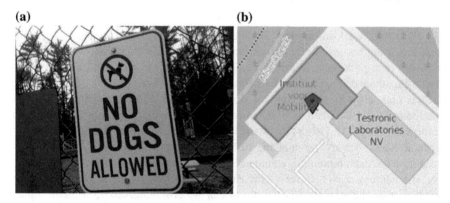

Fig. 2 A 'no-dogs' sign with text and image in New Hampshire, USA (*left*); assignation of the no-smoking SUR, the location of which is shown as a *red marker*, to the building highlighted in *green* (*right*, © OpenStreetMap)

Every classifier issues every possible area for a corresponding SUR a number from −100 (very unlikely) to 100 (very likely). The output of the resulting 'strong classifier' is a weighted sum of these values. The weights of these classifiers are calculated using a genetic algorithm, the starting population of which has equally distributed weights. These classifiers, so-called 'weak classifiers', are: distance from the photo location to the closest edge of an area, distance to the closest vertex of the area, SUR description classifier (e.g., a 'no smoking' sign cannot refer to a water body), photo orientation extracted from EXIF data, point-in-polygon (as SUR area more likely contains the photo location than otherwise) and computer vision-based inside/outside classifier—as such, a 'no skateboarding' sign is expected to be somewhere in the street and a 'no food' sign is most likely to be inside a building.

However, this method has its limitations. The image processing step showed an accuracy of the SUR detection part varies from 82.9 to 91.5 % (depending on the SUR), while the assignation of a SUR to a polygon step has an average of 63.2 % area crossing between ground truth and calculated polygons. Therefore, the results of the method require human correction, which can be performed by either professionals or enthusiasts.

4 Future Applications

Once full up-to-date SUR datasets are gathered and are widely available, we expect various smartphone LBS developers to raise an interest in these. In this section, we discuss possible future applications informing users of SURs or using the information as context data to process their other requests.

4.1 Processing Request 'Can I Perform Activity X Here'?

A straightforward application making use of SUR data could answer user's questions regarding what is legal to do in his or her current location, such as 'can I smoke here?', 'can I swim in this lake' or 'can I walk my dog in this park?' Using the context data about the user, such as local laws regarding whether he or she is old enough to light a cigarette, or whether the user's drone exceeds the allowed weight, keeping in mind the time of the day and weather conditions, the application would be able to answer the question more precisely. In addition, the application could warn the user about additional restrictions (e.g., drone flying requires a permit or dogs should be kept on a leash).

4.2 Using SURs as Context Data in Mobile Assistant Applications

Current context-aware recommendation applications provide a variety of information based on the user's interests and location. We expect these applications to incorporate SUR data once it is widely available, along with the other parameters (Adomavicius et al. 2011), such as distance to a point of interest (POI), weather conditions, time of day, crowdedness etc. Knowledge about the local activity restrictions in the user's location broadens the services that these mobile assistants can offer. Knowing that the user is a smoker, these could suggest cafés or restaurants where smoking is allowed, or those that have a smoking terrace. Undoubtedly, if a user wishes to go swimming, suggesting the closest water reservoir where swimming is forbidden would not suit the user's needs. Another useful feature that could be implemented is displaying local restrictions in the status bar, informing the user of activities that are allowed or forbidden in his or her location.

SUR datasets are useful for the mobile assistants designed for autonomous cars as well. Self-driving cars are expected to drastically decrease deaths in car accidents and make better use of space on the roads (Poczter and Jankovic 2014)—and Google is expecting to release their self-driving car for public use by 2020 (Korosec 2015), while Tesla CEO Elon Musk suggests that their driverless cars will hit the road before 2019 (Thompson 2015). As interaction with the assistants built into driverless cars can be accomplished through voice commands to achieve minimum reaction time, for example, with the recently presented context-aware voice recognition principle embedded in the application Houndify (Deleon 2016), we expect that the development of autonomous cars will also boost research in navigation assistant programmes. Therefore, a SUR knowledge base should be incorporated into these programmes, so that they would be aware not just of the driving rules in the city, but also of the local activity restrictions.

4.3 An Application for Lawmakers

Currently, lawmakers and local landowners who are responsible for rules design do not have any SUR-aware software tools to analyse how introducing or changing a rule actually affects areas on the map. Using an application that visualises a map with a SUR layer and highlighting areas affected by the local and global rules, both professional and amateur lawmakers could adjust the edited rules while observing the map. As such, a rule 'drone flying is prohibited within 2 km from airports' is rather vague from a lawmaker's perspective until the actual area is visualised. Additionally, on a higher level, this application could show the areas where the created rule conflicts with general rules—or others. After a new rule is created or an old one is updated, these automated systems would signal the changes to the existing SUR datasets. Furthermore, the rule creation process in the traditional form has a weak link to people, who do not always check updates on space usage rules, and only get to know about the changes from visual signs and news on websites. We believe that setting up a direct connection between rule or law creators to the people affected by these rules is one of the best possible scenario of SUR development in the future.

5 Conclusion and Discussion

In this article, we have discussed an important limitation of the majority of contemporary location-based services: treating all spatial objects, such as buildings, water bodies or parks, as only geopoints on the map rather than 2D objects. From our survey of 54 popular location-based services, we have found out that context-aware LBS, which resolve disambiguation in users' requests and often process large corpuses of context data, are affected by this issue the most. From the nine suggested questions (see Table 3) requiring various spatial operations that we expect a context-aware application to be able to process, only the three questions that require just point-to-point operations are supported by at least one surveyed context-aware LBS (though one of these is processed imprecisely). Two requests are simplified to point-to-point before processing and provide users with results for a different request, while the other four are just not supported. Therefore, we conclude that the simplification of the request parameters to geopoints does affect user experience with those services. While treating spatial features as geopoints is suitable for the LBS that only show the nearest points of interest on the map, it becomes crucial for services that process custom user requests in a natural language. We strongly believe that for those applications, query processing should be enhanced with deeper understanding of the types of request. As can be seen in Table 3 and Fig. 3, similarly looking questions 'where is the next POI?' (1-2) or 'where is the closest road/area?' (4-5) are actually different requests and therefore should be processed each in their own way. In some cases (question 2), misinterpretation of a request does not significantly affect user experience—a restaurant's entrance is normally next to its centroid on the map. However, for large areas (questions 4-5) this error is important: the user is more likely to be interested in the park entrance or even in a parking next to it, but not in the park centroid, as suggested by current applications.

We have conducted additional analysis on other aspects of LBS, such as their platform: 85.2 % of analysed services have a location-aware mobile app while 57.4 % are available both as a web page (and/or a PC application) and as a mobile app. Out of all abovementioned location-based services 77.8 % cover more than 50 countries or even the whole world. Own private company data as main source is used by 90.6 % of all examined services, while the rest mostly depend on external data sources; on the other hand, all examined LBS use external data, such as a map plugin, weather information etc.

Table 3 Examples of request disambiguation in the context-aware applications. A question is marked as supported (partially supported) if at least one surveyed application processes it correctly (imprecisely, but otherwise correctly)

№	User request example	Actual request	Measurement	Supported
1	Where is the bus stop?	Where is the closest bus stop **marker**?	Point-to-point	Yes
2	Where is the next café?	Where is the closest café **entrance**?	Point-to-point	Yes[a]
3	Any friends nearby?	Are any of the friends' **locations** nearby?	Point-to-point	Yes
4	Where is the next road?	Where is the closest road **polyline**?	Point-to-polyline	No
5	Where is the next lake?	Where is the closest **lake view point**?	Point-to-polygon	Partially
6	Where is the next park?	Where is the closest park **entrance**?	Point-to-point	Partially
7	Am I in Roman Forum?	Am I inside the Roman Forum **area**?	Point-in-polygon	No
8	Can I smoke here?	Do any **area polygons containing my location** have smoking restrictions?	Point-in-polygon	No
9	Can I swim here?	Do any water reservoirs **nearby my location** have swimming restrictions?	Point-to-polygon	No

[a] Usually, a café size is rather small and we consider showing its centroid to be sufficient

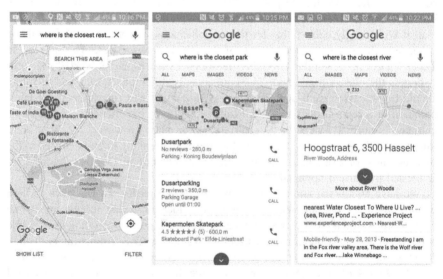

Fig. 3 Three examples of Google Now application processing ,'where is the next …?' requests—searching for restaurants, parks and river bodies. © Google

Additionally, we have pointed out an important feature missing in all but few LBS—space usage rules, lack of which threatens people's safety and rights because of ignorance of the local rules and regulations. We have gathered and analysed existing work on SURs. Lately, a variety of SUR mining methods have been proposed by various researchers, including VGI-based, image-based and text-based methods, while in the paper we also have pointed out the possibility of gathering SUR data in location-enabled games. In our opinion, from the existing methods, the text-based method (possibly combined with image-based and volunteer-based to check the rules in the field) is the most promising as it strengthens the link between lawmakers and the people who are affected by the space usage rules. Furthermore, we have also suggested several use cases of SUR data, which we believe to be very useful however missing in the current app stores. Our ultimate goal is to bring more attention to the SUR unawareness, and induce further improvements to existing location-based applications.

Acknowledgments I want to thank Johannes Schöning for comments on this chapter and Eva Cordery for proof reading.

References

Adomavicius, Gediminas, and Alexander Tuzhilin. 2011. "Context-aware recommender systems." In Recommender systems handbook, pp. 217–253. *Springer US*.

Arno, Christian. 2012. "Mobile search marketing for BRIC countries". *Search Engine Journal*, October 30. https://www.searchenginejournal.com/mobile-search-marketing-for-bric-countries/50545/

Barron, Christopher, Pascal Neis, and Alexander Zipf. 2014. "A comprehensive framework for intrinsic OpenStreetMap quality analysis". *Transactions in GIS 18, no. 6 (2014): 877–895*.

Crook, Jordan. 2012. "Tim Cook apologizes for Apple Maps, points to competitive alternatives". *Techcrunch*, September 28. http://techcrunch.com/2012/09/28/tim-cook-apologizes-for-apple-maps-points-to-competitive-alternatives/

Deleon, Nicholas. 2016. Your next car's dashboard may have voice recognition that's better than Siri. *Motherboard*, January 8. http://motherboard.vice.com/read/your-next-cars-dashboard-may-have-voice-recognition-thats-better-than-siri

Deterding, S., Dixon, D., Khaled, R. and Nacke, L., 2011. From game design elements to gamefulness: defining gamification. *In Proceedings of the 15th international academic MindTrek conference: Envisioning future media environments (pp. 9–15)*. ACM.

Digital News Asia. 2015. "In-store shoppers very interested in location-based services: survey". September 8. https://www.digitalnewsasia.com/digital-economy/instore-shoppers-very-interested-in-location-based-services-survey

Dobson, Mike. 2012. "Google Maps announces a 400 year advantage over Apple Maps". *Telemapics*, September 20. http://blog.telemapics.com/?p=399

Experts Exchange. 2015. "Battle of the virtual assistants". http://pages.experts-exchange.com/virtualassistants.html

Gallen, Christine. 2014. "Third generation of LBS to capitalize on location based advertising, says ABI Research". *Marketwatch*, June 23. http://www.marketwatch.com/story/third-generation-of-lbs-to-capitalize-on-location-based-advertising-says-abi-research-2014-06-23

Haklay, Mordechai. 2010. "How good is volunteered geographical information? A comparative study of OpenStreetMap and Ordnance Survey datasets." *Environment and planning B: Planning and design 37, no. 4 (2010): 682–703.*

Korosec, Kirsten. 2015. "Here's how Google's self-driving car experiment is going in Austin". *Fortune*, September 1. http://fortune.com/2015/09/01/google-self-driving-car-report-austin/

PetaPixel. 2016. "Man arrested for flying drone over Colosseum, could face $100 K + fine". April 13. http://petapixel.com/2016/04/13/tourist-arrested-flying-drone-colosseum-rome/

Poczter, Sharon L., and Luka M. Jankovic. 2014. "The Google Car: driving toward a better future?". *Journal of Business Case Studies (Online) 10, no. 1 (2014): 7.*

Samsonov, Pavel Andreevich, Johannes Schöning and Brent Hecht. 2015. "A user interface for encoding space usage rules expressed in natural language". *In Proc. of CHI Extended Abstracts, pp. 2211–2216).*

Samsonov, Pavel Andreevich, Xun Tang, Johannes Schöning, Werner Kuhn, and Brent Hecht. 2015. "You can't smoke here: towards support for space usage rules in location-aware technologies". *In Proceedings of the 33rd annual ACM conference on human factors in computing systems, pp. 971–974.*

Schiller, Jochen and Agnès Voisard. 2004. "Location based services". *Elsevier.*

Schöning, Johannes, Brent Hecht, and Werner Kuhn. 2014. "Informing online and mobile map design with the collective wisdom of cartographers." *In Proceedings of the 2014 conference on Designing interactive systems, pp. 765–774.*

Soll, Marcus, Philipp Naumann, Johannes Schöning, Pavel Samsonov, and Brent Hecht. 2016. "Helping computers understand geographically-bound activity restrictions". *In Proceedings of the 34rd annual ACM conference on human factors in computing systems.*

Thompson, Cadie. 2015. "Elon Musk says Tesla's fully autonomous cars will hit the road in 3 years". *Techinsider*, September 25. http://www.techinsider.io/elon-musk-on-teslas-autonomous-cars-2015-9

Trippaers, Aäron (2015). "RuleMaDrone: a web-interface to visualise space usage rules for drones". *Master's thesis, Hasselt University.*

Concept Design of #hylo—Geosocial Network for Sharing Hyperlocal Information on a Map

Hanna-Marika Halkosaari, Mikko Rönneberg, Mari Laakso, Pyry Kettunen, Juha Oksanen and Tapani Sarjakoski

Abstract Sharing hyperlocal geospatial knowledge with one's community may foster many positive outcomes, such as increasing a sense of community. However, much of this information travels by word of mouth and, therefore, does not reach everyone in time or even at all. We present the human-centred concept design process of #hylo, a geosocial network in which users can easily share personal, hyperlocal geospatial knowledge about their surroundings. Users may gain hyperlocal information of their own places, filtered with their own interests. We expect #hylo to increase (1) attachment to location, (2) interest in the area, and (3) social participation in local community. We explain the design process and show results from initial user studies that were encouraging, as the participants were interested in this kind of hyperlocal information and found it useful. Even those who are usually hesitant to post in social media, found this kind of community-focused, hyperlocal platform to be a pleasant place to share information.

Keywords Human-centred design · Geosocial · Hyperlocal · Location-based · Map application

---

H.-M. Halkosaari (✉) · M. Rönneberg · M. Laakso · P. Kettunen · J. Oksanen · T. Sarjakoski
Finnish Geospatial Research Institute FGI, Helsinki, Finland
e-mail: hanna-marika.halkosaari@nls.fi

M. Rönneberg
e-mail: Mikko.Ronneberg@nls.fi

M. Laakso
e-mail: Mari.Laakso@nls.fi

P. Kettunen
e-mail: Pyry.Kettunen@nls.fi

J. Oksanen
e-mail: Juha.Oksanen@nls.fi

T. Sarjakoski
e-mail: Tapani.Sarjakoski@nls.fi

© Springer International Publishing AG 2017
G. Gartner and H. Huang (eds.), *Progress in Location-Based Services 2016*, Lecture Notes in Geoinformation and Cartography,
DOI 10.1007/978-3-319-47289-8_16

1 Introduction

People possess large reservoirs of geospatial information about personally important places and routes covering, for example, areas around their homes. This kind of geospatial information is often combined with experiences and emotions, thus creating personal geospatial knowledge in their memory. In many cases, personal geospatial knowledge is not shared, although the information would be interesting and useful to other people as well. Actually, there are groups of people who desperately need this kind of hyperlocal information, like people who just moved to an area or do not speak the language.

Hyperlocal information is relevant to small communities or neighbourhoods (Glaser 2010). Hyperlocal information includes news or other content concerning "a town, village, single postcode or other small, geographically defined community" (Radcliffe 2012). Hyperlocal information can be both functional, telling you what is going on, where and when, as well as emotional, for instance, helping you feel a sense of belonging to a group. Knowing what is around you and what is happening near you raises your community awareness and may also foster a sense of community (Boyd and Ellison 2008). People feel attached not just to their city, town or village, but also to their neighbourhood and street (Radcliffe 2012).

Finding relevant hyperlocal information from, for example, one's neighbourhood can be laborious, as this kind of information is often difficult to obtain from a single source. Some information can be found on social media sites, like local Facebook groups. Official information, such as education or social services, can be found on your city's website. Specific hobby-related news, events at your local school or neighbourhood yard sales all reside on different websites, if on the Internet at all. The information is usually not referenced by location, thus making it difficult to know what is going on near you.

In this paper, we present the early phases of the human-centred design process of #hylo—a geosocial network for sharing hyperlocal information on a map (later referred to just as #hylo). #hylo is a mobile map application where users can easily share personal, hyperlocal geospatial knowledge about their surroundings. Users may gain hyperlocal information of their own places, filtered with their own interests. A geosocial network has been earlier defined as: "a web-based or mobile-based service that allows users to (1) construct a profile containing some of their geolocated data (along with additional information), (2) connect with other users of the system to share their geolocated data and (3) interact with the content provided by other users (for instance by commenting, replying or rating)" (Gambs et al. 2011). We first present the related work and identification of the design opportunities, a need assessment and iterative phase of prototyping. Eventually, we present the conducted user studies. In the end of the paper, we reflect our study on what has already been done and consider future work.

2 Related Work

Previous research on developing geosocial services has been active. Park et al. (2014) studied location-based social questions and answers with the aid of an app called Naver KiN "Here", in which local residents with hyperlocal knowledge answer questions related to the local community, like, for example, local services and places. Pat et al. (2015) studied geosocial search, i.e., geographic search based on user activities in social media. Geosocial search is different from traditional geographical search on which Pat et al. give examples, such as a search for "jogging" can indicate good jogging places or a search for "sunset" helps find popular sunset viewpoints or "romantic walk" that especially reflects how people feel in a place or what they say about the place, and not so much about a geographical property of the entity. A geosocial search can be a great tool for tourists or local residents to discover new places. Espinoza et al. (2001) present a GeoNotes system that seeks to socially enhance physical space, and can be considered as a pioneering work to our research. Gupta et al. (2007) presented how location information of users can be used to identify, for instance, popular meeting places automatically.

Research especially focusing on hyperlocal geosocial media includes, for example, Xia et al. (2014) who present a demo, CityBeat, of a real-time Instagram-focused visualisation of hyperlocal social media content for cities on a map. Hu et al. (2013) present and evaluate Whoo.ly, a web service that provides hyperlocal information based on Twitter posts. Walther and Kaisser (2013) present another Twitter-based event detection map.

There are numerous social networking applications and quite many that are location-based and include maps. Many also allow for hyperlocal use. There are few, such as, *EveryBlock* (http://www.everyblock.com/), where you can browse through news, info and local events happening around you. *Nextdoor* (https://nextdoor.com/) is a private social network for the neighbourhood, where you can stay in the know about what is going on close to your home. *Waze* (https://www.waze.com/) is a community-based traffic and navigation app, where drivers can share real-time traffic and road info. On the map one can also see friends' locations, when connected to Facebook. The Finnish *Nifty Neighbour* (http://nappinaapuri.fi) is a map- and location-based social web service. People can meet each other, by way of their needs and resources. You can ask for help taking your dog out, borrow tools, offer fishing company, help your neighbour and so forth.

Social networking services that allow for location-based user interaction on a map interface also include global networks such as *Foursquare* (https://foursquare.com/), *Airbnb* (https://www.airbnb.com/), *Couchsurfing* (http://www.couchsurfing.com/) and *Yelp* (http://www.yelp.com). However, these global services do not particularly support or promote the viewpoint of local life, but often concentrate on the users' attention towards remote and unknown locations. What sets #hylo apart from the other location-based services, is that it makes finding and sharing hyperlocal information easier.

3 The Human-Centred Design Approach

We utilised a human-centred design (HCD), or user-centred design (UCD), process for the development of #hylo. Defining potential users and understanding their needs by utilising qualitative research techniques is essential for a successful design, as stated e.g. by Cooper et al. (2007). The HCD approach promotes a constant focus on users' needs and limitations throughout the process. The human-centred approach is summed up in the ISO 9241-210 standard (International Organization for Standardization [ISO] 2010) as follows: "The design is based upon an explicit understanding of users, tasks, and environments; users are involved throughout the design and development; the design is driven and refined by user-centred evaluation; the process is iterative; the design addresses the whole user experience; the design team includes multidisciplinary skills and perspectives." The term human-centred design is used in the ISO 9241-210 standard, rather than user-centred design, in order to emphasise the impact on a number of stakeholders, not just those typically considered as users. But as it is stated in the standard, these terms are often used synonymously in practice.

In the field of geoinformatics and location-based services, more attention can be seen to be paid to the HCD approach (e.g. Delikostidis 2011; Haklay 2010; Ooms 2012; Schobesberger 2012), although mainly traditional usability-based methods are used and the potential end-users are mostly only invited to participate in usability tests of the prototypes. Utilising the HCD methods systematically in the entire life cycle of a map-based service is still rare, as has been noticed also by Nivala (2007). In addition to involving users in evaluations, a more holistic view to human-centred design should be taken in order to design engaging map services. A human-centred approach has been recognised as one of the key factors in the successful design of products and services (e.g. Cagan and Vogel 2002; Miettinen 2012).

Our HCD process began with creating design proposals by the multi-disciplinary team, where we utilised the scenario method. We created a mock-up of one of the design proposals. The mock-up was presented to users, who we interviewed and observed trying out the mock-up. We asked them to fill in a questionnaire of related issues. The design was altered in response to the user feedback. The process continued iteratively, including a brainstorming session with a group of teenagers in order to develop useful tasks for them to perform during field tests.

4 Defining the Concept

This study is part of the MyGeoTrust -project (http://mygeotrust.org/) that focuses on the user privacy of location information. The vision of the MyGeoTrust -project is to create an alternative location platform for mobile users, which allows users to enjoy the benefits of location technologies without sacrificing their privacy

(Guinness et al. 2015). In the project, keystone apps were developed to demonstrate a new way of crowdsourcing. Our team was responsible for the app titled 'hyperlocal news and social media' that was described as: The hyperlocal news and social media platform that will allow users to find news, local information, and community members based on common geospatial parameters.

The project plan outlined high-level aims, but it did not set out specific steps. The plan thus encouraged us to explore the possibilities. We started the concept design process of the app by brainstorming and identifying opportunities in a small group of researchers with background from user experience design and geoinformatics.

4.1 Opportunity Identification

As the project plan gave us a relatively wide array of possibilities, we wanted to pay attention to identifying the existing gaps and possible opportunities related to these. During brainstorming, we soon recognised that we wanted to emphasise the value of people's experience as an enabler of meaningful services for users. Cockton (2006) has argued that values unify the design. He points out that designing value, or worth, as he prefers calling it, means designing things that will motivate the users to adopt and use a service as well as preferably recommend it.

After a couple of iterations of brainstorming, we summarised our value propositions. We presented them in a pitch form using NABC (Need, Approach, Benefit and Competition) style (Carlson and Wilmot 2006).

4.1.1 Need

Based on common experience, many neighbourhoods appear to lack a sense of community these days, which has also been shown scientifically (see Francis et al. 2012). People do not know their neighbours, they are not aware of the local area and its services, and they lack personal attachment to local surroundings. Having no sense of community can lead to a fear of one's surroundings and even loneliness. Being part of a local community creates new ties and enriches lives, but still many people do not participate in communal activities.

Not knowing what is around also keeps people from caring about their surroundings. The local park might be in bad condition, but only a few people seem to care about it. The lack of information similarly makes people feel distant from their surroundings. It takes too much effort to find out what is going on, since the information is scattered or it might even be unavailable. Local information usually travels by word of mouth and, therefore, does not reach everyone in time or at all. This means that people do not always have a choice when it comes to participating in activities in the neighbourhood.

4.1.2 Approach

By making community-sourced hyperlocal geospatial information available, and more importantly shareable to a local community, it is much easier for people to feel a sense of community on a larger scale. Knowing what is going on right down the street shortens the bridge to actually participating (Fig. 1). Instead of: "I wonder what is going on over there," people can make a choice: "I think I am going to go to that yard sale gathering money for the new playground sandbox." Meanwhile, sharing knowledge of what is going on in a specific location is not all. "What, where and when?" can be attached to "How does it feel?". Knowing what people experienced in the location is equally, if not even more, important. That local yard sale might have generated a lot of positive feedback from participants, so more people heard about it and joined in. Increased local activity can also be considered more ecological than travelling outside of the local area, e.g., for recreational activities. In addition, a community-sourced hyperlocal geospatial knowledge platform can serve as an interaction channel between the local community and local actors like media, services and administration.

4.1.3 Benefit

Collecting and representing hyperlocal geospatial information has the potential to create a new kind of collective understanding about the environment in light of experiences of people who have the local expertise in their surroundings. Appropriately represented, it may improve people's everyday lives both on a personal and on a community level: individual persons can find new perspectives on their familiar environments and communities can be enlivened by the enriched meanings given to the environment and increased social activity. Community-sourced hyperlocal information may even fight social exclusion by encouraging and

Fig. 1 Locally shared knowledge about what is happening in the neighbourhood shortens the bridge to actually participating

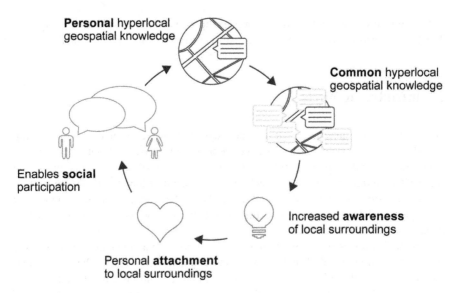

Fig. 2 Hyperlocal geospatial knowledge will be utilised to encourage social participation (Guinness et al. 2015)

inspiring people to explore their neighbourhood and make them acquainted with the community they live in Farrell et al. (2004). With a tailored service platform, people could share their personal tacit hyperlocal geospatial knowledge, thus building a common hyperlocal geospatial knowledge reservoir. This community-sourced hyperlocal geospatial knowledge increases people's awareness of the local area, which enables personal attachment to local surroundings and facilitates social participation (Fig. 2).

4.1.4 Competition

Is it possible to activate and motivate people to explore their local surroundings with a mobile application? And, because of privacy concerns, are people willing to share their hyperlocal information openly?

Many web services have aimed at fulfilling the need for local geosocial networks (see Sect. 2) but seem not to have fully succeeded in motivating people to attach to their surroundings. Privacy concerns of location information and the privacy concerns of sharing information in social media have been recognised both in public discussions and in academic research. Privacy issues of geosocial networks have been studied by, for instance, Gambs et al. (2011) and Alrayes and Abdelmoty (2014). Alrayes and Abdelmoty found that users need means to control and manage their location data, but also to improve their knowledge, access and visibility to their data sets in general. Still, as Gambs et al. conclude, almost no research project is focused on developing a privacy-preserving geosocial network. Using MyGeo-Trust as a location provider, we expected to overcome the privacy concerns.

5 Introducing #hylo

We created three alternative design concepts that we presented at a workshop attended by our colleagues. The concepts were presented with the aid of hand-drawn mind maps. We did a collaborative review of the concepts and stated the positive and negative sides of each one. As a result of the workshop, the amount of design concepts was reduced to two. After the second iteration round, the concept called #hylo was the one that would ultimately become the final design concept (Fig. 3).

The decision was based on the review of the positive and negative opinions, the feasibility, and the level of motivation of the concept design team. The novelty and the possibilities of the idea, as well as the fact that it could improve daily life, were counted for choosing the concept. The greatest risk was stated as how can we find and motivate users and whether it would be dynamic enough for daily use.

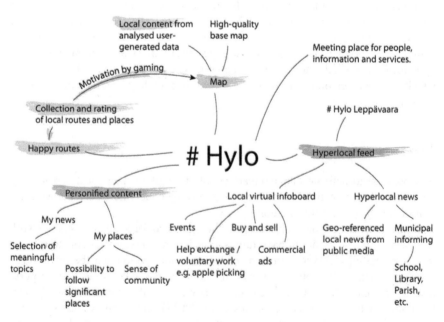

Fig. 3 The initial description of the #hylo concept

5.1 Needs Assessment

To continue the process, we chose a test area that had some communal activity, such as a communal house, an active elderly home, cafés and a library that were all places for many activities. We did some fieldwork and studied the bulletin boards for these places and collected hyperlocal information. The exploration of the chosen test area served us as needs assessment. Collecting the in situ information ensured us that we were on the right track as there was a lot of hyperlocal information that is not available on the Internet referenced by location, if at all. Hyperlocal information included, for example, advertisements from local companies, events and more personal messages, as someone was looking for a lost cat with a note taped to a wall at a train station (Fig. 4).

6 The Design and Prototyping Phase

To get hands-on experience in sharing hyperlocal information on a map, we started creating mock-ups of #hylo early on in the process. The first mock-ups were with simple functionalities created with web-based prototyping tools, in this case Marvel (https://marvelapp.com/). As the map is a crucial part of #hylo, and the user generated content is presented on a map interface, we soon realised that we needed to have a functioning map in order to reflect the idea properly. We found a functioning map essential especially for user studies, as panning and zooming a map is something users are accustomed to. From prototyping the map functionalities in Leaflet (http://leafletjs.com/), we moved on to developing the Android application, as it was to be the final deliverable in the project (Fig. 5). The technical development of #hylo has been published in Rönneberg et al. (2016).

In social networks, the contents are usually presented as a feed, in which the contents are filtered and sorted according to their time of posting, and they are presented as a list that the user can scroll. The location aspects are usually not

Fig. 4 Lots of hyperlocal information is not available on the Internet referenced by location

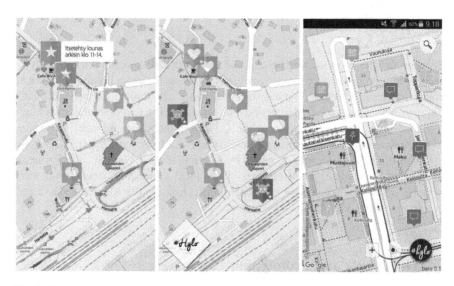

Fig. 5 The mock-up creation evolved from (1) simple prototypes to (2) a functioning map app and finally to (3) an Android app

considered in a feed-style presentation. In #hylo, instead of just sorting the posts according to time, the user-generated content is also sorted by location. This way the #hylo map acts as a multidimensional content feed that sorts and filters the content shared by users according to location and time, while also taking into consideration the users' interests. The users are asked to fill in their interests as hashtags, as well as the locations they are interested in, in their profile information in order to deliver relevant information to them.

The user can share content, called geonotes, by adding them on the map, where they are publicly visible for all #hylo users. Descriptive hashtags and place names can be added to the geonotes, so they are easier to find by other users. The geonotes shown on the map are filtered by default according to the users own interests and own places. This way, the user sees geonotes on subjects and from places they have specified in their profile. Place names are added automatically by fetching them from the OpenStreetMap Nominatim service (Nominatim 2016), but can also be added by the users. In #hylo, the user can view the geonotes on a map (Fig. 6) or from a variety of feeds as newest geonotes, geonotes in my places and geonotes with my tags to name a few. Users can easily search nearby geonotes by pressing anywhere on the map to perform a proximity search. Nearest geonotes in the area are revealed originating from the pressed location. Another similar proximity search is done when the user requests to see his/her own location on the map. Users can also switch between the map and the feed by either opening all the visible geonotes on the map to a feed or showing the geonotes in the feed on the map. The purpose of switching between the feed and the map and the proximity searches functionality is to emphasise the hyperlocal nature of #hylo. Users may also use a more

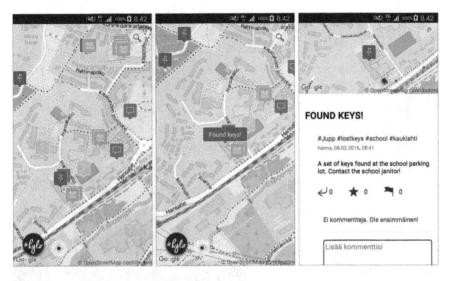

Fig. 6 In #hylo the user-added content is seen on a map

traditional search bar to find interesting content by searching with hashtags or place names.

Some features were eliminated during the process, such as location-dependent sharing and location-dependent visibility, where functionality and content would depend on the current location of the user. One of the initial ideas was that the user could instantly see what others have considered happy or unpleasant places and, based on this, decide, e.g., which way to walk. This was difficult as users' opinions are very subjective and what is negative for some might be positive for others. The first user studies confirmed this and, therefore, this kind of "happy route" idea was left out.

7 User Studies

7.1 Questionnaires

In order to evaluate the concept and the initial ideas of #hylo, we conducted a small user study with eight users (four male, four female). We introduced the concept to the participants, asked them to fill in a questionnaire and let them try the #hylo mock-up (Fig. 7).

The participants were aged from 26 to 41 years old, and they actively explored their neighbourhoods. Seven out of eight users said they do activities, such as jogging or dog walking, close to their home on a daily or weekly basis. However,

Fig. 7 We utilised questionnaires in the first user studies

they were not very active in posting in social media and usually did not include their location in social media posts, if they made any.

At the moment, the user study participants found information of their neighbourhoods from local newspapers, Facebook groups, bulletin boards and word of mouth. When we asked how aware they thought they were of local happenings and news within their neighbourhood on a scale of 1–5 (1 = not at all, 5 = very well), the average was 2.9.

We gave the participants some examples of what the geonotes could be, to give them an idea of the possibilities in #hylo. The examples included an announcement of some roadwork and a closed road, a found unidentified cat, an advertisement of a school bazaar and a reminder of a free, outdoor exercise organised in the local park. We asked if the participants were interested to hear this type of news and if they would be willing to share this kind of information. The participants were, naturally, interested in different things. For example, some would like to hear about yard sales while others would not, or if one does not have children, they are not too interested in hearing about events at a local school. However, all the participants in our study were interested or very interested to hear about local news and official announcements. All the participants said that they would be willing to share this kind of information with others. This was very interesting to learn, as otherwise they did not post very actively in social media.

When we asked if they would use the application, the participants were motivated, but also concerned whether the content would be interesting to them. Especially, in the beginning, when there are not so many users, the content might be unilateral and affected by the type of user groups. For example, if teenagers would be the first group to adopt the service, it might put off some users. To increase the amount of interesting content for different users, geo-referenced content from other social media, such as Twitter, could be included in #hylo. Filtering the geonotes according to the users' own interests should prevent totally unwanted content.

Realising the very different interests among users, we learned that instead of the positive and negative geonotes first initiated, the geonotes should have different themes. The themes were then to be: (1) a general note, (2) a question, (3) an event, (4) a notice, such as lost keys and (5) a more official announcement, such as a city could post information on roadwork. When a geonote is made, the user can freely choose any of these themes based on their own judgement. With the aid of the themes and hashtags, the user can find relevant location-based information.

When users can add content without moderation, there is always the possibility of inappropriate material. We considered if the fact that the geonote shows whether it was created on the location or elsewhere would bring credibility to it. During the user studies, it became evident that the participants did not place high value on knowing whether the geonote was created at the actual location, and they thought that adding geonotes on the go would sometimes be too problematic.

One of the participants suggested that instead of a community created by location, i.e. your neighbourhood, you should be able to create your own community and share content with friends. Most social networks support this kind of maintenance of pre-existing connections (Boyd and Ellison 2008), but in #hylo we especially want to encourage people to get out there, make new acquaintances and gain new experiences.

7.2 Field Studies with Teenagers

To further evaluate #hylo, we conducted field studies with teenagers. As teens are a big user group of social network services, it is important to consider this groups' means of sharing information. A study of how Finnish teenagers experience sharing in social media revealed, not surprisingly, that most of the external content that the teens shared consisted of videos and pictures (Ahonen and Hepolehto 2014). Based on the previous research, we assumed the teens would prefer to create visual geonotes.

In order to create suitable tasks for the teens, we began the process by brainstorming with five students, aged 13 years old. We asked them to study a hand-drawn map around their school and pin meaningful places in it with post-it notes (Fig. 8). We advised the group to think about a scenario, where a new student would join their class and they needed to tell him/her relevant places around the school. We were surprised how fluent map users the teens were, even though the map was a rather rough sketch. They could easily find the right spots for different places. However, the participants seemed to know well only a small, limited area around their home, school or places of hobbies. After the session, we were convinced that they would be suitable #hylo users.

Fig. 8 The teens participated actively in the brainstorming session

In addition to brainstorming with the group of teens, we interviewed two of their teachers about what kind of tasks #hylo could be used for in school. In general, the teachers were very positive about the idea of #hylo and convinced it would be useful in various ways. The teachers were especially pleased that #hylo could be used outside the classroom and in the nature, where the students could also improve their knowledge of and relationship with nature. The possibility of combining different school subjects, such as geography and biology or geography, history and language studies were considered a positive addition.

Based on the brainstorming with the teens and the interviews with the teachers, we created tasks for the students as follows: (1) Look for signs of spring and mark them on the map, (2) Find something historical and make a geonote of it, (3) Describe what you find in nature and mark them on the map, such as what is beautiful, ugly, exciting, amazing and so forth, (4) Look for landmarks and make geonotes of them, (5) Make questions for your classmates using geonotes, (6) Answer the questions your classmates made.

We conducted the field studies with a total of 86 teenagers, aged 13–14 years old, in five different schools. We arranged a one and a half hour session during the school day. In the beginning, we gave a short introduction of #hylo and the coming tasks. The students were divided into pairs or groups of three and they were given the tasks and a mobile phone, where #hylo application was installed and ready to use. The students were allowed to go out to nearby forests or parks. After they had conducted the tasks, we gathered back in the classroom. We had a short discussion and asked the students to fill in a questionnaire about using #hylo.

In the questionnaire, we asked the teens how they felt about using #hylo and what kind of feelings it raised. In general, they were very happy about the assignment and thought it was fun. Going outdoors was thought as a nice change to the school day. The students suggested #hylo could be useful in biology and geography, e.g. for finding plants and getting to know the nearby nature, in physical education for orienteering, and in history for getting to know the local past.

As was expected, the teens took a lot of photos and added them to the geonotes (Fig. 9). Many of the tasks also encouraged taking pictures. The students suggested that #hylo should also allow adding videos and sound. As the geonotes are visible to all #hylo users, we emphasised that the teens should not post photos of each other, and in general they should behave well. However, some insulting photos and comments were added and it created bad feelings. We were able to remove the offensive content quickly, but this highlighted the issue of moderating user-generated content.

As the geonotes in #hylo can be explored both on a map and in a feed, we asked the students which one they preferred. The answers show that the map was the most popular way to look for geonotes. Forty-nine out of 76 answers (64 %) preferred the map to the feed. Four students answered they wanted to use both, the map and the feed.

Based on the previous brainstorming session and our experiences, we expected the students to be rather good in orienteering and map usage. We asked how easy it was to find geonotes in nature that their classmates had made and rate it on a scale of 1–5 (1 = very difficult, 5 = very easy), the average was 3.6. Seeing the geonote on its actual location on the map was thought useful. However, when we asked how interesting it was to look for geonotes made by other users, on a scale of 1 to 5 (1 = not at all interesting, 5 = very interesting), the average was only 3.1.

Fig. 9 The teens took photos e.g. from something historical, signs of spring and added pictures to questions such as 'what is this plant?'

As we expect #hylo to increase the interest of an area and shorten the bridge to participate in local events, we asked the teens whether they would go to a new place after seeing an interesting geonote from there. Thirty-four out of 83 (41 %) answered they would go to a new place if the geonote was interesting to them and close by, 37 % said maybe and 22 % would not go.

The majority (50 out of 81) could use #hylo also for something else other than school assignments. They mentioned announcing events and parties, wayfinding, reporting lost items and exploring new places, e.g. on holidays.

The #hylo suited naturally to the school environment, and it was well received. #hylo could be used for very different kinds of assignments. Further studies should be conducted in schools and different types of tasks should be developed in order to have #hylo be adapted to schools more permanently.

8 Future Work

#hylo is now available in the Google Play Store. During the early release and previous user testing phases we have gained useful knowledge about what the users find important in the application. We identified challenges, such as having relevant content and motivating the users.

Including official information from, e.g., local city administration as well as including geo-referenced content from other social media would increase the amount of meaningful and interesting information for different users in #hylo. However, as was stated by Flatow et al. (2015), social media users share billions of items per year, but only a small fraction of them is geotagged. Many popular social media applications, such as Facebook or Instagram, use only predefined place names for georeferencing, and not the actual coordinate based location. This makes the sharing of content on a map even more problematic.

Motivations for content generation in social media inspire significant amounts of research. For example, a recent study of the Finnish web users identified three motives for content production: (1) development of Web ideology and self—a desire to be involved in the development of the Internet and to develop oneself accordingly; (2) self-expression—people, especially the young, want to act independently and freely on the Web and to share information about their lives; and (3) community—people want to belong to online communities and to interact with one another (Matikainen 2015). Additional motivators unique for local knowledge sharing have been identified to be (1) ownership of local knowledge and (2) sense of community (Park et al. 2014). These motivators could be utilised to encourage user contributions and increase user commitment. Future research on motivation in sharing especially hyperlocal information in a geosocial network is necessary for #hylo to be successful.

Conducting user studies with teens in school was an interesting and thought-provoking experiment. Adjusting #hylo to suit even better to the school environment might be worth investigating. The ultimate goal of #hylo is to increase awareness and personal attachment of people to their local surroundings. We expect #hylo to increase (1) Attachment to location, (2) Interest in users' areas, and (3) Social participation in local community. To see if this is true, more user studies are needed in the longer term. More users need to adapt #hylo and only after some time passes, we can see what the effects really are.

9 Conclusions

In this paper, we presented the concept design process of #hylo, a geosocial network for sharing hyperlocal information on a map. After several rounds of brainstorming and needs assessment, we began the iterative mock-up process. As we were developing a map application, a functioning map was found to be essential even in the prototyping phase, especially for the user studies. We conducted initial user studies first with only eight users; nonetheless, the results were valuable and positive. Then user studies with a total of 86 teens were conducted in five different schools. The results were encouraging, and #hylo was well-suited for the school assignments. As we expected #hylo to increase the interest of an area and shorten the bridge to participate in local events, the teens were asked whether they would go to a new place after seeing an interesting #hylo geonote. The results were surprisingly high, as 41 % answered they would go to a new place if the geonote was interesting to them and close by, 37 % said maybe and only 22 % would not go. The teens were fluent map users, and 64 % of them preferred exploring geonotes on a map instead of a feed.

The user studies showed that this kind of hyperlocal information is interesting and the participants found #hylo useful. Future work requires focusing on motivating the users to ensure #hylo has substantial content for various user groups.

Acknowledgments Support for this research was provided by Tekes (the Finnish Funding Agency for Innovation) through a strategic research grant for the MyGeoTrust-project. We would also like to acknowledge the rest of the project team members and the participants of the user studies for their helpful feedback.

References

Ahonen A & Hepolehto I (2014) How teenagers experience sharing in social media. Thesis in Bachelor's Degree Programme in Business Management. Laurea University of Applied Sciences.

Alrayes FS, Abdelmoty AI (2014) No place to hide: a study of privacy concerns due to location sharing on geo-social networks. International Journal On Advances in Security, 7(3/4), 62–75.

Boyd DM, Ellison NB (2008) Social Network Sites: Definition, History, and Scholarship. Journal of Computer-Mediated Communication 13 (2008) 210–230.

Cagan J, Vogel GM (2002) Creating breakthrough products: Innovation from product planning to program approval. Upper Saddle River, NJ, USA: Prentice-Hall.

Carlson CR, Wilmot WW (2006) Innovation: The Five Disciplines for Creating What Customers Want. New York: Crown Business, 2006.

Cockton G (2006) Designing worth is worth designing. In Proceedings of the 4th Nordic conference on Human-computer interaction: changing roles (NordiCHI 2006). ACM, 165–174.

Cooper A, Reimann R, Cronin D (2007) About face, the essentials of interaction design 3. Indianapolis, IN: Wiley Publishing, Inc.

Delikostidis I (2011) Improving the usability of pedestrian navigation systems. PhD thesis, University of Twente.

Espinoza F, Persson P, Sandin A, Nyström H, Cacciatore E, Bylund M (2001) GeoNotes: Social and navigational aspects of location-based information systems. In Ubicomp 2001: Ubiquitous Computing. Springer Berlin Heidelberg, 2–17.

Farrell SJ, Aubry T, Coulombe D (2004) Neighborhoods and neighbors: Do they contribute to personal well-being? Journal of Community Psychology, 32(1), 9–25.

Flatow D, Naaman M, Xie KE, Volkovich Y, Kanza Y (2015) On the Accuracy of Hyper-local Geotagging of Social Media Content. In Proceedings of the Eighth ACM International Conference on Web Search and Data Mining (WSDM'15). ACM, 127–136.

Francis J, Giles-Corti B, Wood L, Knuiman M (2012) Creating sense of community: The role of public space. Journal of Environmental Psychology, 32(4), 401–409.

Gambs S, OHeen O, Potin C (2011) A comparative privacy analysis of geosocial networks. In: Proceedings of the 4th ACM SIGSPATIAL International Workshop on Security and Privacy in GIS and LBS. ACM, 33–40.

Glaser M (2010) Citizen Journalism: Widening World Views, Extending Democracy. The Routledge Companion to News and Journalism, edited by Stuart Allan. London and New York: Routledge.

Guinness RE, Kuusniemi H, Vallet J, Sarjakoski T, Oksanen J, Islam M, Syeed M, Halkosaari H-M, Kettunen P, Laakso M and Rönneberg M (2015) MyGeoTrust: A Platform for Trusted Crowdsourced Geospatial Data. In Proceedings of the 28th International Technical Meeting of The Satellite Division of the Institute of Navigation (ION GNSS + 2015), 2455–2469.

Gupta A, Paul S, Jones Q, Borcea C (2007) Automatic identification of informal social groups and places for geo-social recommendations. International Journal of Mobile Network Design and Innovation, 7 (3/4), 159–171.

Haklay M (Ed) (2010) Interacting with Geospatial Technologies. London: John Wiley & Sons Ltd.

Hu Y, Farnham SD, Hernández AM (2013) Whoo.ly: Facilitating information seeking for hyperlocal communities using social media. In Proceedings of the SIGCHI Conference on Human Factors in Computing Systems, 3481–3490.

International Organization for Standardization (2010) ISO 9241-201. Human-Centred Design Processes for Interactive Systems. Geneva: ISO.

Matikainen J (2015) Motivations for content generation in social media. Participations, journal of audience and reception studies, 12(1), 41–58.

Miettinen S (2012) Discussion on change, value and methods. In Miettinen S, Valtonen A (Eds) Service design with theory (pp. 5–10). Vantaa, Finland: Lapland University Press.

Nivala A-M (2007) Usability Perspectives for the Design of Interactive Maps. Doctoral dissertation. Helsinki University of Technology. Publications of the Finnish Geodetic Institute, N:o 136, Kirkkonummi, 60 p. + 97 p. App.

Nominatim (2016) OpenStreetMap Nominatim Wiki. Website: http://wiki.openstreetmap.org/wiki/Nominatim

Ooms K (2012) Maps, how do users' see them. PhD thesis Ghent University.

Pat B, Kanza Y, Naaman M (2015) Geosocial Search: Finding Places based on Geotagged Social-Media Posts. In Proceedings of World Wide Web Conference (WWW'15), 231–234.

Park S, Kim Y, Lee U, Ackerman MS (2014) Understanding localness of knowledge sharing: A study of Naver KiN "Here". In: Proceedings of the 16th international conference on Human-computer interaction with mobile devices & services (MobileHCI'14). ACM, 13–22.

Radcliffe D (2012) Here and Now: UK hyperlocal media today. London: Nesta.

Rönneberg M, Halkosaari H-M, Laakso M, Hietanen E, Vallet J, Kettunen P, Sarjakoski T (2016) #hylo – Privacy Preserving Geosocial Network for Sharing Hyperlocal Information on a Map. In Proceedings of the 19th AGILE International Conference on Geographic Information Science.

Schobesberger D (2012) Towards a framework for improving the usability of web-mapping products. Doctoral dissertation. University of Vienna.

Walther M, Kaisser M (2013) Geo-spatial event detection in the Twitter stream. Advances in Information Retrieval. Springer Berlin Heidelberg, 356–367.

Xia C, Schwartz R, Xie K, Krebs A, Langdon A, Ting J, Naaman M (2014) CityBeat: real-time social media visualization of hyper-local city data. In Proceedings of World Wide Web Conference (WWW'14), 167–170.

Multimodal Location Based Services—Semantic 3D City Data as Virtual and Augmented Reality

José Miguel Santana, Jochen Wendel, Agustín Trujillo,
José Pablo Suárez, Alexander Simons and Andreas Koch

Abstract The visualization of cross-domain spatial data sets has become an important task within the analysis of energy models. The representation of these models is especially important in urban areas, in which the under-standing of patterns of energy production and demand is key for an efficient city planning. Location Based Services (LBS) provide a valuable addition towards the analysis and visualization of those data sets as the user can explore the output of different models and simulations in the real environment at the location of interest. Towards this aim, the present research explores mobile alternatives to the visual analysis of temporal data series and 3D building models. Based on the fields of numerical simulation, GIS and computer graphics, this work presents a novel mobile service that allows exploring urban models at different Level of Details (LoDs) using well-known standards such as CityGML. Ultimately, the project enables researchers, city planners and technicians to explore urban energy datasets in an interactive and immersive manner as Virtual Globes, Virtual Reality and Augmented Reality. Using models of the city of Karlsruhe, the final service has been implemented and tested on the iOS platform providing an empirical insight on the performance of the system. In addition, this research provides a holistic approach by developing one application that is capable of seamlessly change the visualization mode.

Keywords Augmented reality · Virtual reality · CityGML energy data · Multimodal visualization · GIS · Temporal datasets rendering

---

J.M. Santana · A. Trujillo · J.P. Suárez
University of Las Palmas de Gran Canaria, Las Palmas de G.C., Spain
e-mail: josemiguel.santana@ulpgc.es

J. Wendel (✉) · A. Simons · A. Koch
European Institute for Energy Research (EIFER), Karlsruhe, Germany
e-mail: wendel@eifer.org

© Springer International Publishing AG 2017
G. Gartner and H. Huang (eds.), *Progress in Location-Based Services 2016*, Lecture Notes in Geoinformation and Cartography,
DOI 10.1007/978-3-319-47289-8_17

1 Introduction

The visualization of urban energy modelling and simulation results is a crucial part in energy research as it is the main communication tool among scientists and decisions makers. Results from energy modelling and simulations are directly linked to spatial objects such as buildings or city furniture. Usually, these results are not aggregated by the building object but at a finer scale such as building surfaces. This requires a higher grade of visualization and therefore 3D becomes a necessity (Nouvel et al. 2014). Interactive 3D visualization of those dataset have been successfully implemented before enabling users to explore and comprehend outcomes from multiple perspectives (Klobe et al. 2015; Prandi et al. 2015; Wendel et al. 2016). However, those applications are mainly targeted as 3D web applications and lacking immersive components to truly emerge, explore and interact on-site in the real environment.

This research focuses on the development of a mobile multimodal visualization application for the display of different energy related simulations and modelling results at the building level. In this regard, "multimodal" refers to a holistic approach that implements seamless transition between a traditional virtual map view to Virtual and Augmented Reality modes in one single application. This is particular of interest to experts and decisions makers that would like to explore results on-site through AR or VR but still would be able to switch to a traditional map display for a better overview and strategic planning capabilities. Furthermore, the user is only required to carry one device instead of specialized hardware with them.

The multimodal application is based on the Glob3 Mobile API. This API, presented in Suárez et al. (2012) is a mobile-oriented framework for the development of map and 3D globe applications, being highly configurable on the camera management and the Level of Detail strategies. Thus, the framework is suitable for the present research, having demonstrated recently the possibilities that mobile devices offer for planning of complex infrastructures and large datasets (Ortega et al. 2016; Fernández et al. 2016).

Furthermore, given the complex nature of multidisciplinary energy related modelling, data sets commonly differ in spatial and temporal resolution, data structure or data storage format that requires an intensive data integration workload. Therefore, in order to maintain flexibility all data sets are stored in an open-source data infrastructure that is based on a PostgreSQL database and PostGIS, for spatial capabilities (Simons and Nichersu 2014). A major requirement of this research is the connectivity to a PostgreSQL databases and the CityGML data structure in which building information and energy models are stored, directly from the mobile application.

In the following section, an overview of the usage of immersive and LBS technology in city application is summarized. Sections 3 and 4 describe the data infrastructure and tools used while Sects. 5 and 6 focus on the methodology and implementation strategies used to develop such a multimodal visualization application. Section 7 addresses the visualization of massive temporal simulation data as a use case for the proposed system. Finally, a discussion and future direction is followed in Sect. 8.

2 Immersive LBS in City Applications

LBS have been widely applied in urban environments, and are frequently used for energy planning and research purposes. Most of these applications are focusing either on providing location services, facilitating data collection, or on the analysis of data that has been generated by users through LBS such as social media (Ahas and Mark 2005; Ratti et al. 2006; Anthopoulos and Fitsilis 2010). In addition, immersive technologies that are using LBS, such as AR and VR in urban environments have a long history and go back to early prototypes such as "MARS" by Höllerer et al. (1999) or the "Touring Machine" by Feiner et al. (1997) where immersive technologies were used for the first time for the exploration of urban features. The development of those applications are tightly coupled with major breakthroughs in technological enhancements. The last decade, with the introduction of smartphones, tablet computers, wearable computing devices and VR Gear boosted research and development of those systems. Most of those systems are based on proprietary systems, such as Oculus Rift (Boas 2013), Google Glass[1] or Samsung's VR Gear[2] that require the usage of specific APIs tailored to the hardware. However, moving from one system to another or seamlessly changing the mode of visualization is still limited. For example, Google Inc. is offering users to explore Google Maps Streetview through its Google Cardboard[3] experience as a VR application that allows users to virtually explore places by changing from a map to a VR Street View mode. In contrast, Google Glasses can only be used for displaying features in AR as transparent layers. In the urban context, multiple case studies have been developed using Google Glasses mainly on the theme of navigation and exploration (Tussyadiah 2013; Rehman and Cao 2015). In the field of urban planning, urban design and energy research, VR applications based on Oculus Rift have been successfully applied (Shen 2012; Celik et al. 2013; Heydarian et al. 2015; Lovett et al. 2015; Tanaka et al. 2015). Novel implementations that tried to used existing mapping APIs for VR purposes have been demonstrated by Paolantonio et al. (2015), who developed a VR application for a virtual representation of drone flights using the popular CESIUM 3D[4] web mapping API on an Oculus Rift.

---

[1]https://www.google.com/glass/start/.
[2]http://www.samsung.com/us/explore/gear-vr/.
[3]https://vr.google.com/cardboard/index.html.
[4]https://cesiumjs.org/.

According to Zamyadi et al. (2013), a review of current and past implementations of AR and VR for 3D data show that most of those systems have only used 3D visualization formats such as CAD, 3DS, and textural markup formats such as KML and X3D. However, not many examples that employ structured 3D data models exist, despite the existence of standards such as CityGML that are crucial for simulation outputs. Other surveys such as Biljecki et al. (2015) point out to a few uses of CityGML in VR and AR showcases, but not in an multimodal fashion. Instances that do use standardized data are for example the visualization of wind flow in CityGML as AR by Heuveline et al. (2011) or the AR visualization of underground structures by Schall et al. (2010). However, to our knowledge, the conjunction of the three visualization modes, map mode, virtual and augmented reality modes in the same mapping API using standardized data structures has not been developed before.

3 Data Infrastructure, Energy Simulations and 3D City Models

The main aim of this research is the multimodal visualization of energy simulation data at the building level, a flexible data representation that can host cross-domain data sets in multiple data formats is required. Furthermore, the data infrastructure needs to be flexible enough not only to run and store urban energy simulation data but also to allow the connectivity to mobile computing devices at the same time.

3.1 Data Infrastructure and Energy Simulations

All energy models and their corresponding geographic abstraction levels (e.g. buildings) are stored in a PostgreSQL database using the CityGML standard. CityGML is an open data model standard developed by the OGC (Open Geospatial Consortium). A benefit of using the CityGML standard instead of other geospatial data formats is that semantic information of each surface or element of a building can be stored. Furthermore, object-modelling specifications can be shown at different Levels of Detail (LoD) (Döllner et al. 2006). Furthermore, the concept of LoDs can be used not only for semantic object abstraction but also for cartographic visualization purposes. For efficient central data storage of information from all energy models and spatial analysis tools, the 3DCityDB,[5] a simplified CityGML database is used (see Fig. 1) (Simons and Nichersu 2014). In addition, the usage of PostGIS within this infrastructure enables the usage and handling of spatial

---

[5]https://github.com/3dcitydb.

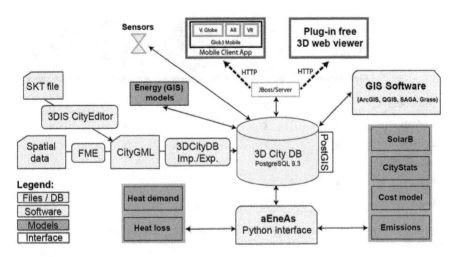

Fig. 1 Infrastructure and workflow

functions that are required for visualization model outputs at different visualization modes.

The open-source data infrastructure (Fig. 1) consists of storage (PostgreSQL, 3D City DB and the PostGIS extension), analysis tools (aEneAs, an energy balance model based on convex hull LoD2 building model approach (Simons and Nichersu 2014). The aEneAs tool includes the energy models whose outputs are used in this research. Those models include SolarB, a Python application developed for city-wide computation of solar radiation received by buildings' wall and roof surfaces based on spatial vector data (Wieland et al. 2015), CityStats, a socio-demographic clustering model for energy usage (Saed and Wendel 2015) and the heat demand and heat loss model (Simons and Nichersu 2014). The here presented research will use results for all spatial energy models to be displayed in a multimodal framework including AR and VR modes as well as mobile 3d mapping using a single API (Wendel et al. 2016). A JBoss[6] web server via HTTP protocols realizes data transmission between the server and the mobile client.

Each of the energy models above produce results that can be directly linked to spatial attributes ranging from point clouds to polygon features. Multiple static and dynamic visualization strategies could be used for each type of result. Figure 2 gives an overview of the model results and its spatial abstraction used in this research.

---

[6]http://www.jboss.org/.

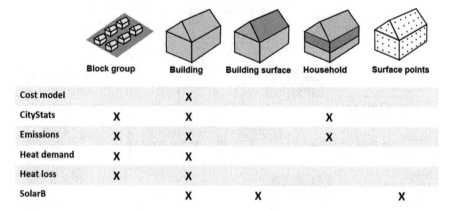

Fig. 2 Visual overview of aggregation scales of the aEneAs energy model outputs used for visualization

3.2 3D City Models

For displaying energy model outputs at the building level, 3D buildings models are required. In this research, the city of Karlsruhe in Germany is used as a case study area. Karlsruhe is a medium sized city of 300,000 inhabitants located near the Franco-German border in the south-western part of Germany. The city does not provide any public access to its 3D spatial data. Although building footprints were available for the whole city, the height information was missing or unusable for the objective of this experiment. Instead, the height information of each building was estimated by the number of floors of each building obtained from areal imagery. In total 86,000 buildings were generated. For each floor, a standard height value in meters, depending on the building type, is used to estimate the total height of a building. Validation of the generated 3D model through a stratified random sampling process showed high accuracy levels of 92.4 % within a 95 % confidence interval for the estimated building heights of the whole 3D city model of Karlsruhe (Saed and Wendel 2015).While the 3D city model was originally created as a LoD1 model, for the purpose of this study all LoD1 buildings have been transformed into LoD2 building models for storage of the surface taxonomy information. Some places of interest (POIs) such as the Karlsruhe Castle are included in the visualization as high detailed models as a proof of concept.

4 Multimodal Visualization with Glob3 Mobile

A main requirement, besides the connectivity to standardized semantic data sources, is the possibility of seamless integration of multimodal LBS visualizations including map, AR and VR display modes. The open-source Glob3 Mobile (G3M)

framework was chosen for this research as no proprietary APIs or mobile hardware is required. The G3M project allows the generation of map applications in 2D, 2.5D and 3D following a zero third-party dependencies approach and provides native performance on its three target platforms (iOS, Android, HTML5) (Trujillo et al. 2013). 3D graphics are supported by the Khronos Group APIs, OpenGL ES 2.0 on portable devices and WebGL (web counterpart of OpenGL) on the HTML5 version. Features of this API include multi-LoD-3D rendering and automatic shading of objects (Suárez et al. 2015; Trujillo et al. 2014). However, up to this date no VR and AR modes for structured data had been implemented in G3M. A major part of this research was dedicated to the development and integration of new features for supporting seamless VR and AR modes to the G3M application core and are describes in the following sections.

4.1 Virtual Reality Support

The VR mode replicates the map scenario as a real environment in which the presence of the user is simulated. In order to achieve a realistic feeling of immersion, the device must respond to the changes in position and attitude of the device, showing the 3D scene that is theoretically behind it.

G3M includes multiplatform support for tracking the orientation of the mobile device, as well as, its location in the geographical space, using the API for accessing accelerometers, magnetometers and GPS antennas.

At every frame, the final attitude and location of the virtual camera is achieved by a combination of three transformations.

- Rotation of n * 90° of the 3D viewport (M_{UI}) within the device's inner coordinate system.
- Relation of the device's inner coordinate system with the global coordinate system. This transformation (M_{DEV}) is a free rotation that tracks the attitude of the device in its local space. Both iOS and Android system provides this value as a 4 × 4 matrix, even though their inner coordinate systems HTML5 Orientation API (Bolstad et al. 2011) generates an attitude description based on Euler angles, which G3M combines on a single matrix.
- Geolocation on the 3D scene. The camera needs to be placed on a particular latitude, longitude and height. These coordinates are a translation (M_G) of the global coordinate system of the device in scene space. The Z axis is oriented towards the terrain's normal direction through a general rotation matrix that depends on the kind of virtual globe that we are using (flat, spherical, ellipsoidal…).

These three matrix transformations are linearly combined in a single matrix and applied to the virtual camera before every frame for high responsiveness of the system. Nonetheless, the attitude and location tracking is affected by typical elements of urban environments such as walls or electronic equipment. The limitations of these measurements, although sufficient for our case study, have been analyzed in other works (Blum et al. 2013).

4.2 Stereo Rendering

By tracking the location and attitude of the device the user can consider the screen as a small window through which is possible to explore the 3D map. However, the experience is still not completely immersive. Therefore, stereoscopic images will cover the field of view of both eyes. In the present work, the client implements stereo rendering that creates on a single screen separated images for both eyes. This stand-alone stereo rendering makes the app compatible with any Cardboard-like VR set by generating two sub cameras with parallel view directions that are separated by the interaxial separation. As the stereo is intended to run on a realistic VR environment such separation matches the natural human interocular distance, that is about 6.3 cm (Rosenberg 1993). In this research, the visibility of the objects in the scene from a central point of view using a frustum that includes both sub-frustums, as seen in Fig. 3, is calculated. In addition, the user position is used for the terrain LoD calculation, computing the scene as a view independent rendering target (Gateau 2009).

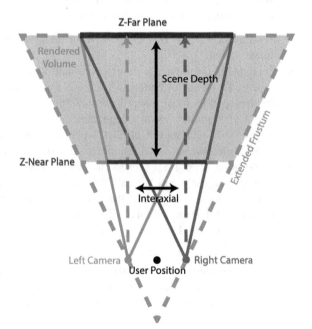

Fig. 3 Composite frustum used for visibility checking in stereo rendering

4.3 Augmented Reality Setup

In this implementation, the AR mode uses the rear camera of the mobile device and adds to the captured image the elements present in the urban scenario as well as all visible buildings and data (e.g. point clouds) from the energy model outputs. The AR image fusion has been implemented in iOS by using a two viewport strategy. The rear viewpoint contains an iOS *AVCaptureVideoPreviewLayer* that updates frame by frame the image captured by the camera, covering the background of the whole UI. In addition, the G3M widget shows a frame buffer of 32 bits, which allows the inclusion of an alpha channel and the use of transparencies for merging the rendered image with the underlying camera viewport. In order to combine real image with synthetic additions the parameters that define the camera of the virtual scene (field of view, sensor offset, distortions), must match the physical configuration of the real camera (Azuma et al. 2001). In this research, the vertical and horizontal fields of view of the virtual 3D camera have been adjusted to the values present on the iOS device camera. These values vary even among devices of the same family and vendor, which normally would require measurements of the captured image of a known scene. However, since iOS v7.0, the operating system informs the application about the horizontal field of view of the current device through the property *AVCaptureDeviceFormat::videoFieldOfView* (Apple Inc. 2014a). With this value, and the aspect ratio of the captured image, the vertical FOV can be derived by:

$$\alpha_V = \arctan\left(\tan(\alpha_V)\frac{I_H}{I_W}\right) \quad (1)$$

where α_V descries the horizontal field of view and I_H and I_W the aspect ration of the captured image in pixels (see footnote 2).

5 Client Design

5.1 Multimodal Visualization

The spatial data and energy model outputs stored as CityGML in the PostgreSQL database can be applied in different environments. In this research, at any given moment, the user can switch between four different modes (1. Map mode, 2. VR mono, 3. VR stereo, and 4. AR mode) allowing the user to explore the dataset in the most suitable way and in different environment using the same mobile device (Table 1).

Fig. 4 Map mode displaying CityGML buildings on G3M

5.1.1 Map Mode

The default visualization mode displays a classical virtual globe scenario, based on the WGS84 earth model, in which the 3D model of city of Karlsruhe is loaded at the beginning of the execution. Once the 3D model is visible, the camera is animated from whichever place it is to a position where the whole city model is visible. The main goal of this visualization mode is to display the user the city model as a whole for familiarization of the study area and to explore for example urban planning scenarios or the outputs from energy models at city level scales. Different coloring and aggregation schemes for building models (described in Sect. 6.3) enables to study different model and simulation outputs seamlessly. Web Map Service (WMS) layer support has been implemented for base map displays such as OpenStreetMap (OSM), which offers a non-commercial free-to-use WMS service (Wiki 2013). During the execution of the map mode, the user can navigate freely the map using the navigational multitouch control. These gestures allow the user to control the position of the camera in geographical space, as well as its pitch and heading (Fig. 4).

5.1.2 Virtual Reality Mono Mode

The virtual reality mode aims to create an interactive experience in which the user can use his mobile device to explore in first person the surroundings of its actual location. The VR mode has been successfully tested and implemented for exploration of building energy performance model outs such as heat loss and emissions visualizations. The VR and AR modes, rely heavily on tracking the device's location and attitude as described in Sect. 4.1. Transitioning to and from the map mode to VR Mono mode is realized through a camera animation that lets the user assess its current location within the virtual scenario. Once there, holding the mobile device at a natural distance the user can explore the different building

Fig. 5 Virtual reality mode with mono rendering

models that populate the area. At such low height, the user is unable to read any label placed on the map surface. Thus, the OSM data labeled imagery is replaced with Bing Maps areal imagery WMS for optimized visualization. While any kind of satellite image is blurry at this scale level, it still shows more detail than a web map display and furthermore mediate a more realistic feel that is required for VR applications (Fig. 5).

5.1.3 Virtual Reality Stereo Mode

Similarly, to the VR Mono mode, the virtual reality stereo mode presents an interactive experience to the user that allows exploring the map and the city model in first person. However, this time the experience is fully immersive through the usage of a VR set such as Google's Cardboard that isolate the user from the rest of the environment. This mode uses the techniques described in Sect. 4.2 to generate two stereoscopic images of the 3D scenario. These images are then projected through the lenses of the VR headset to the eyes of the user, generating a realistic depth perception. All other elements of the user interface are hidden so they do not interfere with the binocular perception (Fig. 6).

Fig. 6 VR stereo rendering and Google's Cardboard headset

5.1.4 Augmented Reality Mode

This final mode of the viewer combines the image captured by the camera of the mobile device and the 3D image rendered by Glob3 Mobile. These two images combined generate a scenario in which the actual buildings are covered by their 3D models that display relevant semantic information to the user (Fig. 7). The possibilities and limitations of different AR systems have been extensively discussed in the literature (Krevelen and Poelman 2010). In our particular use case, this setup allows the user to enhance the perception of their environment by adding to it meaningful information about data sets describing the urban environment. In this research a solar radiation model was used as a case study (Sect. 7). Notably, in this last viewing mode the terrain is no longer rendered. The view of the terrain and other nearby objects is therefore replaced by the actual image of the objects, and only the semantic CityGML building model and data from the energy model remain.

The following table summarizes the main differences of the different modes of the viewer

Fig. 7 Augmented reality scene recorded in Karlsruhe

Table 1 Attribute table of different visualization modes

Feature/mode	Map	VR mono	VR stereo	AR
Terrain rendering	x	x	x	
Multitouch navigation	x			
Touch building selection	x	x		x
Attitude tracking		x	x	x
Location tracking		x	x	x
Rear camera usage				x
VR set needed			x	

5.2 User Interface

The client application is designed to present a single-view UI containing all the elements necessary to visualize the 3D scene and interchange the current mode of visualization. This main view is composed of the following elements, as presented in Fig. 8.

- *Header*. Contains the logos of the University of Las Palmas de G.C. and the European Institute for Energy Research (EIFER). During the rendering of a time dependent data series, a label shows the current timestamp.
- *Wait message*. During the loading process the app allows the interaction so the user can navigate through the virtual globe. However, a message indicating the progress of the processed buildings is shown. Once the whole data set is loaded the virtual camera centers on the map, offering a view of the data set.
- *Glob3 Mobile widget*. The background of the UI is composed by the 3D viewer, following the premise of "content deference" (Apple Inc. 2016a). It shows the virtual scenario and changes reactively to the commands performed in the menu. If no other element responds to a touch event it is captured by the G3M engine, which may use it to trigger actions such as navigation or objects selection.

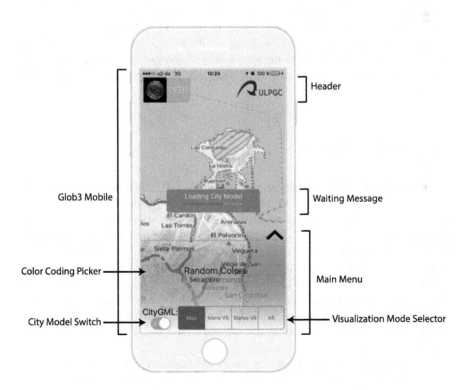

Fig. 8 User interface layout design

- *Slide-up menu.* This menu is placed at the bottom of the screen and becomes active by clicking on an arrow button. It includes the following widgets:
  - A scrollable picker menu featuring a list of data sets available for display. Selecting one of them triggers a recoloring of the 3D buildings according to the selected value, as described in Sect. 6.3.
  - A switch that enables/disables the rendering of the city model, which allows the user to see the underlying terrain map or real image.
  - A scope bar (Apple Inc. 2016b) that allows to switch between the different visualization modes of Sect. 5.1.

6 iOS Client Implementation

As a proof of concept, a first implementation of the application has been developed and tested on iOS and is aimed at any iOS device (iPhone, iPad and iPod Touch) for versions of iOS 9.0 or newer. However, some features, such as the rendering capabilities and the location tracking accuracy may vary from device to device. All the experiments and numerical tests regarding the software performance have been obtained by conducting experiments on an iPhone 6 smartphone (64 GB), running iOS 9.3.2. Due to the multiplatform nature of the G3M API portability to Android or HTML5 is possible. A major contribution of this research is the transformation and visualization of CityGML for multimodal visualizations. The following sections will focus on the processing steps necessary for displaying CityGML in VR and AR.

6.1 CityGML Parsing

In order to visualize the values of the energy model outputs from the CityGML database, the values have to be fetched, stored and parsed first. For intelligent processing, the data set of the whole city has been split in manageable data sets ranging from 10 to 40 MB. To avoid bottlenecks in data transfer and processing, a caching strategy is applied to improve the download speed from the JBoss server. Synchronous execution would have the undesirable effect of blocking the app, making it unresponsive and impeding any kind of progress notification to the user. Hence, the client application has been designed to delegate these blocking operations to different threads using iOS Grand Central Dispatch (Apple Inc. 2014b). The multithreaded sequence of fetching, parsing and tessellating the model is depicted in Fig. 9.

Even when the parsing of the city model has been delegated to a non-blocking thread, the memory and time consumption had to be minimized to make it feasible for a mobile device. As a consequence, an implementation study was performed to

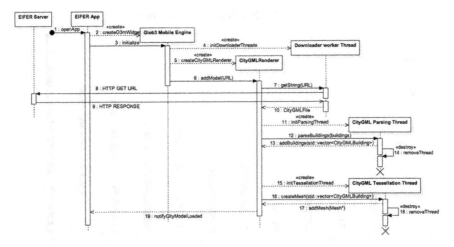

Fig. 9 Sequence diagram of the multithreaded loading of a CityGML model

compare the two main parsing strategies for XML data formats, known as Document Object Model (DOM) (World Wide Web Consortium 2006) and Simple API for XML (SAX) (Megginson 2004) respectively. Both strategies have iOS implementations using LibXML (Veillard 2009) and NSXML respectively. DOM allows parsing the whole CityGML document using barely four XPath (Clark and DeRose 2010) queries per building, whereas SAX implies a lower level and harder to maintain parsing strategy. However, benchmarks showed a critical advantage of SAX over DOM regarding the processing time (DOM: 2.27 ms/Building in contrast to SAX: 0.78 ms/Building), as seen in Fig. 10. Moreover, DOM is unable to parse a document with more than 23,000 buildings due to the limited heap memory on an iPhone 6.

This analysis confirmed that SAX was needed for parsing large amounts of data. Furthermore, SAX features other desired properties such as stream parsing, scalability and lower garbage generation on *JVM.CityGML* building object model. In order to visualize the CityGML data structure a phrasing of the data is required. The parsing phase extracts the basic features of these models that are of use and converts them to an object representation. Each building is represented by a *bldg:Building* node, containing several *gml:Multisurface* instances that represent the boundary surfaces of the structure. The UML diagram of the parsed buildings is depicted in Fig. 11.

As can be seen in Fig. 11 the *CityGMLBuidling* class is a container of three kinds of information: surface data, property data and tessellation data. The most prominent information in these data is the surface information extracted from the CityGML model. The class *CityGMLBuildingSurface*, contains geometrical information used by the tessellation process. The coordinates are extracted from the *gml::Surface* (defined in the EPSG:4326 coordinate system) from its *gml::posList*

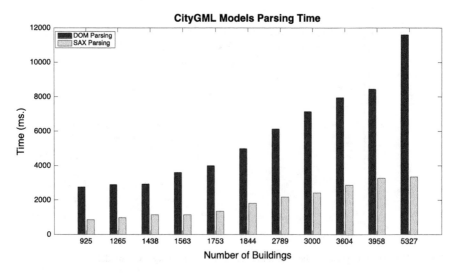

Fig. 10 DOM versus SAX parsing performance for CityGML documents

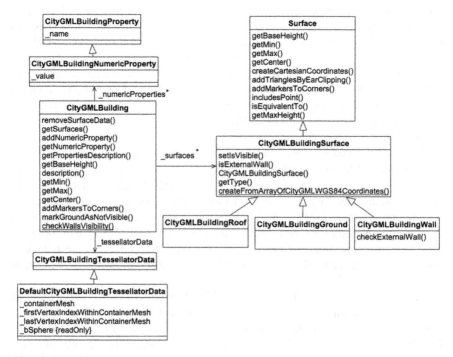

Fig. 11 CityGMLBuilding object diagram

attribute. Just as in the CityGML model, surfaces are classes modeled as Roof, Ground and Walls.

The differentiation at object level of each kind of surface allows to implement different behaviors at the tessellation stage. For instance, Ground is considered to be a non-external wall so it can be dismissed at the tessellation of external 3D models. In a similar way, Roofs are always visible and Walls may be checked against neighboring buildings. Different strategies might be also implemented in the future, regarding the coloring, selection and manipulation of such surfaces.

Property data are values that can be associated to a specific instance of building. These values are extracted from the CityGML model itself or in subsequent data acquisitions. In this research, all the stores properties have numerical values, which are used in the recoloring of the city model as described in Sect. 6.3. Finally, after the tessellation, each building is attached to an instance of *CityGMLBuildingTessellatorData* that is implemented by the *CityGMLTessellator*. The current implementation includes a pointer to the part of the city model that represents the building that is accessed during the recoloring task. In addition, a bounding sphere is associated to the building that is used for the building selection.

6.2 CityGML Buildings 3D Tessellation

One of the main challenges in this research is the inclusion of a client-based tessellation technique that transforms the CityGML LoD2 models into a triangle mesh usable by the OpenGL ES 2.0 API. The applied LoD2 schema description is well defined in the CityGML specification (Gröger et al. 2008) and it is sufficient for the purposes of this work. However, the definition is sufficiently vague for allowing many representations of the same LoD and several changes have been proposed (Biljecki et al. 2016).

For optimized tessellation, the external visibility of the building surfaces is checked eliminating ground surfaces and shared facades from the 3D model. The first step in the tessellation task is the transformation of coordinates to Cartesian space. This transformation is performed using the current planetary model that shapes the terrain model on G3M. The terrain model must also include a DEM displacement with no vertical exaggeration, in order to fit the altitude of the building model. The Cartesian coordinates of each wall are then represented as a 3D polygon that has to be triangulated. However, the triangulation problem is inherently a 2D problem. Hence, the 3D polygon is firstly represented in the 2D subspace of its containing plane.

The ear clipping strategy has been implemented as the tessellation algorithm. This technique offers good results on mobile devices, using an implementation based on a doubly linked chain of vertices, similar to the one described in Eberly (2002). This method, based on the two ears theorem (Meisters 1975) simplifies iteratively the polygon by extracting outer triangles from the shape. The order of this algorithm is $O(n^2)$ whereas some other approaches might offer a better performance. For instance, the expected running time of the Seildel's algorithm is O (n log * n + k log n). However, considering the simplicity of most surfaces, a lighter implementation such as the ear clipping method may be more performant. Another key point for improving the tessellation performance is the rectangular nature of buildings. In fact, in the 3D model of Karlsruhe, 91 % of the city surfaces are rectangular. In those cases, checking the convexity of the polygon allows us to derive the mesh without running the ear-clipping algorithm.

A by-product of the tessellation process is a bounding sphere, attached in the process to each building that is used as a selection mechanism for enhanced touch screen support.

6.3 City Model Coloring

The generated mesh model of the city each building is colored according to the selected energy simulation output. Each option corresponds to a particular numerical value stored in the building set. For optimal color usage ColorBrewer and Jenks Natural breaks for optimal class, selection has been implemented (Harrower and Brewer 2003; Jenks 1977). Once the color and class schema is determined, it is integrated on the VBO of the mesh for every vertex of the building.

6.4 Rendering

6.4.1 CityGML Building Mesh Rendering

Subsequently, to the tessellation and coloring described in Sects. 6.2 and 6.3 the mesh data has to be packed for GPU processing. In our current implementation, a non-photorealistic light model has been applied for the rendering of the city model, based on the Phong shading (Phong 1975). The implemented shading strategy applies to each surface a combination of the ambient light color with a diffuse component coming from a directional source. Hence, each vertex must contain the normal of the surface to which it belongs. The tone of each fragment is calculated as described by a simplified version of the Phong reflection model:

$$I_P = k_a i_a + k_d (\hat{L} \cdot \hat{N}) i_d \qquad (2)$$

where k_a and i_a are the reflection constant and intensity of the ambient light, k_d and i_d represent the same parameters of the directional light, $\hat{L}$ the vector from the rendered fragment to the light source and $\hat{N}$ the surface's normal vector at that point.

The rendering of the building models of higher LoDs is performed using a flat shading, as the textures associated to such models normally include a natural look illumination. Figure 12 shows screenshots of the high definition model of the castle of Karlsruhe in Map and AR mode as proof of concept.

6.4.2 City Model Rendering Performance

In order to minimize the drawing overhead due to the use of G3M and the underlying OpenGL ES 2.0 API, the city mesh has been grouped into groups of buildings of up to 3000 vertices each. This batch rendering of building models avoids CPU-induced bottlenecks as described in Trapp (2004).

An experiment comparing the rendering times of isolated buildings to the batch rendering shows that, for the used hardware, the latter outperforms the first by a factor of $\times 49$ (0.078 ms per batched building vs. 3.827 per isolated buildings). With a performance goal of 30 FPS, the system would not be able to render isolated buildings for a city bigger than 250 buildings, as seen in Fig. 13. Other strategies could be followed in order to improve the performance of the drawing commands, such as distance ordering (Trapp 2004) which leads to a better performance of the fragment shader or the usage of interleaved Vertex Buffer Object (Civil 2008). However, the video memory of the mobile device seems to be the main restriction limiting the possible size of our city model. A clever LoD strategy or a more compact modeling of the city objects are, in this regard, promising lines of future work.

7 Dynamic Point Clouds from Energy Models

As a proof of concept for dynamic multimodal data visualization, outputs from a vertical solar radiation energy model (SolarB) have been implemented. The solar radiation data model consists of massive sets of points associated to a time series that describe the intensity of the received sunlight at a 1 m resolution (Wieland and Wendel 2015). The time series covers a whole year and is composed of a radiation value per point averaging an hour of sunlight. This form of visualization is particularly useful as AR as the user can explore solar gains and façade shadowing of potential solar installations as animations on site.

In order to visualize and animate the time series data, the user selects the building of interest. Once the selection occurs, the user is prompted with a command that triggers the visualization of the point cloud associated with the selected building (Fig. 14).

To display this solar radiation model, the geometry model of the building becomes invisible, being replaced by a point cloud. The vertices of such point clouds are stored server-side and retrieved by the client on demand. With the geometry in place, a periodical task scans the data series converting the radiation

Fig. 12 High definition model of Karlsruhe's castle view on map mode (*left*) and AR mode (*right*)

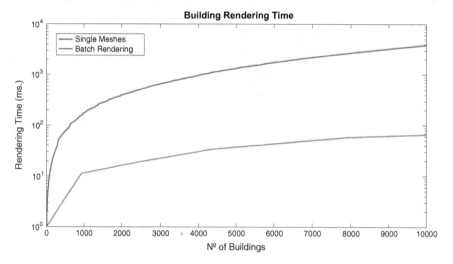

Fig. 13 Rendering time of individual building meshes versus batched buildings

values to colors using a linear interpolation. This evolution is accelerated by a factor of ×3600, so an hour of data is displayed during one second. The point cloud mesh contains the color values for the previous timestamp and the next. At every frame, the rendering time is sent to the GPU and a linear interpolation between the two values is applied to the point colors. Each second, the data for an hour slot is fetched from the server. A preprocessing stage has identified the nighttime values, eliminating the need for processing them for the solar energy model visualization. The fragment shader given the point position and size, discards all the fragments beyond a given point radius enabling a rapid circle-shaped point rendering. Regarding the rendering of massive point clouds, an interesting addition to this pipeline would be an adaptive circle radius in order to simulate a perspective rendering and enhance the depth perception.

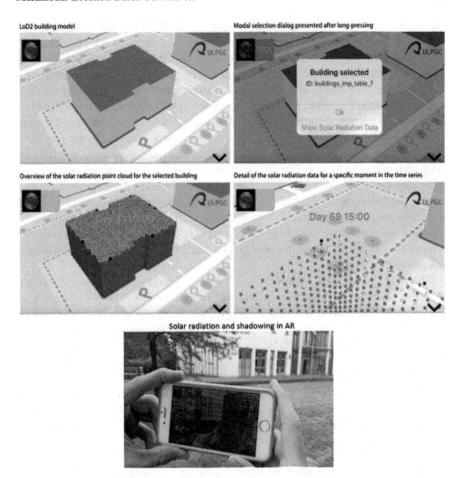

Fig. 14 Dynamic solar radiation point cloud visualization process

8 Discussion

This research has developed a novel showcase of multiple immersive location based mobile modes of energy modelling for the city of Karlsruhe. In this proof of concept, the client application fetches, tessellates and integrates data up to a LoD2 CityGML urban model. This virtual scenario serves as the basis for the visual analysis of massive temporal datasets, in which the simulation of solar radiance models has been represented. The use of location based multimodal visualization techniques enables intuitive access and interaction to complex energy simulations that are highly comprehensible to a wide audience through the real-life camera view as AR visualizations and immersive VR renderings. The results of numerical simulations will be available to decision-makers and citizens, raising the impact and improving the communication of scientific results. While other studies have, implemented 3D mapping technologies such as CESIUM for AR representations (Paolantonio et al. 2015) before, to our knowledge the seamless integration of AR and VR and virtual globe displays has not been undertaken so far.

The performance analysis of this system has revealed the importance of using a lightweight, multithreaded and error resilient approach to the phases of parsing and tessellation of the CityGML model. In this regard, SAX offers a feasible alternative, in contrast to more traditional DOM parsing strategies. Besides, the "Ear Clipping method" proved to be a good choice for the tessellation. Ultimately, the rendering of these massive models seems to be limited by the graphics memory available on modern mobile hardware. The major limitations of these systems on VR and especially on AR modes are the accuracy of GPS, accelerometers and magnetometer sensors embedded in the device. However, upcoming advances in hardware and other positioning techniques might reduce these drawbacks. Nevertheless, the accuracy and available memory of current devices seemed sufficient for the exploration of static and dynamic energy simulation models as the ones presented in this research.

The usage of VR and AR for the exploration of energy simulation outputs has been successfully demonstrated for multiple energy models. The VR modes have been successfully implemented for energy planning purpose where users could explore the heat energy loss and gains model in an immersive environment. However, as only a LoD2 CityGML model was implemented as a proof of concept a higher more detailed abstraction level is necessary in future versions for a more realistic display. The AR mode was successfully tested for the dynamic visualization of point clouds generated by a vertical solar radiation model (SolarB). The user was able to select a building and visualize solar gains and the shadowing effect of neighboring buildings as animations on-site. This use case sparked great interest, as it is a visual way to communicate the benefits and drawbacks of potential future solar panel installations. Furthermore, the seamless switch from map mode to VR or AR mode was beneficial for users when exploring the study site in the real environment. Without changing from one device or application to another, the user could rapidly change the modes seamlessly and in consequence could more efficiently analyze the data.

Future research will focus on expanding the applicability of the system, integrating new urban models and simulation outputs to the service. A focus will be on the integration of new extensions in data standards such as the CityGML Energy ADE[7] (Application Domain Extensions) that includes energy specific attributes or the OGC SOS[8] (Sensor Observation Service) for standardized processing of data received from sensors.

Moreover, other visualization techniques such as multi-LoD rendering could be applied in order to process bigger datasets. The VR and AR modes could also be enhanced by improving the position tracking (for instance by using visual

---

[7]http://en.wiki.energy.sig3d.org/index.php/Main_Page.

[8]http://www.opengeospatial.org/standards/sos.

landmarks) and adding camera effects such as lens distortions to the 3D rendering. In future projects, it would be also of great interest to include other data sources such as Internet of Things (IoT) real-time data or on-demand simulations in the system.

Acknowledgments The first author wants to thank *Agencia Canaria de Investigación, Innovación y Sociedad de la Información*, and the *European Social Fund*, for the grant "Formación del Personal Investigador-2012" that made possible this work.

References

Ahas, R. and Mark, Ü. (2005). Location based services—new challenges for planning and public administration? *Futures, 37* (6), 547–561. doi:10.1016/j.futures.2004.10.012
Anthopoulos, L. and Fitsilis, P. (2010). From Digital to Ubiquitous Cities: Defining a Common Architecture for Urban Development. *2010 Sixth International Conference on Intelligent Environments* (S. 301–306). IEEE. doi:10.1109/IE.2010.61
Apple Inc. (2014a). AVCaptureDeviceFormat Class Reference.
Apple Inc. (2014b). Grand Central Dispatch (GCD) Reference.
Apple Inc. (2016a). iOS Human Interface Guidelines. *IOS Developer Library*. http://developer.apple.com/library/ios/#documentation/UserExperience/Conceptual/MobileHIG/Introduction/Introduction.html. Accessed 10 June 2016
Apple Inc. (2016b). iOS Human Interface Guidelines: Bars. https://developer.apple.com/library/ios/documentation/UserExperience/Conceptual/MobileHIG/Bars.html#//apple_ref/doc/uid/TP40006556-CH12-SW1. Accessed 10 June 2016
Azuma, R., Baillot, Y., Behringer, R., Feiner, S., Julier, S. and MacIntyre, B. (2001). Recent advances in augmented reality. *IEEE Computer Graphics and Applications, 21* (6), 34–47. doi:10.1109/38.963459
Biljecki, F., Ledoux, H. and Stoter, J. (2016). *An improved LOD specification for 3D building models. Computers, Environment and Urban Systems* (59). doi:10.1016/j.compenvurbsys.2016.04.005
Biljecki, F., Stoter, J., Ledoux, H., Zlatanova, S. and Çöltekin, A. (2015). Applications of 3D City Models: State of the Art Review. *ISPRS International Journal of Geo-Information, 4* (4), 2842–2889. Multidisciplinary Digital Publishing Institute. doi:10.3390/ijgi4042842
Blum, J.R., Greencorn, D. and Cooperstock, J.R. (2013). Smartphone sensor reliability for augmented reality applications. *Mobile and ubiquitous systems: computing, networking, and services*, 127–138. doi:10.1007/978-3-642-40238-8_11
Boas, Y. (2013). Overview of virtual reality technologies. *Interactive Multimedia Conference 2013*.
Bolstad, L.E., Jackson, D., Nilsson, C., Percivall, G. and Turner, D.M. (2011) DeviceOrientation Event Specificaiton. W3C Working Draft. https://www.w3.org/TR/2011/WD-orientation-event-20110628/. Accessed 22 August 2016.
Celik, B., Karatepe, E., Gokmen, N. and Silvestre, S. (2013). A virtual reality study of surrounding obstacles on BIPV systems for estimation of long-term performance of partially shaded PV arrays. *Renewable Energy, 60*, 402–414. doi:10.1016/j.renene.2013.05.040
Civil, A. (2008). Best Practices for Working with Vertex Data. *Data Management*.
Clark, J. and DeRose, S. (2010). XML Path Language (XPath). Engineering. https://www.w3.org/TR/xpath/ Accesed 27th September 2016.
Döllner, J., Kolbe, T.H., Liecke, F., Sgouros, T. and Teichmann, K. (2006). The Virtual 3D City Model of Berlin-Managing, Integrating, and Communicating Complex Urban Information. *Proceedings of the 25th Urban Data Management Symposium UDMS*, (May 2006), 15–17.
Eberly, D. (2002). Triangulation by ear clipping. *Magic Software, Inc*, 1–13.

Feiner, S., MacIntyre, B., Höllerer, T. and Webster, A. (1997). A touring machine: Prototyping 3D mobile augmented reality systems for exploring the urban environment. *Personal Technologies*, *1* (4), 208–217. Springer-Verlag. doi:10.1007/BF01682023

Fernández, P., Santana, J.M., Ortega, S., Trujillo, A., Suárez, J.P., Domínguez, C. et al. (2016). SmartPort: A Platform for Sensor Data Monitoring in a Seaport Based on FIWARE. *Sensors*, *16* (3), 417. Multidisciplinary Digital Publishing Institute. doi:10.3390/xx

Gateau, S. (2009). 3D Vision Technology - Develop, Design, Play in 3D Stereo. *Acm Siggraph 2009*.

Gröger, G., Kolbe, T.H., Czerwinski, A. and Nagel, C. (2008). OpenGIS City Geography Markup Language (CityGML) Encoding Standard, Version 1.0.0. *OGC Document No. 08-007r1*, 234.

Harrower, M., and Brewer, C. A. (2003). ColorBrewer.org: an online tool for selecting colour schemes for maps. *The Cartographic Journal*, *40*(1), 27–37.

Heuveline, V., Ritterbusch, S. and Ronnas, S. (2011). Augmented Reality for Urban Simulation Visualization. *INFOCOMP 2011 : The First International Conference on Advanced Communications and Computation*, (c), 115–119.

Heydarian, A., Carneiro, J.P., Gerber, D., Becerik-Gerber, B., Hayes, T. and Wood, W. (2015). Immersive virtual environments versus physical built environments: A benchmarking study for building design and user-built environment explorations. *Automation in Construction*, *54*, 116–126. doi:10.1016/j.autcon.2015.03.020

Höllerer, T., Feiner, S., Terauchi, T., Rashid, G. and Hallaway, D. (1999). Exploring MARS: developing indoor and outdoor user interfaces to a mobile augmented reality system. *Computers and Graphics*, *23* (6), 779–785. doi:10.1016/S0097-8493(99)00103-X

Jenks, G. F. (1977). Optimal data classification for choropleth maps. University of Kansas. Dept. of Geography.

Klobe, T. H. Burger, B., and Berit, C. (2015). City GML goes to Broadway. Photogrammetric Week '15, Wichmann, 2015, 343–356

Krevelen, D.W.F. van and Poelman, R. (2010). A Survey of Augmented Reality Technologies, Applications and Limitations. *The International Journal of Virtual Reality*, *9* (2), 1–20. doi:10.1155/2011/721827

Lovett, A., Appleton, K., Warren-Kretzschmar, B. and Von Haaren, C. (2015). Using 3D visualization methods in landscape planning: An evaluation of options and practical issues. *Landscape and Urban Planning*, *142*, 85–94. doi:10.1016/j.landurbplan.2015.02.021

Megginson, D. (2004). SAX. http://sax.soursforge.net. Accessed 15 June 2016

Meisters, G.H. (1975). Polygons have ears. *American Mathematical Monthly*, *82* (6), 648–651. Mathematical Association of America.

Nouvel, R., Zirak, M., Dastageeri, H., Coors, V. and Eicker, U. (2014). Urban Energy Analysis Based on 3D City Model for National Scale Applications. *IBPSA Germany Conference*.

Ortega, S., Trujillo, A., Santana, J.M., Suárez, J.P. and Gómez-Deck, D. (2016). Rendering large datasets of georeferenced markers in mobile devices, In Press. doi:10.1145/2948628.2948638

Paolantonio, M. Di, Fernández, C.G., Latorre, M.J. and Pedrera, F. (2015). 3D virtual representation of drones' flights through Cesium. js and Oculus Rift. *Geomatics Workbooks n 12 – „FOSS4G Europe Como 2015"*, 577–578.

Phong, B.T. (1975). *Illumination for Computer-Generated Images. Communications of the ACM*. The University of Utah.

Prandi, F., Staso, U. Di, Berti, M., Giovannini, L. and Amicis, R. De. (2015). Hybrid approach for large-scale Energy Performance estimation based on 3D city model data and typological classification, (November), 329–330.

Ratti, C., Frenchman, D., Pulselli, R.M. and Williams, S. (2006). Mobile Landscapes: Using Location Data from Cell Phones for Urban Analysis. *Environment and Planning B: Planning and Design*, *33* (5), 727–748. doi:10.1068/b32047

Rehman, U. and Cao, S. (2015). Augmented Reality-Based Indoor Navigation Using Google Glass as a Wearable Head-Mounted Display. *Systems, Man, and Cybernetics (SMC), 2015 IEEE International Conference on* (S. 1452–1457).

Rosenberg, L.B. (1993). The effect of interocular distance upon operator performance using stereoscopic displays to perform virtual depth tasks. *Virtual Reality Annual International Symposium, 1993. IEEE* (S. 27–32). doi:10.1109/VRAIS.1993.380802

Saed, S. and Wendel, J. (2015). Estimating heating energy consumption and CO_2 production - A novel modeling approach. *Proceedings of the 14th International Conference on Computers in Urban Planning and Urban Management.*

Schall, G., Schmalstieg, D. and Junghanns, S. (2010). Vidente - 3D Visualization of Underground Infrastructure using Handheld Augmented Reality. *Integrating GIS and Water, 1* (4), 1–17. doi:10.1.1.173.3513

Shen, Z. (2012). *Geospatial Techniques in Urban Planning* (Advances in Geographic Information Science). Springer Berlin Heidelberg.

Simons, A. and Nichersu, A. (2014). Development of a CityGML infrastructure for the imple-mentation of an energy demand method with different data sources. *GIScience 2014.*

Suárez, J.P., Trujillo, A., De La Calle, M., Gómez-Deck, D. and Santana, J.M. (2012). An open source virtual globe framework for iOS, Android and WebGL compliant browser. *Proceedings of the 3rd International Conference on Computing for Geospatial Research and Applications.*

Suárez, J.P., Trujillo, A., Santana, J.M., de la Calle, M. and Gómez-Deck, D. (2015). An efficient terrain Level of Detail implementation for mobile devices and performance study. *Computers, Environment and Urban Systems, 52,* 21–33. Elsevier. doi:10.1016/j.compenvurbsys.2015.02.004

Tanaka, E.H., Paludo, J.A., Cordeiro, C.S., Domingues, L.R., Gadbem, E. V. and Euflausino, A. (2015). Using Immersive Virtual Reality for Electrical Substation Training. *International Association for Development of the Information Society.* International Association for the Development of the Information Society.

Trapp, M. (2004). OpenGL-Performance and Bottlenecks. *Hasso Plattner Institut - Universität Potsdam,* (January 2004), 15 pages.

Trujillo, A., Suárez, J.P., La Calle, M. De, Gómez, D., Pedriza, A. and Santana, J.M. (2013). Glob3 mobile: An open source framework for designing virtual globes on iOS and android mobile devices. *Lecture Notes in Geoinformation and Cartography* (S. 211–229). Quebec, Canada: Springer. doi:10.1007/978-3-642-29793-9-12

Trujillo, A., Suárez, J.P., Santana, J.M., De La Calle, M. and Gómez-Deck, D. (2014). An efficient architecture for automatic shaders management on virtual globes. *Proceedings - 5th International Conference on Computing for Geospatial Research and Application, COM. Geo 2014* (S. 38–42). doi:10.1109/COM.Geo.2014.23

Tussyadiah, I. (2013). Expectation of Travel Experiences with Wearable Computing Devices. *Information and Communication Technologies in Tourism 2014* (S. 539–552). Cham: Springer International Publishing. doi:10.1007/978-3-319-03973-2_39

Veillard, D. (2009). The XML C parser and toolkit of Gnome. http://xmlsoft.org/. Accessed 10 June 2016

Wendel, J., Murshed, S.M., Sriramulu, A. and Nichersu, A. (2016). Development of a Web-Browser Based Interface for 3D Data—A Case Study of a Plug-in Free Approach for Visualizing Energy Modelling Results (S. 185–205). Springer International Publishing. doi:10.1007/978-3-319-19602-2_12

Wieland, M. and Wendel, J. (2015). Computing Solar Radiation on CityGML Building Data. 18th AGILE International conference on Geographic Information Science 2015. Lisbon, Portugal. 9th June 2015

Wiki, O. (2013). Tile usage policy — OpenStreetMap Wiki. http://wiki.openstreetmap.org/w/index.php?title=Tile_usage_policy&oldid=947750. Accessed 28 May 2016

World Wide Web Consortium. (2006). W3C Document Object Model. http://scholar.google.com/scholar?hl=en&btnG=Search&q=intitle:W3C+Document+Object+Model#2. Accessed 10 June 2016

Zamyadi, A., Pouliot, J. and Bédard, Y. (2013). A Three Step Procedure to Enrich Augmented Reality Games with CityGML 3D Semantic Modeling (S. 261–275). Springer Berlin Heidelberg. doi:10.1007/978-3-642-29793-9_15

Part V
User Studies, Privacy and Motivation

Ephemerality Is the New Black: A Novel Perspective on Location Data Management and Location Privacy in LBS

Mehrnaz Ataei and Christian Kray

Abstract Location information is essential to location-based services (LBS), but also has the potential to reveal sensitive information about the users of LBS to malicious agents. Therefore, location privacy is an important issue to address for both users and providers of LBS. In this paper, we investigate how location privacy can be realized in the context of a location-based service. Based on a review of architectures for LBS and key issues related to location privacy, we discuss several measures to integrate location privacy into LBS. In order to address privacy threats associated with the storage of location information, we propose an approach based on privacy-by-design principles and introduce a conceptual model to facilitate the implementation of those principles. In addition, we investigate the role of location data management in the context of privacy preservation, and propose the concept of temporal and spatial ephemerality to improve location privacy in the context of a location-based service.

Keywords Location-based services · Privacy by design · Location privacy · Ephemerality

1 Introduction

The defining feature of location-based services (LBS) is that they respond to the requests of users according to their physical location, which is not the case for other types of services. This dependency on positional information enables new and more user-friendly services but also entails issues regarding location privacy (Junglas and

M. Ataei (✉) · C. Kray
Institute for Geoinformatics (Ifgi), University of Münster, Münster, Germany
e-mail: m.ataei@uni-muenster.de

C. Kray
e-mail: c.kray@uni-muenster.de

Watson 2008; Barkuus and Dey 2003; Fodor and Brem 2015). Striking a balance between providing a service based on the user's location while protecting their (location) privacy is thus a key challenge in this area. In principle, the location privacy of users can be compromised in two ways: (1) using real-time location information enables an attacker to find you *right now* and carry out different attacks; (2) using past data facilitates the discovery of who you are, where you live, and what you do. It can be used, for example, to predict your behavior *at any time in the future* (Krumm2009). Ideally, issues related to location privacy are considered at design time, i.e. when a location-based service is developed. The 'Privacy by Design' approach (PbD) has been applied in other domains to *"prevent privacy invasive events before they happen"* (Cavoukian 2010). It thus constitutes a good starting point for developing a process for building LBS that actively considers location privacy during the design process rather than tinkering with the service after location privacy has been compromised. The work presented in this paper proposes a new model to realize location privacy by design and an approach to tackle location privacy by focusing on the *management* of location information in LBS. We also introduce the concept of ephemerality of location data and demonstrate how it can help to address privacy threats resulting from the retention of location data. The remainder of the paper is structured as follows. We first discuss different models and architectures that have been proposed to describe the structure and inner processes of LBS. Section three reviews different approaches to location privacy. The main part of the paper (section four) outlines the basic model underpinning our approach and then reviews in detail each element and strategy for location privacy protection. The penultimate section discusses the limitations and implications of our approach. The final section summarizes our key findings and provides an outlook on future research.

2 The Anatomy of LBS

Location-based services (LBS) cover a broad range of application scenarios, from navigation support (Ran et al. 2004) over local recommender systems (Foursquare 2016) to intelligent transport services (Uber 2016) and games (O'Hara 2008). Such services are different from more conventional services as they are aware of the context in which they are being used and can adapt their contents and presentation accordingly (Steiniger et al. 2008). While a traditional service usually only relies on networking and computing resources to "collect, process, filter, transmit, and disseminate data that represents information useful for a specific purpose or individual" (Schiller and Voisard 2004), a location-based service also intrinsically considers positional information. This enables a location-based service to deliver "information to its users in a highly selective manner, by taking the user's past, present, or future location and other context information into account" (Schiller and Voisard 2004). Consequently, a location-based service is subject to additional requirements compared to standard services (Chow and Mokbel 2009) and its architecture may also differ to accommodate those requirements. In the following, we therefore review several architectures and models that have been proposed for LBS and analyze some examples of LBS with respect to how they function.

Kido et al. (2005) proposed a location-based service model that consists of a geographic information system (GIS), a service provider and a database. In their model, a user of a location-based service obtains their location through a positioning device and then sends the position data to a service provider. The service provider, in turn, creates a response after communicating with the database and the GIS. Spiekermann (2004) developed a general communication model, which includes three layers: the positioning layer, the application layer, and the middleware layer. The positioning layer calculates the position of a user. The application layer comprises all services that request location data to integrate it into their offering. The middleware layer sits between the positioning layer and the application layer in order to reduce the complexity of service integration. All layers access the GIS directly. Strassman and Collier (2004) also discuss the development of a location-based service, a commercial friend finder application. The application is built around a location engine, which encapsulates the 'intelligence' of the service. It includes functionality such as geocoding, reverse geocoding, and routing, and retrieves data from both database and server. Deep Map (Malaka and Zipf 2000) was an early and complex location-based service providing intelligent guidance to tourists. The underlying architecture was agent-based, and components such as the routing agent or the presentation planner communicated over a shared message bus.

On a more abstract level, Hightower et al. (2002) introduced a layered approach for different positioning systems, which they termed the 'location stack'. It is inspired by similar models in the networking domain and consists of a set of layers that build upon one another. From the bottom to the top, the sensor layer deals with low-level hardware and raw data values. The measurements layer combines sensor data to derive location information such as distances or angles. The fusion layer determines the location of objects, and the arrangements layer provides information about spatial relationships between objects. The contextual fusion layer combines location information with other contextual information, e.g. to detect states. The activities layer is concerned with semantics and application-specific states, while the intentions layer deals with user needs and goals.

The example systems and the abstract architectures for LBS discussed above cover a broad range of perspectives and propose different models to conceptualize and build a service that takes into account location. One aspect that is not covered much (if at all) is the question of how location information is managed after the position of the user/device has been determined (e.g. by a set of sensors such as a GPS receiver). Few, if any of the proposed approaches consider how this information is stored and retrieved, how it can be accessed and what should happen with it 'over the long run'. This aspect is however quite central, in particular when considering privacy, which we will discuss in the following section.

3 Location Privacy

In order to receive the full benefits of a location-based service, users have to share location data, i.e. where they are or where they have been. Such location data is quite sensitive as it reveals the current physical location of users, and if disclosed would thus pose a serious threat to their privacy and safety. For example, attackers could use this information to either track them down or to exploit their absence, e.g. to break into their home while they are away. Historic location data incurs further privacy threats: attackers can, for example, use it to predict behavior (e.g. to waylay victims) or to infer information about people (e.g. where they live and work or who they know). Even though not all users are aware of these issues, the sensitivity of the location information incurs challenges and difficulties in the process of LBS adoption by users (Xu et al. 2009 and Zhou 2011).

Privacy as a concept has many facets (Waldo et al. 2008), and different definitions have been proposed—from the classic "the right to be left alone" (Warren et al. 1890) to "choose freely under what circumstances and to what extent" people share information about themselves with others (Westin 1968). Location privacy can thus be understood as privacy relating to the location information of a person, i.e. "a special type of information privacy which concerns the claim of individuals to determine for themselves when, how, and to what extent location information about them is communicated to others" (Duckham and Kulik 2006). Beresford and Stajano (2003) define location privacy along similar lines as "the ability to prevent other parties from learning one's current or past location" (Beresford and Stajano 2003).

In order to appreciate the importance of location privacy, it is important to understand the risks and threats associated with leaked location data. This is also the first step for exploring possible countermeasures to the identified threats and risks. The rapid proliferation of LBS has resulted in the collection of large amounts of location data, which, in turn, has enabled the *analysis of movement patterns*. This analysis, if applied by an attacker, is one of the most discussed threats associated with leaked location information (Krumm 2009). It has been shown that a broad range of sensitive user-related information can be extracted from analyzing movement patterns. This includes the identity of the user, their (home) address, individual (points of) interests as well as significant events (e.g. strikes or protests) that a user participated in (Hoh et al. 2006; Patterson et al. 2003).

A related issue resulting from large-scale collection of location data is *dataveillance*, "the systematic use of personal data systems in the investigation or monitoring of the actions or communications of one or more persons" (Clarke 1988). Key privacy risks associated with dataveillance (Abbas et al. 2015) are the loss of control, (continuous) monitoring, identification, social sorting, and profiling. In general, threats linked to location data have the potential to "disclose a great deal about the movements of entities, and hence about individuals associated with those entities" (Clarke and Wigan 2011). When exploited in attacks, these threats may cause psychological, social, and economic harm (e.g. loss of control over one's life, social embarrassment, financial damage) to individuals (Clarke and Wigan 2011). Although many attacks depend on access to recorded (past) location data, the way in which location data is managed has not received a lot of attention.

In order to neutralize these and other threats to (location) privacy and to counter attacks, research has identified a number of general methods to protect privacy. One of the most common methods to secure data in general (and thus location data in particular) is encryption. Encryption is platform and service agnostic and can be applied to secure data. As a key area of cryptography, encryption provides data security through hashing and secret communication (Balogun and Zhu 2013). While cryptography is considered as an essential and necessary aspect to secure communication, it is not sufficient by itself unless its deployment and implementation are managed adequately (Kessler 2016).

In the context of location data and the associated threats, Duckham and Kulik (2006) discuss further measures for privacy protection. Regulatory strategies are a promising approach, where the government defines rules on the use of personal information, for example by passing laws that are binding for LBS providers. A second option is the use of privacy policies, which are trust-based agreements between individuals and whomever they are sharing their location data with. Another generally applicable method is to rely on anonymity. For example, a user might use a pseudonym instead of their real name or create ambiguity by grouping with other people. Finally, it is also possible to use obfuscation, which reduces the quality of location data and thereby prevents attackers from easily learning where exactly a user is located. When applied sensibly, all these methods as proposed by Duckham and Kulik (2006) can be implemented without compromising the quality of the LBS.

As a practical example of a privacy through data management implementation, Stroeken et al. (2015) developed a privacy preserving location-based service called *Zone-it,* a virtual notice board which permits users to have location-based interaction in self-zoning areas and under certain categories. The service places offers and requests with their exact coordinates on a map. Users can find offers and requests based on their interest and location (zone). After a match is found, the message disappears. Zone-it is a social media service, which shifts the focus from person-based (e.g. Facebook) to goal-oriented communication (Stroeken et al. 2015).

On a more technical level, several of the approaches listed above have successfully been implemented. Examples in this area include work by Krumm (2007, 2009), where computational countermeasures to mitigate threats are discussed including anonymity, spatial-temporal degradation, specialized queries, spatial cloaking, noise, and rounding. Other countermeasures proposed at this level are the use of a trusted third party, which improves location privacy by serving as an intermediary between providers and users of a location-based service (Mokbel et al. 2006). This intermediary can then employ various strategies, for example dynamically adjusting location quality based on the number of nearby users. A similar approach is the use of mix zones (Beresford and Stajano 2003), which are spatial areas inside which all clients of a location-based service stop sharing their location with the service provider and also change the pseudonym they are using. This makes it difficult to track individuals when they leave a mix zone. Most of the countermeasures discussed above work based on the assumption that location data is perpetually stored. The role of location data management and its impact on location data privacy are not considered explicitly in the cited papers. Due to the increasing importance and practical relevancy of privacy, Cavoukian (2010) proposed to consider privacy from the start, i.e. the design stage. Their 'Privacy by Design' (PbD) approach describes general principles and essential steps towards realizing better privacy protection in all type of information systems. The goal of PbD is to secure the privacy of individuals by providing them with control over their information (Cavoukian 2010). For this purpose, the author defines seven basic principles that should be followed when designing an interactive system to ensure that the resulting system respects the privacy of its users:

1. **proactive not reactive**: rather than wait for privacy risks to occur, such risks should be anticipated and prevented from materializing.
2. **privacy as the default setting**: the default behavior of a system should be such that the privacy of its users are automatically protected—no prior user action is required.
3. **privacy embedded into design**: rather than 'patching' a system with some privacy-protection measures, privacy-related functionality should be considered as an integral part of the system and be realized without interfering with its overall purpose.
4. **full functionality**: unnecessary trade-offs (e.g. security vs. privacy) should be avoided and all legitimate requirements should be realized ("win-win").
5. **end-to-end security**: all data collected in the system should be protected by strong security at all stages of its life cycle (from creation to deletion).
6. **visibility and transparency**: all parties involved in the provision of a service and the running of the corresponding system, should expose their practices, policies and technologies so that they can be independently verified.
7. **respect for user privacy**: the interests, needs, and preferences of users should be considered first and foremost to ensure a user-friendly privacy-preserving system.

While the Privacy by Design approach in principle can be applied to LBS, it is not clear how it could be folded into a location-based service and how it can be used to make existing LBS more privacy-aware. In addition, the issue of managing location data is only implicitly covered and deserves a more thorough analysis due to the role historic location data plays in enabling different types of attacks. In the following section, we therefore propose a conceptual model to facilitate location Privacy by Design, and we introduce the concept of ephemerality of location data as a fundamental approach to realize Privacy by Design in the context of LBS.

4 Location Privacy by Design

Service and content providers of LBS are collecting location data from users and are usually storing it for a substantial period of time (Sathe et al. 2014). The rationale for storing the data is manifold. Depending on the country, there may be legal requirements to keep the data for at least a certain amount of time. Being able to analyze historic location data might also provide insights that can help to improve the service. Finally, historic location data also allows for deep profiling of the users, and such profiles constitute a commercial value, such as targeted advertising. From a user's perspective, in particular, the latter use can be perceived as an unwanted intrusion of their privacy.

By default, many LBS rely on a number of different databases for retaining and maintaining various types of data such as service-specific content data, digital map data, or user location data (Lee et al. 2005). These databases frequently are accessed remotely on an as-needed basis and are usually under the control of the service provider. Based on a sample of commercial LBS, the number of LBS that are self-contained on a mobile device is relatively small (e.g. navigation systems with local map databases to avoid roaming charges while traveling abroad). Research investigating how location data is stored is mostly focusing on technical challenges relating to, for example, handling large amounts of spatio-temporal location data or increasing system performance by optimizing access to location data (Mokbel et al. 2003). In the light of the various privacy threats discussed above, it makes sense to look at location data management not only from a technical perspective but also from the perspective of how it affects privacy. This aspect, however, has not received much attention in literature. When looking at existing architectures of and models for LBS such as Kivera (Schiller and Voisard 2004) or the location stack (Hightower et al. 2002), we can observe that privacy protection for location data is not an inherent part of these models. As discussed in the previous sections, there are a number of approaches to protect location privacy but these are frequently either external to the LBS, e.g. as a trusted third party (Mokbel et al. 2006), or not integrated into the architecture of a location-based service, e.g. the mix zones proposed by Beresford and Stajano (2003).

In order to describe more clearly how location privacy protection can be integrated into a location-based service, we propose a conceptual model (see Fig. 1) that facilitates applying existing methods for privacy protection as integral parts of a location-based service. In addition, the model provides means to explicitly consider how location data is managed and how strategies for privacy protection in this context can be realized. It also captures how the configuration of location privacy settings can be exposed to users of a location-based service without requiring thorough modifications of the internal core logic of a service.

The model describes how a location-based service interacts with the world and provides a user with a service while explicitly considering location privacy. A set of *sensors* observes the world and provides information about it, in particular, location data and context data. While the former refers mainly to the position of a user, the

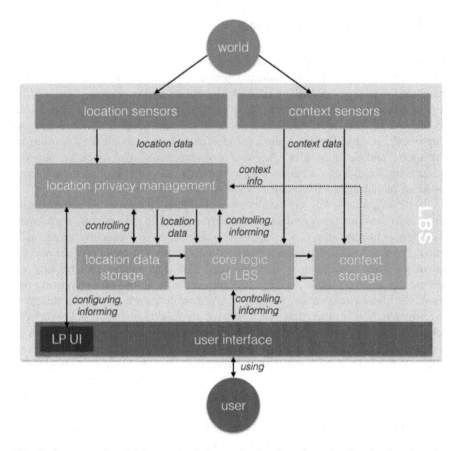

Fig. 1 A conceptual model for seamlessly integrating location privacy in a location-based service

latter includes aspects such as environmental factors, the time of the day or the current task of the user. Both types of information are usually stored for later perusal by the service (in a *location data storage* and a *context storage*). They are also needed for processing by the *core logic* of the location-based service. This part encapsulates the main functionality of the location-based service, for example, routing algorithms for a navigation service, or means to retrieve real-time traffic data. This component also interacts with both the location data storage and the context storage, i.e. to retrieve information (e.g. historic location data to carry out dead reckoning) or to update it (e.g. to set the current task of the user to navigation after directions have been requested by the user).

The *location privacy management component* (LPM) is strongly connected to the location data storage in order to implement various privacy protection measures. It observes and controls the location data storage according to the rules and procedures defined by the designers, developers and/or users of the location-based service. In order to address the location privacy issues, it can actively control the data storage process. For example, it can reduce the quality of location data received from the sensors prior to passing it on to the location data storage or the core logic of the location-based service along the lines of the concept underpinning Caspar (Mokbel et al. 2006). From the perspective of the Core Logic, only the provider of the location data is different compared to a more traditional architecture, where it receives location data directly from the sensor component. Users of the location based service rely on a *user interface* (UI) to control the location-based service and receive information from it. The user interface can also incorporate a *location privacy user interface* (LP UI), which enables users to directly access the LPM in order to inspect how location privacy is managed and to configure it according to their preferences and needs. Providing a component which separately addresses the user interface design and options for LBS users with the goal of increasing their control over their location privacy can be a suitable approach to realize standardized privacy controls at the UI level.

The integration of the LPM and its interaction with the other components of a location-based service facilitates different ways to build a service that takes into account location privacy. In a legacy system, the LPM basically corresponds to a forwarding mechanism that forwards all location data directly to the location data storage and the core logic. A first step towards more location privacy would be to introduce a set of simple rules that the LPM uses when deciding what information to pass on to which component and at what granularity. An example of a rule is to reduce the quality of the location data to improve location privacy if the user specifies this or when the service does not require completely accurate location information to function. A more sophisticated set of rules could also take into account contextual information such as the time of the day and automatically stop providing location information after the working hours of a user are over. Such a rule set could also facilitate the realization of user-driven preferences with respect to location privacy protection (Toch et al. 2010).

The simple approach described above could be integrated into a location-based service without the need to modify the core logic (beyond changing which component provides it with location data). An alternative and complementary strategy that would also not require any changes to existing components is for the LPM to take more detailed control of the location data storage. In this case, the LPM could directly access the location data storage (e.g. using the same means that the core logic employs) to apply various strategies to recorded historic location data. For example, it could continuously monitor the stored location data to ensure k-anonymity (Sweeney 2002) (e.g. by accessing social networking sites where other people publish their location). This approach also forms the basis for the temporal ephemerality approach introduced in the following section.

More sophisticated strategies for protecting location privacy might require more involved interaction between the core logic and the LPM, and thus entail changes to the former. For example, location privacy could be negotiated on a case-by-case basis with the core logic providing a rationale why positional information of a certain quality is required. Conversely, the LPM might inform the core logic component about new location privacy settings requested by the user so that the core logic might change its behavior in response to this. A complementary strategy is to consider the way in which location data is stored. Beyond technical considerations, there are also different options regarding how the system stores location data, where the data resides in the physical world and whom it is shared with. These aspects play a key role in the realization of spatial ephemerality as a means to protect location privacy (as discussed in Sect. 4.2).

In respect to the privacy-by-design principles, the introduction of the location privacy management component thus facilitates addressing location privacy issues from inside the LBS architecture and it supports realizing this in different ways. In addition to simplifying the integration of location privacy protection into legacy systems, the overall model of course also allows for the creation of privacy-aware LBS from scratch. The following two subsections will demonstrate the usefulness of our model further by first introducing the concepts of temporal and spatial ephemerality for location data and then highlighting how these can be realized using the LPM and the model in general.

4.1 Temporal Ephemerality

When location data is stored in LBS databases indeterminately, the window of attacks is substantially enlarged: malicious agents have an unrestricted amount of time to obtain access to the location data and carry out attacks on the user's location privacy. As discussed above, historic location data is particularly sensitive as it allows for very deep inferences on users and their behavior. In order to address this key issue of location privacy, it makes sense to consider *how long* location information is stored.

Rather than assuming that location information is stored, a more fine-grained consideration of the *temporal ephemerality* of such information can contribute towards better protecting the location privacy of the user. The basic idea of this concept is that all location information can be assigned an expiration date, after which it is deleted. By defining an expiration time and then discarding location data once it has passed, attackers will be unable to use this data for further attacks in the future (assuming that the location data storage was not breached while location data was still in the storage).

The temporal ephemerality of location data can be specified in different ways. It is possible to assign an overall expiration time for all location data, either in relative (e.g. "delete 24 h after recording") or absolute terms (e.g. "delete today at midnight"). Alternatively, a more fine-grained control is possible as well using a set of rules that determines for each individual piece of location data when it should expire. Such rules could take into account various factors such as context (e.g. "delete all location data when I leave my work place") or user preferences (e.g. "delete very precise location data immediately after recording"). The exact way in which temporal ephemerality is realized can be specified by the designer and developers of a location-based service (while designing and building the service) and by the users while interacting with the location-based service.

In our model, the temporal ephemerality of location data can be easily realized by encapsulating the corresponding rule sets inside the LPM. For legacy LBS, this could be achieved in a completely transparent way as outlined above. The only component that would need to be modified slightly is the location data storage as each entry would get an additional attribute (expiration date) to facilitate the timely removal of expired location data. The LPM could then use this attribute to periodically query the location data storage for all entries with an expiration time in the past and to then delete the returned entries.

In addition to better protecting the location privacy of users in general, implementing temporal ephemerality of location data in this way also realizes several basic principles of the privacy-by-design approach for location data. Supporting the idea of being preventative and not remedial by discarding the data from database, the risk of inference attacks will be reduced as there will be no record of data available for attackers after the expiration date. This follows the 'proactive, not reactive' principle, where "Privacy by Design comes before-the-fact, not after" (Cavoukian 2010). With regards to the approach and implementation described above, the discarding of location data occurs automatically once the expiration time has passed, thereby realizing the 'privacy as the default setting' principle (This behavior could be changed by users, e.g. via the LP UI, should they wish to keep location data forever). Finally, the approach outlined above very strongly connects with the 'privacy embedded into design' principle of PbD. By encapsulating the functionality for temporal ephemerality in the LPM, designers can easily design systems that realize the location-based service while respecting location privacy as the other components are largely unaffected. They can focus on the location-based service functionality and relegate considerations about location privacy to LPM and (to some degree) the location data storage. The LP UI then provides an easy way to expose issues related to location privacy to users. The model also facilitates reuse of components: designers can create generic LPM, location data storage, and LP UI components and then use them to create different location privacy-aware location-based service.

4.2 Spatial Ephemerality

In the discourse of (location) privacy, many aspects are discussed but the issue of *where* location information is stored (and accessible) has received little attention. Usually, the underlying assumption is that stored location data can be accessed from anywhere. If, however, such data is only accessible inside a well-defined spatial area, then attackers or their proxies have to be co-present in order to carry out an attack. In analogy to temporal ephemerality for the time domain, the concept of *spatial ephemerality* refers to location data having a spatial 'expiration' zone: location data is stored in a particular area, and only accessible for users who reside inside this area. More specifically, spatial ephemerality entails that all location data is assigned a spatial expiration zone, and once a user leaves the zone for a particular location entry, it is deleted. To put it differently, location data would not leave a particular geographic area (e.g. the area where the location-based service is most relevant or where the user intends to use it) so that an attacker could not use it once a user has left that area - assuming the data was not retrieved while the user was still inside the area.

Similar to the temporal case, the spatial ephemerality of location data can also be defined in relative (e.g. "delete location data that is further than 2 km from the current position") or absolute terms (e.g. "delete location data that is more than 2 km from the city center"). In addition to specifying general rules for all location data, it is possible to define this for individual pieces of location data. The rules encoding spatial ephemerality can also consider various other factors such as context (e.g. "delete all location data inside a 2 km radius around locations that are visited only by few people") or user preferences (e.g. "delete all location data inside a 2 km radius around my home"). As with the temporal case, the exact way in which spatial ephemerality is realized can be specified by the designers and developers of a location-based service (during design and development) and by the users (during usage of the location-based service).

In our model, spatial ephemerality could be realized via the LPM to encapsulate the rule set defining the spatial ephemerality of location data. This approach has the advantage of being completely transparent and thus would lend itself easily to making legacy LBS more location privacy-aware. As with the temporal case, it would be necessary to introduce an additional attribute for location data. Consequently, the location data storage component would have to be modified accordingly. This attribute would hold the spatial expiration area of an entry, for example, in the form of a polyline corresponding to the boundary of the area. The LPM could then periodically query the location data storage component with the current location to obtain all entries, which do not contain this location within their expiration areas. The returned entries could then be deleted.

Spatial ephemerality can contribute towards location data privacy by deleting location data based on spatial conditions, and thereby reduce the risk of inference attacks. The proposed model and approach to realize spatial ephemerality of location data also facilitates the application of PbD principles to location data. By geographically limiting the storage of location data and encapsulating the corresponding rules with default values inside the LPM, the 'privacy as the default setting' principle can easily be realized. Similarly, this approach supports the 'privacy embedded into the design' principle. The 'proactive not reactive' principle of PbD applies as well, as location data is systematically deleted before an attack occurs. The considerations regarding the design and development of privacy-aware LBS (ease of improving legacy LBS, concentration of location privacy concerns in the LPM, reuse of components) we discussed for temporal ephemerality (in Sect. 4.1) hold true for spatial ephemerality as well.

In order to further investigate how ephemerality can be implemented and used in everyday life, we have started to develop an initial prototype[1] based on our proposed model. The prototype is a service designed to enable users to share their experiences while visiting or exploring a city (e.g. special events). The application provides a means to share short messages anonymously with people in the same geographic area. In addition, it empowers users to define an expiration time for each message. The system design is implemented to not store any location data of the users or their messages over time. The location data of users is discarded from the system (website[2] or app) as soon as the user leaves the geographic area or when the messages expire. Our next step is to carry out user studies based on this prototype to gain a deeper understanding on how users act when they are given increased control over their location privacy.

5 Discussion

The proposed location-Privacy by Design approach and the corresponding model for LBS as well as the concept of ephemerality offer benefits and are also subject to a number of limitations. The key benefit of the PbD approach in combination with the proposed model is facilitating the realization of location-privacy-preserving LBS. In Sect. 4, we discussed in detail how this can be achieved both for existing LBS that should be made more privacy-aware and during the design of a new LBS from scratch. The benefits of the ephemerality concept include facilitating sophisticated privacy-protection without having to substantially modify all of the components of a location-based service. In addition, ephemerality of location data reduces the amount of storage needed to hold historic location data, and it provides a unified and simple approach to implement legal requirements (e.g. via expiration dates corresponding to the legally required duration of storing data). Considering that LBS can produce a large amount of privacy sensitive data every day, which requires a secure storage and proper treatment to comply with existing law, the ephemerality approach will also not require the system to obtain more servers over time, which may incur financial savings. From the user's point of view, key benefits of the proposed approach and the model as well as ephemerality include an increased level of privacy and a fine-grained control over the user's location privacy.

These benefits also come with a number of drawbacks and challenges. While there are no inherent technical issues preventing the implementation of the proposed ideas, there are potential business-related implications. Location data can have a commercial value for advertisement partners to LBS companies as the collected location data can provide deep insights into the behavior and habits of

---

[1]https://github.com/heinrichloewen/SC-App.
[2]https://github.com/chack05/sc16-ephemeral-lbs-server.

users. For example, providing tailored advertisements to users based on those insights can be a viable revenue stream for LBS companies, which ephemerality could negatively affect. A key question in this context is if users would be willing to pay for a service with increased privacy and control to compensate for reduced revenues of service providers due to this. With the non-permanent storage of location data also comes the challenge of maintaining the functionality of LBS that rely on forward predictions based on past behavior. Time-limited data storage could pose substantial challenges when making advance analysis of user data. Selecting expiration conditions (both spatial and temporal ones) carefully to ensure optimal service provision would be one way to address this challenge. Another limitation or drawback of the proposed LPM component is the fact that it is still vulnerable to attacks, and may also be subject to new kinds of attacks. While in principle it can reduce the severity of successful attacks aimed at retrieving historic location information (by reducing the amount of data being stored), the attacks can still be applied. In addition, the component may become a target by itself, for example, by introducing rules into the LPM component that counteract user-specified rules.

A consequence and potential drawback of using the ephemerality feature is the loss of data. This can be discussed from a provider and a user perspective. From the provider perspective, data storage allows the information they gather to be used for profiling or categorizing their users for purposes such as targeted advertising. Historical tracks of location data is a commodity that can be sold to other companies to be used for the same purposes. From the user's perspective, the loss of data can also have consequences. By not storing location information, it may not be possible to get user-adapted or a localized service provision. For applications that strongly rely on recorded location data (e.g. Foursquare), the ephemerality feature may severely affect service quality.

In section three, we have listed a number of approaches and solutions developed and proposed to protect (location) privacy in LBS. It is crucial to mention that LPM as a solution is a complementary approach. Our solution can be combined with other approaches such as encryption or anonymization. Privacy is regarded as a multifaceted problem that is challenging to solve with one single solution. Due to this, combinations of different methods and approaches can be advisable and/or necessary in order to protect the privacy of users.

In our discussion, we mainly focused on the management of location data due to its importance and potential in the context of realizing location privacy in a location-based service. We did not analyze contextual aspects in detail, which can also have a severe impact on privacy in general. One option to deal with this issue could be the introduction of a context data management component into our model that would operate on contextual data in a similar way as the LPM deals with location data. Another area we did not discuss relates to users and their understanding of location privacy. The model foresees a subcomponent of the user interface, the LP UI, as a means for users to configure settings related to location privacy and to access information about it. In order to build LBS that facilitate proper protection of the users' location privacy, these user-facing parts need to be further investigated. In particular, there is a lack of knowledge about the user's understanding of location privacy and related concepts and options, and it is also not clear how to best communicate this to users.

6 Conclusion

In this paper, we investigated how location privacy can be realized in the context of LBS. In particular, we looked into the role of location data management in the context of privacy preservation. Based on privacy-by-design principles, we then proposed an approach tailored to LBS and defined a conceptual model to facilitate the implementation of those principles. We showed that this model supports the realization of different privacy protection mechanisms and enables an explicit and fine-grained control of location data management in the context of privacy preservation. In addition, we proposed the concept of temporal and spatial ephemerality as a means to improve location privacy in the context of a location-based service, which can both be realized using the proposed approach and model. The conceptual model and ephemerality concept are complementary to existing methods to protect location privacy such as encryption or obfuscation.

Though the proposed approach is subject to some limitations, there are several promising options for further research. One interesting and underexplored area relates to the understanding users have of location privacy, related concepts, and options, and to how to effectively communicate these aspects to them. We are planning to carry out user studies to compare different systems to communicate threats and countermeasures and to gain a deeper understanding of (mis)conceptions about location privacy. The LP UI component will serve as a platform to facilitate this line of research. A complementary direction for future research relates to the concept of spatial ephemerality. Here, we plan to investigate how opportunistic information sharing can enable spatial ephemerality at the level of the location data storage and/or the core logic of a location-based service. This line of work will rely on the LPM component to realize and to test the prototypes in realistic settings.

Acknowledgment The authors gratefully acknowledge funding from the European Union through the GEO-C project (H2020-MSCA-ITN-2014, Grant Agreement Number 642332, http://www.geo-c.eu/).

References

Abbas, R., Michael, K. and Michael, M.G., 2015. Using a Social-Ethical Framework to Evaluate Location-Based Services in an Internet of Things World. *International Review of Information Ethics*, 22(12).

Balogun, A.M. and Zhu, S.Y., 2013. Privacy Impacts of Data Encryption on the Efficiency of Digital Forensics Technology. *arXiv preprint* arXiv:1312.3183.

Barkuus, L. & Dey, A., 2003. Location-Based Services for Mobile Telephony: a Study of Users' Privacy Concerns. In *Proceedings of the International Conference on Human-Computer Interaction (INTERACT).* pp. 1–5.

Beresford & Stajano, F., 2003. Location privacy in pervasive computing. *IEEE Pervasive Computing*, 2(1), pp. 46–55. Available at: http://ieeexplore.ieee.org/lpdocs/epic03/wrapper.htm?arnumber=1186725.

Cavoukian, A., 2010. Privac y by Design. *Identity in the Information Society*, 3(2), pp. 1–12.

Chow, C.-Y. & Mokbel, M.F., 2009. Privacy in location-based services: a system architecture perspective. *Sigspatial Special*, (2), pp. 23–27. Available at: http://dl.acm.org/citation.cfm?id=1567258.

Clarke, R., 1988. Information technology and dataveillance. Communications of the ACM, 31 (5), 498–512.

Clarke, R. and Wigan, M., 2011. You are where you've been: the privacy implications of location and tracking technologies. *Journal of Location Based Services*, 5(3–4), pp. 138–155.

Duckham, M. and Kulik, L., 2006. Location privacy and location-aware computing. *Dynamic & mobile GIS: investigating change in space and time*, 3, pp. 35–51.

Fodor, M. & Brem, A., 2015. Do privacy concerns matter for Millennials? Results from an empirical analysis of Location-Based Services adoption in Germany. *Computers in Human Behavior*, 53, pp. 344–353. Available at: http://www.sciencedirect.com/science/article/pii/S0747563215300066.

Foursquare. 2016. *foursquare*. [ONLINE] Available at: https://foursquare.com. [Accessed 15 June 2016].

Hightower, J., Brumitt, B. & Borriello, G., 2002. The location stack: A layered model for location in ubiquitous computing. *Proceedings—4th IEEE Workshop on Mobile Computing Systems and Applications, WMCSA 2002*, pp. 22–28.

Hoh, B. et al., 2006. Enhancing security and privacy in traffic-monitoring systems. *IEEE Pervasive Computing*, 5(4), pp. 38–46.

Junglas, I. & Watson, R., 2008. Location-based Services., 51(3), pp. 65–70. Available at: http://pewinternet.org/Reports/2013/Location.aspx.

Kessler, G.C., 2016. An Overview of Cryptography (Updated Version, 3 March 2016).

Kido, H., Yanagisawa, Y. & Satoh, T., 2005. An anonymous communication technique using dummies for location-based services. *Proceedings—International Conference on Pervasive Services, ICPS '05*, 2005, pp. 88–97.

Krumm, J., 2007. Inference Attacks on Location Tracks. *Pervasive Computing*, 10(Pervasive), pp. 127–143. Available at: http://www.springerlink.com/index/TG64551RW2716103.pdf \nhttp://research.microsoft.com/en-us/um/people/jckrumm/publications2007/inferenceattackrefined02distribute.pdf.

Krumm, J., 2009. A survey of computational location privacy. *Personal and Ubiquitous Computing*, 13(6), pp. 391–399.

Lee, D.L., Zhu, M. & Hu, H., 2005. When Location-Based Services Meet Databases. *Mobile Information Systems*, 1(2), pp. 81–90. Available at: http://iospress.metapress.com/index/d3fy03rg5vvvbc29.pdf\nhttp://www.hindawi.com/journals/misy/2005/941816/abs/.

Malaka, R. & Zipf, A., 2000. Deep Map: Challenging IT research in the framework of a tourist information system. *ENTER 2000: 7th. International Congress on Tourism and Communications Technologies in Tourism, Barcelona, Spain, 26–28 April 2000*, pp. 1–11. Available at: http://195.130.87.21:8080/dspace/handle/123456789/581.

Mokbel, M.F., Ghanem, T.M. & Aref, W.G., 2003. Spatio-Temporal Access Methods. *IEEE Data Engineering Bulletin*, 26(2), pp. 40–49. Available at: http://dl.acm.org/citation.cfm?id=1458187.

Mokbel, M.F., Chow, C.-Y. & Aref, W., 2006. The new Casper: query processing for location services without compromising privacy. *Vldb'06*, (1), pp. 763–774. Available at: http://dl.acm.org/citation.cfm?id=1164193.

O'Hara, K., 2008. Understanding geocaching practices and motivations. *Proceedings of the SIGCHI Conference on Human ...*, p. 1177. Available at: http://portal.acm.org/citation.cfm?doid=1357054.1357239\nhttp://dl.acm.org/citation.cfm?id=1357239.

Patterson, D.J. et al., 2003. Inferring High-Level Behavior from Low-Level Sensors. *UbiComp 2003 Ubiquitous Computing*, 2864, pp. 73–89. Available at: http://www.springerlink.com/index/k3x004g773qj80kg.pdf.

Ran, L., Helal, S. & Moore, S., 2004. Drishti: An integrated indoor/outdoor blind navigation system and service. *Proceedings - Second IEEE Annual Conference on Pervasive Computing and Communications, PerCom*, pp. 23–30.

Sathe, S. et al., 2014. Enabling Location-Based Services 2.0: Challenges and Opportunities. *Mobile Data Management (MDM), 2014 IEEE 15th International Conference on IS–SN -*, 1, pp. 317–320.

Schiller, J. & Voisard, A., 2004. *Location-based Services*, Morgan Kaufmann.

Spiekermann, S. 2004. General Aspects of Location-Based Services. In: Schiller, J. & Voisard, A., 2004. Location-based Services. Morgan Kaufmann, pp.15–33

Steiniger, S., Neun, M., Edwardes, A. and Lenz, B., 2008. Foundations of LBS. *CartouCHe-Cartography for Swiss Higher Education. Obtido em*, 20, p. 2010.

Strassman, M. and Collier, C. 2004. Case Study: Development of the Find Friend. In: Schiller, J. & Voisard, A., 2004. Location-based Services. Morgan Kaufmann, pp.34–48.

Stroeken, K., Verdoolaege, A., Versichele, M., Backere, F.D., Devos, D., Verstichel, S. and Weghe, N.V.D. 2015. Zone-it before IT zones you: A location-based digital notice board to build community while preserving privacy. *Journal of Location Based Services*, 9(1), pp. 16–32.

Sweeney, L., 2002. k-ANONYMITY: A MODEL FOR PROTECTING PRIVACY. *International Journal on Uncertainty*, 10(5), pp. 557–570.

Toch, E. et al., 2010. Empirical models of privacy in location sharing. *Proceedings of the 12th ACM international conference on Ubiquitous computing—Ubicomp '10*, (April 2016), p. 129. Available at: http://portal.acm.org/citation.cfm?doid=1864349.1864364.

Uber. 2016. https://www.uber.com. [ONLINE] Available at: https://uber.com. [Accessed 15 June 2016].

Waldo, J. et al. (2008), Engaging Privacy and Information Technology in a Digital Age. National Academies Press.

Warren, Samuel D., and Louis D. Brandeis. "The right to privacy." Harvard law review (1890): 193–220.

Westin, A.F., 1968. Privacy and Freedom. *American Sociological Review*, 33(1), p. 173.

Xu, H. & Gupta, S., 2009. The effects of privacy concerns and personal innovativeness on potential and experienced customers' adoption of location-based services. *Electronic Markets*, 19(2–3), pp. 137–149.

Zhou, T., 2011. The impact of privacy concern on user adoption of location-based services. *Industrial Management & Data Systems*, 111(2), pp. 212–226.

Classes for Creating Location-Based Audio Tour Content: A Case of User-Generated LBS Education to University Students

Min Lu, Masatoshi Arikawa and Atsuyuki Okabe

Abstract A course named *Seminar of Culture Studies* is given to undergraduate students in the School of Cultural and Creative Studies of Aoyama Gakuin University in Tokyo, Japan, in which a new mobile application named *Manpo* was applied for the students to create location-based audio tour content in 2013, 2014 and 2015. With little background knowledge on geography and cartography, the students tried to draw maps and create georeferences for positioning on the maps using *Manpo*'s functions. They also focused on recording audio guides for the POIs and walking routes, and bundled them to the maps together with photos and texts. The procedures of the classes and the students' performances were observed, in which the problems encountered by the students showed the difficulties, such as achieving appropriate georeferencing for accurate positioning, for non-professional users. However, the classes and students provided hints to the further development of *Manpo*, which also improved the classes and the students' works. As the course was proved to be a success, it inspired future research plans of a platform that can involve researchers, developers, mapmakers, local communities, and ordinary users to provide user-generated content including diverse maps for more attractive location-based services.

Keywords User-generated content · Hand-drawn maps · Location-based audio tours

M. Lu (✉) · M. Arikawa
Center for Spatial Information Science, The University of Tokyo, Tokyo, Japan
e-mail: lu@csis.u-tokyo.ac.jp

M. Arikawa
e-mail: arikawa@csis.u-tokyo.ac.jp

A. Okabe
School of Global Studies and Collaboration, Aoyama Gakuin University, Tokyo, Japan
e-mail: atsu@csis.u-tokyo.ac.jp

1 Introduction

In today's location-based services (LBS), user-generated content (UGC) has become an important component (FitzGerald 2012; Yap et al. 2012). Apart from the content generated in all kinds of web mapping based applications, location-related content are created in different forms and styles, and published in different media with more and more new tools and platforms. How to generate, collect, manage, share and ultimately use such content becomes a very important research issue (Mooney et al. 2013).

In a course named *Seminar of Culture Studies*, which is set for undergraduate students in the School of Cultural and Creative Studies of Aoyama Gakuin University in Tokyo, Japan, the new trends and technologies of location-related multimedia content are introduced. An important goal of the course is to extend the students' sight of spatial information representation and dissemination, and inspire their creativity of location-related information design. For this purpose, most of the classes in this course are carried out to let the students create location-based audio tour content by themselves, and experience the content using smartphones at the real places.

As the course is collaboration between Okabe Lab in Aoyama Gakuin University and Arikawa Lab in Center for Spatial Information Science, the University of Tokyo, newly developed applications and prototypes of Arikawa Lab, such as *pTalk* (Kaji and Arikawa 2013), *maPodWalk* (Arikawa et al. 2007) and *Manpo*, are introduced in the course as tools for creating the content. As most students have little background knowledge on Geography, Cartography or Information Science, it becomes a valuable opportunity to test and improve the usability of these applications or prototypes for ordinary users.

Among the above tools, *Manpo* was introduced to the course since 2013. It provides functions of georeferencing hand-drawn maps for positioning on them using smartphones. Thus, the students in the course could have opportunities to create their audio content on their own hand-drawn maps, which made an obvious change to the course. Before *Manpo* was introduced, most students considered less about the importance of maps in LBS, and they tended to believe that the Google-style web mapping is already enough for all the usages. Therefore, they concentrated more on creating the multimedia content, such as audio clips and photos, and then just plotted them to certain places on web mapping. However, when they were drawing maps by themselves, they have to consider more about the visual representation and communication of the spatial information according to the topic of their content. Their self-designed maps also enabled the students to realize the importance of the diversity of maps, as well as the necessity and possibility of human-centered mapping.

This paper introduces the course since *Manpo* came into use in 2013 and was carried on in the next two years. It will start from an overview of the classes, including the purposes, plans, and the students' background, Followed by a brief introduction of the software *Manpo* used as a tool in the course. Students' performances in each of their main tasks are introduced in details. Finally we discuss on the key points proved to be crucial or difficult for the students in the classes, which include positioning, guidance and storytelling. The plans of future activities and researches inspired by the classes are discussed with the conclusions.

2 Overview of the Classes

2.1 Purposes

The overall purpose of the course *Seminar of Culture Studies* is to enable the students to create location-based audio tour content with hand-drawn maps to introduce the surrounding areas of their campus. During the classes, the students are supposed to extend the sights of spatial information representation and dissemination, and inspire their creativity of location-related information design.

As the students' major is a bit far from spatial information sciences, the course was supposed to provide necessary knowledge and open their minds before they start creating their own works. The knowledge to be gained by the students was mainly on the following three aspects:

- *Basic knowledge on spatial information sciences*: concepts of LBS and UGC, basic principles of smartphone positioning, showcases of latest trends and applications and so on were supposed to be learnt.
- *Diverse of maps and spatial information representation*: in order to break the students' stereotype of maps, modern and historic maps of different designs for different purposes were shown. Various location-related UGC in different media, including podWalks and travel vlogs were also introduced.
- *Knowledge about the areas surrounding their campus*: brief history of the area near the campus was introduced, and then the students were supposed to investigate using Internet and visit the area to find more interesting places and stories by themselves.

As a seminar-style course, the creativity inspired in the practice is considered more important than passively accepted knowledge. We want the students to gain new experiences from the following aspects.

- *Creating hand-drawn maps according to the decided topics or stories*
- *Georeferencing analog maps for mobile positioning*
- *Storytelling using multimedia tools and locations on the maps*
- *Recording narrations and dialogs for the audio tours from original scripts created by themselves*
- *Role division in teams and work in cooperation*
- *Organizing the created content for mobile LBS applications*

An implicit purpose of this course is to practice English. In the classes of 2013 and 2014, it was optional to create English or Japanese audio tours. In 2015, the students were required to use English in all their content.

2.2 Background of Students

All the students in the course were from *School of Cultural and Creative Studies, Aoyama Gakuin University*. The number of students in the course was from more than ten to around thirty in each year. A majority of the students were from Grade 2, and the other few were from Grade 3. The specialties of the students were mainly related to media, communication and public policy, but a bit far from spatial information sciences. Table 1 shows some basic information of the students in 2015 as an example.

As audio tour content created by the students was supposed to be used on smartphones, and the tools were developed on Apple Inc.'s iOS, surveys were conducted to the students at the beginning of the classes to learn the ownership of smartphones and the usage of maps. The results showed that all of the students were smartphone users, and the ownership of iPhone was usually more than 50 % (and even more than 90 % in 2015).

In 2014, a survey using questionnaires was conducted to the students. Their answers showed that most of the students prefer to use Google Maps than other web mapping services on their smartphones, and more than 60 % of them often use maps with smartphones (as shown in Table 2). The result also showed that when traveling in unfamiliar places, students would like to use both web mapping and conventional paper maps (as shown in Table 3).

2.3 Procedure of the Classes

The course is held every year, usually from the middle of April to the middle of July. There is one class for 90 min every week, and totally twelve to thirteen classes in the semester.

Usually the first three to four classes are lectures for introduction and knowledge preparation, in which one class is carried out for the students to experience their

Table 1 Students in *Seminar of Culture Studies* in 2015

Grade	Gender	Number
Grade 2	Female	6
	Male	4
Grade 3	Female	0
	Male	3
Total		13

Table 2 Students' preference of web mapping and the frequency of using maps on smartphones in the course of 2014

Gender	Preferred web mapping service on smartphone	Frequency of using maps on smartphone
Female: 25 (73.5 %) *Male*: 9 (26.5 %)	*Google Maps*: 31 (91.2 %) *Apple Maps*: 3 (8.8 %) *Others*: 0 (0 %)	*Almost every day*: 3 (8.8 %) *Several times a week*: 17 (50 %) *Several times a month*: 10 (29.4 %) *Seldom*: 4 (11.8 %) *Never*: 4 (0 %)

Table 3 Students' preference of web mapping and the frequency of using maps on smartphones in the course of 2014

When traveling in unfamiliar places, how do you prefer the following maps?		
Maps in guidebooks	Maps in leaflets from local information desk	Maps in my smartphone
Always: 9 (26.5 %)	*Always*: 8 (23.5 %)	*Always*: 20 (58.8 %)
Often: 13 (38.2 %)	*Often*: 13 (38.2 %)	*Often*: 7 (20.6 %)
Sometimes: 3 (8.8 %)	*Sometimes*: 9 (26.5 %)	*Sometimes*: 4 (11.8 %)
Seldom: 9 (26.5 %)	*Seldom*: 4 (11.8 %)	*Seldom*: 2 (5.9 %)
Never: 0 (0 %)	*Never*: 0 (0 %)	*Never*: 1 (2.9 %)

seniors' works in previous years. The students are supposed to basically understand what they are going to achieve in the course, and we also explain the general procedures of their work.

After the lectures, the students start to work in groups. We usually provide three classes for them to discuss and decide their topics, investigate the walking route and the places to introduce, and prepare some raw materials. The next three classes are used for creating the content. They have to finish drawing the map and recording the audio, and then assemble all the materials using *Manpo*, and make georeferences. Tools and devices are provided for the students, which include watercolor pens, paper, scanners, computers, audio recorders and iPhones. Students can use their own devices as well.

After the first draft is finished, we encourage the students to spend enough time to test, modify and finalize their content. They are also advised to experience other groups' works and give comments. In this stage, each group is supposed to give a brief oral report (in one minute) of their progress at the end of the classes.

Finally the students should submit their works and reports, and give final presentations in the final class. Table 4 shows the schedule of the classes in 2015 as an example.

Table 4 Schedule of the course *Seminar of Culture Studies* in 2015

No.	Date	Content of the class	
1	April 21	Introduction of the class	*Lectures*
2	April 28	Try *Manpo* (seniors' works) off campus, questionnaires	
3	May 12	Fundamentals of location-based services and *Manpo*	
4	May 19	Template of scenario design, discussion of subjects	
5	May 26	Discussion on the details of content in groups	*Discussions, fieldworks and material preparation*
6	June 2	Decision of details—map drawing, scripts, role division	
7	June 9	Materials collecting, audio recording (indoor/onsite)	
8	June 16	Content making, test and modification	*Content making and Testing*
9	June 23	Assemble content with *Manpo*, make georeferences	
10	June 30	Test on site, refining and finishing *Manpo* content	
11	July 7	Experience and compare the results with other groups	*Finalizing, experiencing and conclusion*
12	July 14	Finalize and submit *Manpo* content and final reports	
13	July 21	Final presentation	

2.4 Expected Results

Each group in the class should create a set of audio tour content, which is designed for a walking excursion for 10–30 min in the surrounding area of the campus. As such content is created and used in *Manpo*, we call it *Manpo Content* for short. The basic components in a set of *Manpo Content* are shown in Fig. 1. We have the following requirements to the expected results of the students' work.

- **Theme**: Each group should decide a specific theme different form other groups before they start working. We assumed the target users of the content would be *the new students to the university* in 2013 and 2014, and changed the setting to *foreign visitors* in 2015. To help the students make decisions, we divided several areas and periods of the history for them to pick up.
- **Map**: Each group should at least create one map according to the theme. They can either use watercolor pens to draw maps on paper and scan it into computers, or make digital ones on computers using software like Photoshop or Illustrator.

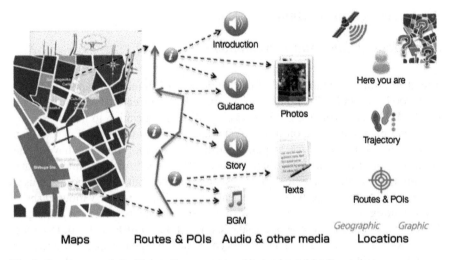

Fig. 1 Components of the *Manpo Content* supposed to be created by the students

- **Point of interest (POI)**: Each group has to make at least four to six POIs on the map to be introduced in details. They should investigate the places and create audio guides, photos and text information for them. They must confirm the locations of the POIs correctly, both on their maps and in the geographic world (as longitudes and latitudes). It is recommended that the POIs may have an order to visit.
- **Walking route**: Each group should design a route for the users to walk along to visit all the POIs. The walking route can either be drawn directly on the map, or be inputted by *Manpo*. For the latter case, we recommend making audio clips and binding them to the routes. Such audio can be guides of the tour or background music for walking. Audio of the walking route is optional.
- **Multimedia content**: The students have to use original audio clips and photos created all by themselves. They must make sure they have the copyright if they use others' resources, for example the background music. We recommend making lively conversations in the audio guides, instead of only narrations.

We suppose that the assumed target users can use the *Manpo Content* on their smartphones, locate their positions correctly on the hand-drawn maps, follow the route to visit all the POIs, get useful information and appreciate the students' stories, and enjoy the walking tour. Therefore, at the end of the course, the teachers and teaching assistants experience all the content to score the quality, usability and creativity. The best groups are presented with some special awards.

3 Introduction to *Manpo*

3.1 Methodology and Development

The ideas of *Manpo* mainly came from the theory of Ubiquitous Mapping (Gartner et al. 2007; Morita 2007) and the framework of Human-Centered Mobile Mapping (HCMM) (Lu and Arikawa 2013; 2015). HCMM pays attention to the diversity of maps and map representation in the digital and mobile environments. The current outstanding commercial web mapping and mobile mapping products have a phenomenon of *Googlization* (Muehlenhaus 2014) that the machine-generated multi-purpose digital maps in similar visual styles are dominating the maps users can access. At the same time, in conventional media, there are various maps of diverse designs for different specific purposes. Especially some well-designed tourist maps concern more about aesthetics, which has a historical link to cartography (Rees 1980; Kent 2005), but is less concerned in digital maps. However, they are not accessible in the mobile environments and cannot take the advantages of the location-aware mobile devices to provide Cartographic Interaction (Roth 2013). The HCMM framework proposes a solution to geo-enable the analog maps with georeferences and enrich them with multimedia content to provide LBS applications with a variety of maps.

As a prototype of HCMM, *Manpo* is designed for the usage of the digital guides of walking tours with local tourist maps. The name of *Manpo* came from a Japanese (also Chinese) word, which means taking a stroll. The development of *Manpo* was started in 2012, during the Ph.D. course of the first author. It is operated on Apple Inc.'s iOS platform. By the time the paper is written, *Manpo* is still a prototype

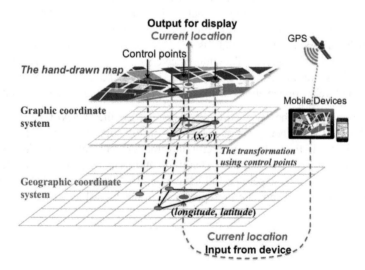

Fig. 2 Processes of positioning on a hand-drawn map using nearby control points

under improvement, but several iOS apps of hand-drawn maps and illustrated maps based on *Manpo*'s browse functions are published in Apple Inc.'s App Store.

Manpo provides functions including positioning on georeferenced maps, multimedia POI information invoked by the user's location, audio narration along walking routes, and so on. It applies control point-based transformation algorithms to calculate the user's position on the analog map with the control points that have geographic coordinates and corresponded graphic coordinates on the map (as shown in Fig. 2). In the early versions of *Manpo*, a *Nearest-Two-Point Based Similarity Transformation* was applied; and the later versions applied a *Nearest-Three-Point Based Affine Transformation*. It also provides edit functions for creating *Manpo Content* by georeferencing map images and bundling multimedia content to maps.

3.2 Browse Functions of **Manpo**

Figure 3 shows the map view and the POI view as two main interfaces of *Manpo*'s browse functions of the iPhone version, which is used by the students of the classes. The main functions are listed as follows.

- **Map browsing**: With the touchscreen, users can use gestures and virtual buttons to pan, zoom in, and zoom out the map image. More than one map can be included in one set of *Manpo Content*, and the user can switch among the maps.
- **POI information**: By tapping a POI button on the map image, users can view its information, including photos, texts, and audios. Users can take photos and store with the POI as their memories.
- **Audios on routes**: Polylines with arrows indicates the recommended walking routes on the maps. Audio clips like narrations and background music can be bundled to the routes.

Fig. 3 Main interfaces of *Manpo*'s browse functions of the iPhone version used in the classes of 2015. (*Left* the map view; *right* the POI view)

- **Current location indicator**: The user's current location obtained from the device is transformed through the georeferences and displayed on the map. The user can switch to the Egocentric Mode to keep the current location at the center of the screen.
- **Trajectory recording**: The user's trajectories are recorded with a certain frequency and shown as a series of dots on the map. A gradient effect is applied to the dots to show the moving direction.
- **Hints of nearby content**: The POIs or route audios can be invoked by the user's location in a certain range. The invoked content will be highlighted to call the user's attention.
- **Playback trajectories**: In Playback Mode, the recorded trajectories can be displayed in the order of time to show the user's movement. Functions like pause, forward, rewind, and speed control are available.

3.3 Edit Functions of Manpo

In the edit mode of *Manpo*, users can create new *Manpo Content* or modify the existing ones. The main processes of creating a new set of *Manpo Content* are listed as follows.

- **Importing maps**: The maps can be imported to *Manpo* from the camera of the device or through the photo album of iOS. Files of the map images can also be imported from a computer.
- **Creating POIs**: A POI can be created at a selected point on the map, and its corresponded geographic coordinates as longitude and latitude can be inputted using Apple Maps. The POIs also act as control points for positioning. The user can input the POI's title, text, photos and audio of the POI with the keyboard, camera and microphone of the devices. They can also be imported from multimedia files. Figure 4 shows the steps and main interfaces of creating and editing a POI.
- **Creating routes**: A polyline with arrows showing the direction as a part of the walking route can be inputted with the interface shown in Fig. 5 *(left)*. A split view showing both the hand-drawn map image and the Apple Maps are used for inputting the graphic coordinates and geographic coordinates of the polyline. Audio bundled to the polyline can be recorded using the microphone of the device, or imported from files.
- **Inputting additional control points**: In addition to the POIs, control points that are invisible to the end users of the *Manpo Content* can be inputted with the interface shown in Fig. 5 *(right)*. Pairs of the graphic coordinates and geographic coordinates can be inputted in a split view showing both the hand-drawn map image and the Apple Maps.
- **Packaging the content**: After inputting the maps, POIs, routes, control points and bundled media files, the content are packaged. The packaged content can be

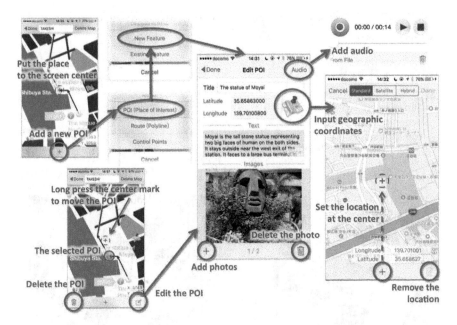

Fig. 4 Steps and interfaces of inputting a POI on a hand-drawn map using *Manpo* (the iPhone version used by the students in 2015)

viewed in the browse mode of *Manpo*, and can be exported to *Manpo* in another device with the help of Apple Inc.'s iTunes.

4 Performances of the Students

After the lectures, the students in each group decided their topics and started to work on creating their own *Manpo Content*. In 2015, the thirteen students formed four groups, as shown in Table 5. The students' roles and performances in the processes of content creating are introduced in this chapter, which will mainly take the classes in 2015 as examples. Figure 6 shows the scenes of the students working in groups in the classes.

4.1 Role Division

Students in each group worked as different roles in many aspects. As each group only had three or four members, each student worked as multiple roles.

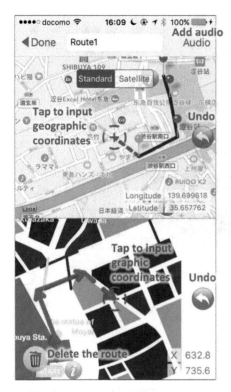 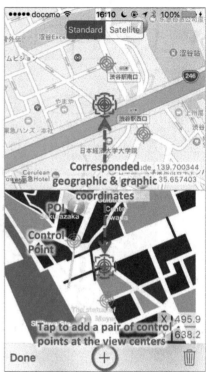

Fig. 5 Interfaces of inputting routes (*left*) and additional control points (*right*) on a hand-drawn map using *Manpo* (the iPhone version used by the students in 2015)

Table 5 Groups of the students and their chosen topics in the course *Seminar of Culture Studies* in 2015. (* The group number four is avoided because it is considered unlucky in Japanese culture)

Group no.	Students	Female	Male	Grade	Topic of the *Manpo* Content
1	4	2	2	2	Traditional Japanese local scenes in Yoyogi-Hachiman
2	3	2	1	2	An urban legend of a mysterious boy near Shibuya Station
3	3	2	1	2	"Cool" places near the Omotesando area
5*	3	0	3	3	Slope streets near Shibuya Station

- **Overall coordination**: Similar as creating a short movie, students in each group require comprehensive arrangement of their tasks. They need roles like *directors* and *producers*.
- **Voice casting**: As the content is designed for an audio tour, voice casting is one of the main tasks. Lively conversations for storytelling are recommended instead

Fig. 6 Students working in groups in the course *Seminar of Culture Studies* in 2015

of only third person narrations. Roles including at least one *main character*, one *supporting character*, and one *narrator* are needed.

- **Map Drawing**: The *mapmakers* play an important role in presenting the topics with correct geographic information. The mapmaker should be in charge of, or cooperate with georeferencing.
- **Content organizing**: The *designers* of the POIs and routes decide the number, locations and topics of the POIs and control the durations of the audio guides on the routes. The *scriptwriters* create the lines for voice casting.
- **Media editing**: The *audio editors* clip the recorded voices and can add background music. The *visual editors* design icons, logos, cover images, and modify the photos of POIs taken by the *photographer*.
- **Manpo editing**: The *Manpo editors* assemble the materials using *Manpo*, and create georeferences. The *testers* test the functionality and positioning indoor and outdoor for modifications.

4.2 Map Drawing

Although most of the students had little knowledge or no experiences about mapmaking before, they tried their best to present the geographic information related to their topics. Actually, very few instructions about how to draw maps were given, as the course was not on cartography. Instead, hand-drawn maps in tourist guidebooks and leaflets were shown as examples. As the walking tour maps are usually large-scale for small areas, students were not required to draw every object precisely as topographic maps. Lively and vivid presentations according to the topics were recommended, for example the focused places can be drawn exaggeratedly in easy-to-understand shapes.

Most of the groups draw their maps using watercolor pens and color pencils. A few of them use computers to draw or modify the maps. Some of the students' maps in the three years are shown in Fig. 7. The styles of the maps are various, although some of them look a bit naïve. Many interesting representations can be found, for example, Group 5 in 2015 used shades to present the altitude changes of the slope streets.

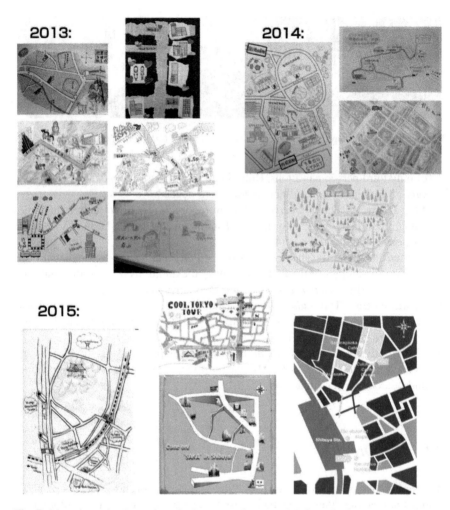

Fig. 7 Maps drawn by the students in the course *Seminar of Culture Studies*

It can be found that the maps of 2013 were more rough and contained more mistakes, as the students cared less on the geometric accuracy. In the later course, we advised the students to care more about the roads, and make sure the topologies of roads are correct. Actually, we found the students in 2014 and 2015 tended to draw their maps based on Google Maps or Apple Maps, as they at first made sketches of the important roads copied from the accurate maps, and then draw other features created by them.

4.3 Script Making

The students were supposed to create scripts at the beginning as detailed guidelines for making content on the maps. Each group should at first define the *title*, *theme*, *purpose*, *target users*, *geographic region* and *temporal range* of the content in their scripts. After that, they were recommended to organize the scripts of the POIs and routes in a list in the order of the tour, in the form shown in Table 6.

4.4 Audio Recording

In the scripts of the POIs and routes, the students wrote lines for recording the audio guides. They were suggested to use storytelling to provide tourist information to the users in a more attractive and lively way, and build communications with the user. Thus, conversations among different characters are considered better than neutral third-person narrations. The following is an example of the script for the conversation created by Group 5 of 2015, in which introduction of the origin of the strange name of a slope street are embedded in a dialogue of questions and answers.

POI 5: Ma Saka (Slope Street)
Mr. Slope *"Here is Masaka."*
Slope Jr. *"Masaka? That's strange name."*
Mr. Slope *"I thought you'd say that. Literally, Masaka means 'really', 'unbelievable' and 'that can't be true!' in Japanese."*

Table 6 The form of the script list for the POIs and routes

POI/Route	Title	Description	Coordinates	Photo	Audio script
POI 1 (Start Point) <Duration>	<string>	<string>	(x, y) (latitude, longitude)	<images>	Lines for the conversations or narrations
Route 1 (POI 1 to POI 2) <Duration> <Distance>	–	–	{(x, y)} {(latitude, longitude)}	–	Lines for the conversations or narrations
POI 2 <Duration>	<string>	<string>	(x, y (latitude, longitude)	<images>	Lines for the conversations or narrations
...					
Route N-1 <Duration> <Distance>	–	–	{(x, y)} {(latitude, longitude)}	–	Lines for the conversations or narrations
POI N (Ending Point) <Duration>	<string>	<string>	(x, y) (latitude, longitude)	<images>	Lines for the conversations or narrations

***Slope Jr.** Masaka, who thought out such a strange name?"*
***Mr. Slope** "It is Shibuya LOFT. It is the origin of Japanese chain stores that sells everyday commodities. The store faces to this slope street. LOFT has collected naming ideas and decided the name MASAKA, because the word can give a deep impression. The word "Masaka" also has a meaning of 'the slope in the space among the buildings', if we split it into two Japanese words 'Ma' and 'Saka'. Ma means between and Saka means a slope street."*
***Slope Jr.** "Hmm, I see"*

With the scripts, the students casted the characters and recorded audio guides with voice recorders. A major number of the groups recorded indoor to make sure of the quality of the audio. Some of the groups recorded outdoor to create a sense of presence, although there may be noise from the streets. Recording audio outdoor also helped matching the duration of the audio to the time for walking. Background music was recommended as it can develop certain atmosphere while walking, but the students were asked to take care of the copyright issues.

4.5 Content Assembling and Testing

With all the maps, scripts, audio clips and images prepared, the students imported all the materials to the iPhones, and assembled them using *Manpo*'s edit functions introduced in Sect. 3.3. This task was considered easy and could be completed in one hour. However, making correct georeferences of the POIs were proved to be difficult for the students, which affected the accuracy of positioning on the hand-drawn maps.

In the courses of 2015, outdoor testing of positioning and the functionality of all the POIs and routes became a required task after the content had been completed. A tool for testing the positioning was developed in *Manpo*, which can input simulated walking trajectories on Apple Maps and display them on the hand-drawn maps. With the actual or simulated trajectories, students were supposed to find out the mistakes of georeferences through the inaccurate positioning results.

4.6 Final Presentation

Finally, the students experienced their audio tours outside and made final reports. In the final presentations, each group introduced their works to the whole class in ten to fifteen minutes. Each presentation included an overall introduction to the *Manpo Content*, the roles of each student in the group, the most worked out points, the difficult points, and the future improvements. Good comments and suggestions to the future development of *Manpo* were also welcomed. Figure 8 shows the scenes of the students experiencing their content and giving final presentations.

Fig. 8 The scenes of the students experiencing their content and giving final presentations in 2015

Fig. 9 Awards and prizes for the students in the course *Seminar of Culture Studies* in 2015

The professors and teaching assistants of the course experienced the audio tours of all the groups and gave marks to them on the strong and weak points. Awards and prizes were given to the best groups for encouragement. Figure 9 shows the prizes in 2015, which are customized mugs with the title, map and icon of the content, and the names of the group members.

After the classes, the *Manpo Content* was also experienced by the students in the next year as a review their seniors' works, as well as an evaluation of Manpo's browsing functions. In 2014, a survey of the usability of Manpo and user satisfaction has been conducted with questionnaires and interviews. The results can be found in (Lu and Arikawa 2015).

5 Discussions

The students' performances in the classes provide hints for improving *Manpo* and the future development of HCMM. Although *Manpo* is proved to be functional when used by the students, who can be considered as non-professional users, some unexpected problems and experiences in the classes are worth in-depth discussions.

5.1 The Difficulties of Georeferencing and Positioning

A major number of the students had difficulties in achieving accurate positioning on their hand-drawn maps. Especially in 2013, all the groups had obvious errors in positioning. Therefore, in 2014 and 2015, extra short lectures on the brief principles of georeferencing and positioning with good and bad examples were presented to the students before they were going to start making georeferences. However, still many groups could not get expected results. The following points that influenced the positioning qualities are discussed.

(1) **Reliability of Smartphones' Location Information**

Although it is a common sense to the researchers that the accuracy of the location sensors, such as GPS, of smartphones is easily influenced by the circumstance, the students did not have appropriate understandings of the possible errors that their smartphones may bring to the positioning on their hand-drawn maps. Figure 10 shows the results of a survey to the students in 2014 using questionnaires about their awareness of the errors of the location information from their smartphones, in which only very few students had correct understandings.

The target areas of the students' *Manpo Content* are mainly downtown areas in Tokyo, which have dense tall buildings and narrow streets. For example, Fig. 11 shows the errors of location information when walking around the area near Shibuya Station, which is the target area of the students' content. A simple program was developed in iPhone to record and show the locations and the accuracy data, which was obtained from the device, on Apple Maps. Although the data shows that most of the points can reach an accuracy of ten meters, it is still sometimes confusing as the streets are narrow and dense. It can be found that in some places even the data show an accuracy of less than five meters, the actual error is more than ten

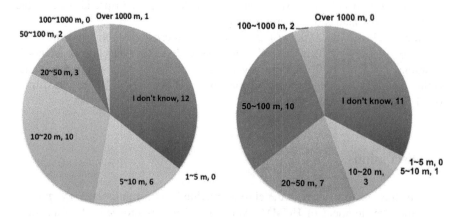

Fig. 10 Results of questionnaires on the awareness of the accuracy of smartphones' location information to the students in 2014. (*Left* the best accuracy when the smartphone is in a good circumstance; *right* the worst accuracy when the smartphone is in a very bad circumstance.)

meters. Such examples were shown to the students in the classes to help them understand the reliability of the location information of smartphones.

(2) **Mistakes and Distortions of Map Drawing**

In the class of 2013, the students drew the maps in an arbitrary way as they had little experience and were not well informed. Some maps had obvious mistakes that some areas and roads were missing or in wrong topologies. On such mistaken maps, accurate positioning can never be achieved. Some other maps were not mistaken, but had distinct distortions that were too large to be handled by the control point based positioning, as it became too difficult to find appropriate control points.

In 2014 and 2015, good and bad examples of hand-drawn maps were shown to suggest the students taking more care on the roads, landmarks and their topologies. As the students tended to draw their maps based on sketches of roads copied from Google Maps or Apple Maps, fewer obvious mistakes were found.

(3) **Improper Placement of POIs and Control Points**

Even the students avoided mistakes in map drawing, they could not achieve accurate positioning without appropriate placement of georeferences. Many of the students were found making mistakes in placing the correctly corresponded points on their hand-drawn maps and Apple Maps for the graphic and geographic coordinates of the POIs. The typical mistakes of the improper POI placement are categorized into the following three types.

Fig. 11 Accuracy of an iPhone's location information when walking around Shibuya Station in Tokyo. (Photos on the *left*: the scenes of the streets near Shibuya Station; maps on the *right*: location information and its accuracy obtained from the iPhone on Apple Maps)

- **Mistaken geographic locations**: Such mistakes were usually due to the insufficient geographical knowledge of the target areas and the careless fieldworks of investigating the POIs. However, sometimes the deficiency of detailed information of some local area (e.g., the narrow paths in the woods of a park) in the web mapping (i.e. Apple Maps) used for inputting geographic coordinates in *Manpo* also brought difficulties to the students. In this case, walking trajectories recorded in the fieldworks can be utilized for confirming the geographic locations.
- **Improper placement on exaggerated drawings**: In the students' hand-drawn maps, some drawings of the geographic features are significantly exaggerated, which cannot easily be matched to their original shapes and locations in the web mapping (i.e. Apple Maps). Because of the limitations of the two point based positioning algorithms by the early versions of *Manpo*, the placement of the control points should take care of their relative spatial relationships to make sure that such relationships are kept geometrically similar in the hand-drawn map and in the geographic space. However, it proved to be difficult for the students as non-professional users. After the three point based algorithms had been applied, the results could be better, but careful POI placement is still necessary for the exaggerated drawings to avoid destroying the relative spatial relationships of the control points.
- **Mismatched placement on nonpoint features**: It often occurs when a nonpoint geographic feature is represented by a POI as a point object, under the scale of the hand-drawn map. For example, one group in 2014 tried to use a POI to give introductions of a street, however, they put it at the center of the street on Apple Maps, but at the intersection on their hand-drawn map, which caused distinct errors. Therefore, the students were suggested using coherent and easy-to-identify points of the large features on both maps. For exam*ple,* Fig. 12 shows the improvement of positioning results after moving the POIs of the large stadiums from their centers to their entrances near the roads.

5.2 Guidance for Improving Positioning

After analyzing the problems that prevented the students from achieving accurate positioning in their content, additional guidance was added to the classes with optimized authoring functions in *Manpo*. Table 7 lists up the short lectures, suggestions and requirements as additional guidance giving to the students in each year.

As the students were lack of background knowledge on positioning, they were found to have no sense about how their input of control points could influence the final positioning result. From 2014, a short lecture on positioning of smartphones was given to make students understand the accuracy and limitation of the location information in their smartphones. When they were going to make georeferencing, a

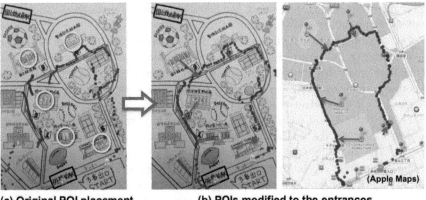

(a) Original POI placement   **(b) POIs modified to the entrances**

○ Original places of the POIs on the hand-drawn map
❘ Original locations of the POIs on Apple Maps
→ Modification of the POIs from the original locations to the new ones
••••• The user's moving trajectory as positioning results on the hand-drawn map

Fig. 12 The improvement of positioning results after modifying the placement of POIs. (The original content is created by a group of students in 2014.)

Table 7 Short lectures, suggestions and requirements as additional guidance giving to the students in each year for improving positioning of *Manpo Content*

Additional Guidance	2013	2014	2015
Short lecture of positioning of smartphones	None	Applied	Applied
Short lecture of *Manpo*'s positioning methods	None	Only poor examples shown	Applied
Suggestions on map drawing	None	Applied	Applied
Additional control points	None	Optional	Required
Outdoor testing	Optional	Optional	Required
Indoor testing	None	None	Tools provided

short lecture was given for making proper placement of POIs, with poor examples of the previous years. In 2015, additional control points other than POIs became required in the students' *Manpo Content*, as the numbers of POIs were usually not enough to achieve acceptable accuracy all over the maps.

Most groups in 2013 and 2014 ignored testing, and failed in finding out some obvious mistakes in map drawing and POI placement before they had submitted their content. Actually it could be easy to find such problems if they could use their content in the actual places. For example, the comparison of the recorded walking trajectories on the hand-drawn map and Apple Maps will show the mistakes of the control points; the POI not invoked by the location just in the annotated place will show the improper POI placement; the audio guide of the route finished too early will show the failure in time control of the content.

One reason of ignoring testing is the limited knowledge of the principles of georeferencing and positioning, for which the students was not aware of the possible large errors that could be brought by their unconscious or arbitrary mistakes. Another possible reason is that the target areas of some groups were too far from the classroom for the students to conduct testing in the classes.

For the latter reason, a test function of *Manpo* using simulated walking trajectory was developed for the students to test the positioning results of their georeferences indoor. The simulated trajectory can be inputted by adding key points (i.e. the turning points) on Apple Maps, while the inner points between the key points are linearly calculated with certain recording time interval and walking speed. In the edit functions of *Manpo*, the simulated trajectory can be imported to the georeferenced hand-drawn maps to check the results of positioning calculation.

With the testing function, the groups in 2015 were required to conduct indoor and outdoor test, and modify their content before submitting. Obvious improvements of the quality compared to the previous years could be found from the results of the classes in 2015.

5.3 Storytelling in Students' Manpo Content

Story is always an efficient way for human to convey knowledge since ancient times. Stories can connect the isolated events or information using sequences or causal links. Such sequences and links can connect the unknown facts or things to the existing knowledge, and stimulate the curiosity to further exploration, which can make the new knowledge easy to memorize and spread. Storytelling is an important factor to the success of the students' *Manpo Content*, as the audio tours have to attract the users and convey knowledge of the target areas to them in a comfortable way. If the POIs and their audio guides are just isolated points with no relations, the content is easy to be considered boring. Therefore, the students were suggested embedding the facts or the descriptions of the places they wanted to introduce to a story, in which they should organize the POIs in an appropriate order and infer the connections between them in their narrations or dialogues. It can be found that some of the groups tried to connect the current places to the certain events or persons in history and legends, which made the audio guides of their POIs and routes more attractive.

In recent years, tools for storytelling with digital maps are brought to the users, such as Story Map of Esri (2016) and StoryMap JS of Knight lab (2016). In these applications, digital maps are used for illustrating the slides or as base maps for attaching multimedia materials. As map is considered as a *graphic communication vehicle* (MacEachren 1995) for conveying certain spatial knowledge, map itself also often contains stories in a broader concept. In the students' *Manpo Content*, apart from the stories in the audio content, storytelling is also important in their excellent hand-drawn maps. Firstly, the exaggerations and simplifications of the depictions can easily attract readers' attention. Secondly, the highlighted spots (e.g., landmarks) can be anchor points that are connected to readers' existing knowledge, which makes it easier to extend cognitions to unfamiliar regions (Golledge 1999). Thirdly, on some maps, routes connecting the checkpoints for readers to follow and synchronize can help find the destinations when used on site (Etienne et al. 1999). Finally, maps can also have characters, while some are explicitly depicted on the maps; some are implicitly referring to the readers.

6 Concluding Remarks

The course in the three years was proved to be a success, as the most students have shown passions in working on the interesting content, and they were satisfied and inspired after the classes. The course also showed the functionality of *Manpo* and the feasibility of the framework of *Human-Centered Mobile Mapping* in utilizing various analog maps, including hand-drawn maps, for tourism and storytelling with smartphone applications. One of the students even applied *Manpo* to her undergraduate project in 2015. Besides, the classes revealed some deficiencies of *Manpo* and the required new functions and features by the students as non-professional users. Especially the difficulties that the students encountered in creating appropriate georeferences on their maps are very important hints for improving *Manpo*, as the functionality and accuracy of positioning is crucial in a LBS application (Raper et al. 2007). With the help of the classes and students, *Manpo* has evolved a lot during the three years, as well as the guidelines for ordinary users to create georeferenced analog maps.

The classes encouraged the authors that, new mapping services and applications can be created with the approach of *Manpo* to increase the diversity of maps in mobile environments. In that case, platforms need to be established to involve cartographers, illustrators, ordinary users and mapping communities, to contribute well-designed maps and accurate georeferences cooperatively. After the classes, a collaborative project named *Manpo Project* has been started to collaborate the powers and resources of researchers, developers and local communities to provide local map applications with *Manpo*'s approach. Excellent sets of *Manpo Content* (*TAKESHI* and *Come on! "SAKA" in Shibuya*) that created by the students in the course have already been developed to independent applications under *Manpo Project*, and published in Apple Inc.'s App Store. Cooperation with local tourist organizations for developing applications with their tourist maps are undergoing smoothly. It is predictable that, user-generated content in LBS can be extended to a broader range, which include not only media associated to locations but also user-created maps and georeferences, and will contribute to more diverse, practical and attractive LBS applications.

References

Arikawa M, Tsuruoka K, Fujita H, Ome A (2007) Place-tagged Podcasts with Synchronized Maps on Mobile Media Players. Cartography and Geographic Information Science 34(4): 293–303. doi:10.1559/152304007782382945.

Esri (2016) Story Map. https://storymaps.arcgis.com/en/. Accessed 27 August 2016.

Etienne A, Maurer R, Georgakopoulos J, Griffin A (1999) Dead Reckoning (Path Integration), Landmarks, and Representation of Space in a Comparative Perspective. Wayfinding Behavior: Cognitive Mapping and Other Spatial Processes. The Johns Hopkins University Press, Baltimore: 197–228.

FitzGerald E (2012) Creating User-generated Content for Location-based Learning: An Authoring Framework. Journal of Computer Assisted Learning 28(3): 195–207. doi:10.1111/j.1365-2729.2012.00481.x.

Gartner G, Bennett D, Morita T (2007) Towards Ubiquitous Cartography. Cartography and Geographic Information Science 34(4): 247–257. doi:10.1559/152304007782382963.

Golledge R (1999) Human Wayfinding and Cognitive Maps. Wayfinding Behavior: Cognitive Mapping and Other Spatial Processes. The Johns Hopkins University Press, Baltimore: 5–45.

Kaji H, Arikawa M (2013) Blog Based Personal LBS. Distributed, Ambient, and Pervasive Interactions - Lecture Notes in Computer Science, Springer, Berlin Heidelberg: 122–127. doi:10.1007/978-3-642-39351-8_14.

Kent A (2005) Aesthetics: A Lost Cause in Cartographic Theory?. The Cartographic Journal 42(2): 182–188. doi:10.1179/000870405X61487.

Knight lab (2016) StoryMap JS. https://storymap.knightlab.com. Accessed 27 August 2016.

Lu M, Arikawa M (2013) Location-based Illustration Mapping Applications and Editing Tools. Cartographica: The International Journal for Geographic Information and Geovisualization, 48 (2): 100–112. doi:10.3138/carto.48.2.1835.

Lu M, Arikawa M (2015) Creating Geo-enabled Hand-drawn Maps: An Experiment of User-generated Mobile Mapping. International Journal of Cartography, 1(1): 45–61. doi:10.1080/23729333.2015.1055110.

MacEachren A (1995) How Maps Work: Representation, Visualization, and Design. The Guilford Press, New York.

Mooney P, Rehrlb K, Hochmairc H (2013) Action and Interaction in Volunteered Geographic Information: A Workshop Review. Journal of Location Based Services 7(4): 291–311. doi:10.1080/17489725.2013.859310.

Morita T (2007) Theory and Development of Research in Ubiquitous Mapping. Location Based Services and TeleCartography - Lecture Notes in Geoinformation and Cartography, Springer, Berlin Heidelberg: 89–106. doi:10.1007/978-3-540-36728-4_7.

Muehlenhaus I (2014) Web Cartography: Map Design for Interactive and Mobile Devices. CRC Press, Boca Raton.

Raper J, Gartner G, Karimi H, Rizos C (2007) A Critical Evaluation of Location Based Services and their Potential. Journal of Location Based Services 1(1): 5–45. doi:10.1080/17489720701584069.

Rees R (1980) Historical Links between Cartography and Art. Geographical Review 70(1): 60–78. doi:10.2307/214368.

Roth R E (2013) Interactive Maps: What We Know and What We Need to Know. Journal of Spatial Information Sciences 7: 59–115.

Yap L F, Bessho M, Koshizuka N, Sakamura K (2012) User-generated Content for Location-based Services. Virtual Communities, Social Networks and Collaboration. Springer, New York, NY: 163–179. doi:10.1007/978-1-4614-3634-8_9.

Gamification as Motivation to Engage in Location-Based Public Participation?

Sarah-Kristin Thiel and Peter Fröhlich

Abstract In the last decade there have been various attempts to foster location-based public participation. Researchers as well as municipalities have explored several methods, among them the implementation of digital participation tools. However, the actual level of participation has remained low. This gives cause to both analyze the reasons for the ineffectiveness of the explored methods as well as to "think outside the box" and try novel approaches. One of such approaches is gamification. This research investigates the effects adding game-inspired elements can have on participation. In particular, we focus on how motivational factors differ for gamified participation and whether motivations influence citizens' level of engagement. To do so, we conducted an experiment with a location-based mobile participation prototype. Our results suggest that participation in a gamified application was higher than in one without, but also decreased intrinsic motivation, which was found to influence activity in location-based public participation. The strongest reported motivation was reporting issues regarding urban topics and stating one's opinion.

Keywords Gamification · Motivation · Public participation · Location-based

1 Introduction

Many democracies are nowadays facing a growing alienation of citizens from institutional politics, which results in low engagement and decreasing voter turnout. This poses a problem for both citizens as well as governance. Governments worry about their position as en-actor of democracy as by definition democracies get their

---

S.-K. Thiel (✉) · P. Fröhlich
Innovations Systems Department, AIT Austrian Institute of Technology GmbH, Giefinggasse 2, 1210 Vienna, Austria
e-mail: sarah-kristin.thiel@ait.ac.at

P. Fröhlich
e-mail: peter.froehlich@ait.ac.at

© Springer International Publishing AG 2017
G. Gartner and H. Huang (eds.), *Progress in Location-Based Services 2016*, Lecture Notes in Geoinformation and Cartography,
DOI 10.1007/978-3-319-47289-8_20

legitimacy from citizens. Citizens on the other hand miss their chance to influence policies and politics, which means that other people will decide for them, potentially causing dissatisfaction.

Reasons for a decreasing involvement include decreased levels of trust, ignorance (cf. Theory of rational ignorance; Krek 2005) or lack of interest. In contrast to the reported alienation, it has been argued that citizens do want to be heard, especially if it concerns aspects that affect the quality of their life and their experience of place (Torres 2007). However, due to the long-term decline in civic involvement (cf. Putnam 2011), either citizens' civic skills needed to articulate requests have decreased (Vigoda 2002) or they have already grown accustomed to the "easy chair of customer" (Vigoda 2002). Next to a dearth of necessary resources (such as skills) and lack of interest, Brady et al. (1995) add another reason for non-engagement by stating that citizens refrain from politics because nobody asked (them). It is hoped that the significance of some of these factors will be reduced by recent technological advancements and the emergence of digital location-based public participation tools. In fact, there is evidence that participants are not ignorant to public issues and do want to engage or at least share their views (e.g. Stuttgart 21, Occupy movements in New York). Citizens currently seek different forms (such as social media or bottom-up movements) than traditional and official ways to participate. A relevant question hence is how to design novel participation methods that will be accepted?

This research focuses on exploring strategies to increase people's motivation to engage. In particular, we analyze the impact gamification can have in this respect. Gamification has been proven to increase engagement with usage of systems in a variety of domains (e.g. Itoko et al. 2014, Thom et al. 2012, Ueyama et al. 2014). This strategy encompasses incorporating elements and mechanics inspired from games into applications. It is anticipated that this will make the experience of using the system more "gameful" and thus more enjoyable (e.g. Flatla et al. 2011), which if successful in turn increases users' motivation to engage. Up until now, research combining the domains of digitally mediated participation (= e-participation) with gamification or game research has mostly focused on the design of full-fledged (*@Stake, ParticiPécs*) or serious (e.g. Poplin 2014) games aiming to leverage from the assumed positive influence of game elements on motivation. The rationale behind turning public participation processes into games lies in the hope of evoking learning processes (Gordon and Baldwin-Philippi 2014). By playfully exploring options and engaging with others (*collective reflection*, Devisch et al. 2016) it is anticipated that citizens deepen and enrich their knowledge and understanding of issues, roles of key actors and processes (Shkabatur 2011). Gamifying participation—as opposed to creating games—follows a different approach. In the first instance the integrated game elements ought to evoke the interest of previously non-engaged (*onboarding*; Bowser et al. 2014). Secondly, incorporating game mechanics (e.g. competition, achievement) should make participation a more engaging experience (*fun*, Eveleigh et al. 2013, Flatla et al. 2011). Together these effects are posited to foster public participation.

Instead of focusing on the design of a platform, this research concentrates on analyzing the actual benefits of gamifying digital location-based public participation tools. While this is a first step on our research agenda, our overall goal is to explore how different game elements affect participation. This will ultimately contribute to work related to gamification frameworks, game research in general and the design of gamified applications. This paper explores the questions of whether gamification has an impact on location-based public participation and specifically whether the motivations of participants differ when using a gamified engagement platform opposed to one without game elements. We report findings on whether people motivated by certain aspects engage more (= quantity) than others. To do so, we evaluated two versions of a sophisticated mobile participation prototype in two field studies, where one group received a version with and another one without game elements. The specific questions that we address with this research are: *RQ1*: Do motivations differ when using a gamified participation application? *RQ2*: Are people with certain motivations more active in location-based public participation?

2 Background

This research is placed at the intersection of two broad research domains: gamification and e-participation. The following section provides some background information on both domains explaining main concepts and discussing the relevance of combining the two.

2.1 E-Participation

E-participation has been defined as "the utilization of information and communication technology in order to extend and deepen the political participation of citizens" (Macintosh 2004). Its objective is to "motivate and engage wider citizens through diverse modes of technical and communicative skills to ensure broader participation in the policy process, real-time qualitative and accessible information, transparent and accountable governance" (Islam 2008). Having seen a shift toward greater direct citizen involvement in the latter part of the 20th century in response to the perceived "political apathy", digital platforms that allow citizens to raise their voice and discuss urban topics among themselves but also with city officials have been setup by a large number of municipalities around the world (Coronado and Vasquez 2014).

While evaluations are yet scarce and research in this area still being in its infancy, several scholars have noted that the vast majority of these e-participation tools are not suited to establish an efficient participation as they do not actively involve citizens (e.g. Åström and Grönlund 2012). Among the critics is the point that these tools often only provide a one-way communication channel and are designed as dissemination rather than an ideation tool (Thomas and Streib 2003). Using Macintosh's (2004) framework for categorizing levels of participation, they are merely implementing the first level, enabling. Only if the participation method allows citizens to actively become involved and citizens are treated as partners of authorities, they become empowered and a sustainable participation is possible. In recent years, also mobile applications aiming to increase location-based public participation have been developed. They are still few in number and apart from reporting apps (e.g. *FixMyStreet*) they are mainly being created by academia (e.g. *Mobile Democracy, Täsä*). Furthermore, a recent study has shown that also those applications do not achieve a two-way communication between citizens and authority (Ertiö 2015), a shortcoming that limits the amount of influence citizens can exert on decision-making and agenda-setting (Nam 2011).

2.2 Gamification

Gamification has been broadly defined as the use of game-elements in non-game contexts (Deterding et al. 2011). While several scholars use Deterding's definition, many researchers have come up with their own definitions. Likewise, there is no standard on how to do gamification. Somewhat loose definitions of the strategy leave it up to the designers or developers to interpret what those "game-related elements" actually are and how they should be applied to a system.

Bringing gamification and location-based public participation together is arguably not a new approach. Especially in the related domain of citizen science, many projects exist that make use of game elements (e.g. *BioTracker* (Bowser et al. 2014), *OldWeather* (Eveleigh et al. 2013)). A recent review of projects regarding gamified participation has found that the majority of developed applications utilize achievement systems, which foster extrinsic motivation (Thiel 2016). Some scholars, including supporters of gamification have criticized that gamification has become a synonym of adding points and badges for interactions within an application (Nicholson 2012). Criticism includes that by adding rewards or incentives to an activity, people might feel that the activity is not valuable enough to be done without the help of rewards. Moreover, seeing rewards as a call for action by system operators, users might feel like their actions are being controlled and it is expected of them to perform these activities. They argue that pleasure is not additive and rewards can backfire (e.g. Kohn 1999). For instance Kohn's studies have shown that the quality of the activity will decrease, if users are offered rewards or incentives. In addition, participants stated to not have enjoyed the activity as much as before. This indicates that their intrinsic motivation was reduced and replaced by extrinsic motivation (i.e. the longing for the reward). Other scholars reported similar findings, which were verified by Lepper et al. (1973) who refers to this mannerism as the "overjustification" effect. Yet, there is more to the strategy than incentivizing participation. In contrast to incentive systems, which are a common method in business and marketing (e.g. collecting loyalty credits), gamification can also mean that the interaction within an application becomes more "gameful". This can be achieved by employing mechanics such as scarcity (e.g. time constraint), competition (e.g. comparisons with other users) or/and meaning (e.g. adding a narrative). While several studies have reported on the success of gamification (i.e. by increasing the usage) (cf. Bowser et al. 2013a, Poplin 2012), there are also a notable number of works that claim the inclusion of game-like features did not have any significant effects (cf. Flatla et al. 2011). In light of these both positive and negative effects of gamification, it is highly relevant to investigate the effect of gamification and especially the impact of individual elements in various contexts.

2.3 Gamified Participation

Although not entirely new, academia has only recently started to explore the affordances of adding game-inspired elements to location-based public participation platforms. Most of these gamified applications are intended for urban planning or citizen science. Evaluating their gamified citizen science application *OldWeather* where users can transcribe historic ship logs, Eveleigh et al. found that while some were motivated by competitive game mechanisms (such as collecting points), these turned out to be a reason for discontinuing participation for others (Eveleigh et al. 2013); echoing earlier findings that not everyone shares a competitive nature (Preist et al. 2014). Yet again the results of Bowser et al. show that gamifying citizen science does have the potential to engage new groups (Bowser et al. 2013a). Their study participants indicated to be motivated by social motivations (e.g. community membership), personal benefits (e.g. fun and education) as well as competition and rewards (here badges). On a similar note, analyzing the impact of added game elements in the civic engagement mobile app *DoGood*, Rehm found that competition was the least effective element, while social aspects (i.e. the feeling of being part of a group) supported participation (Rehm 2015). These opposing findings indicate that the effects of gamification are highly dependent on both the implementation and context.

There are also some examples of gamification of location-based public participation (complying with our definition of the term). *Love Your City!* for instance is an interactive platform using augmented reality that allows users to annotate places with suggestions or organize community gatherings (Stembert and Mulder 2013). Recognizing limited authority levels, in this application there are roles with varying responsibilities. *Love Your City!* implements game mechanics such as scarcity (time constraint), achievement (usage statistics and heart points) and personalization (profile). As the evaluation of *Love your city!* did not focus on the impacts of game elements, we cannot report on any implications of this study for gamified participation.

Another example targeting urban planning is *B3- Design your marketplace* (Poplin 2014). This application can be classified as a whole game rather than implementing a number of game elements. Users can design their own vision of a marketplace from a toolbox by dragging it on the game board. Designs can be rated and discussed by other users. Further game-like aspects within *B3* are a little helper that guides users through the system. The evaluation of this system showed that especially young people appreciated the game elements whereas elderly did not seem to particular notice of them.

In order to facilitate the process of deciding on a new location for the university campus in Hamburg, a non-competitive simulation game was created (Poplin 2012). Within the game users plan the new campus as a team while trying to balance constraints such as available budget. Results from a user study revealed that for the game to be effective in motivating usage, the structure of the game needs to match the objectives of all stakeholders (citizens and authorities). This also entails designing the game environment as close to the represented real-world situation as possible. Another finding was that the game was perceived as too complex, motivating the question not only which game elements to integrate but also how many.

3 Studying Citizen's Motivation to Engage

While there already is a significant body of work in the domain of gamified participation (Hamari et al. 2014), studies on whether and more importantly how incorporated game elements work to increase citizens' motivation to become involved in political decision-making processes are still scarce. The objective of this research is to explore the effectiveness of gamification in location-based public participation. As a central aspect of gamification is to influence users' motivation towards a desired behavior (Hamari et al. 2014), this paper focuses on what motivated users to engage. We conducted two field trials (between-group design) to get more profound insights into how the addition of game-inspired elements to digital public participation tools would influence participation. Participants of both trials used the same application: in the one trial, we deployed a version of our mobile participation prototype that contained game elements (called *Game trial* hereafter) and in the second one, we deployed a version where we had removed all such elements (called *Non-Game trial*). As both versions used the same backend, we decided to run two separate studies. By this we wanted to avoid that participants' activity of the one trial influenced the activity of participants from the other trial. Furthermore, this design ensured that participants would not meet each other in the real world and discuss different versions/elements of the application. Regarding the basic structure of the study design and the given tasks, the Non-Game trial and the Game trial were identical. A previous evaluation of these two studies had shown that levels of engagement did not differ between the two studies (Thiel and Lehner 2015). Having focused on the quantitative aspect of participation (i.e. how many contributions or comments were posted) and themes addressed by participants, the results led us to conclude that gamification did not impact the level of participation nor thematic focus. In this paper, we want to explore whether gamification has an impact on motivations to engage. In particular, we investigated whether the motivations of using a gamified participation tool differed compared to one without game elements. We further contrast these insights with levels of participation in order to test for dependencies between motivation/intention and action.

3.1 Study Design

For both studies we recruited participants through volunteer sampling. Making use of the research institute's user database, we sent out e-mail invitations asking them for participation in a one-day field study on location-based public participation with mobile devices.

Both trials took place during one afternoon in the same district in Vienna, Austria. Participants were instructed to walk around a pre-defined area in the city and post about aspects they came across and found worth sharing. To help frame participation, we provided a list of missions that participants were free to contribute to. This missions had been proposed by the urban renewal group[1] of the district the study took place in. We told them to interact with this application as if the city administration was really reading their input. In fact, for both trial we had researchers located in our office building and thus invisible to the participants, who would respond to participants' input. Participants knew that they were not actually city officials, but were told that someone is reading their contributions. We further planned unobtrusive observations of participants during their walk, but due to logistic reasons (the area being an entire city district) and us not wanting to make participants uncomfortable by obviously shadowing them, we let them wander off. Occasionally, we crossed individuals path and took notes on their current activities (e.g. of what they were taken pictures, around what areas they stopped).

After each trial, participants filled out a survey reflecting on their experience during the trial. Apart from the measures described in a following section, participants answered demographic questions and questions about their experience with the application and the trial in general.

3.2 The Prototype

The prototype, called *b-Part*,[2] used in both studies is a location-based mobile application and has been developed in an iterative user-centered design process. As

---

[1]Urban renewal groups are institutions representing a link between the city administration and citizens. Their main function is to support citizens with challenges arising from living in cities (e.g. moving to new apartments) as well as organizing small events that serve community building in neighborhoods.

[2]*b-Part* has been introduced in Thiel et al. (2015), where the application's concept is described in more detail.

such, we conducted several workshops with both citizens and city officials in preparation of the prototype. The application's concept is loosely based on the participatory sourcing approach common for urban planning applications (e.g. *SeeClickFix, FixMyStreet*). In contrast to existing issue-focused reporting apps (Ertiö 2015), this participation platforms aims to support both top-down as well as bottom-up initiatives. Central to the application's concept is the process of posting and discussing contributions. Contributions are geo-references pieces of content that are visible to all users and be commented and voted upon. By having three types of contributions (Idea, Poll and Issue) citizens are encouraged to not only report problems but propose own ideas and engage in discussions with other users. Officials such as urban planners can create missions to address specific topics and to receive input on them.

We included a number of game elements into the Game trial version of the platform in order to be able to study the effects of a variety of components. These game elements are listed and described in Table 1, a selection of the elements can further be viewed in Fig. 1 (from top: points, progress overview; highscore and top list). Previous to both field trials described in this paper we conducted an objective user study with the prototype where we evaluated the overall concept of the system as well as explored the acceptance and validity of the integrated (game) elements.

Besides implementing the game mechanic scarcity or time constraint, the lifetime elements was included in the application as a measure against spamming and to avoid posting irrelevant content. This is based on the assumption that users would not comment on or interact with a contribution that does not meet public or individuals' interest. Interacting with a contribution adds to the lifetime and at the same time earns the respective user points (measured in square meters). These points are meant to serve both as additional motivation to respond to contributions

Table 1 Summary of included game elements in the prototype

Game element	Description
Lifetime	All contributions start with a lifetime that decreases over time and runs out if there is no activity with this contribution
Mission	Tasks that can be created by both citizens and city officials to get input on particular topics
Square meter	Users are awarded points for in-app activity (e.g. posting, commenting). These points are measured in square meter to reflect the area of influence a user has gained with his or her actions
Leaderboard and Top list	In both lists users can compare their performance within the game with other users
Profile	In this view users can check their activity within the app (e.g. number of comments posted). It also includes the leaderboard, an overview of their lasted game progress
Emoticons	To every posts, users could add emoticons to show how they felt about the topic
Social interaction	By commenting contributions and replying to comments, users have the opportunity to socially interact with other users

(or post new ones) and as means to compare one's own progress (level of activity) with fellow citizens. Allowing for better comparison and potentially also competition, we further included a highscore list and a top list. While the top list always shows the highest ranking users, the highscore lists gives also low-achieving users the chance to for comparison. All three elements (points, highscore and top list) are displayed in the profile which is visible to all users. This enables participants to learn more about other contributors rather than just their usernames.

Apart from required details such as description and title as well as optional information like nearby point-of-interest (POI) and a category tag, users can further choose to attach an emoticon to their contribution. The idea of indicating one's mood about the topic is that it can to some extent serve as prioritization (e.g. how urgent one believes it to be) as well as lets fellow citizens and city administration get a glimpse of the opinions and attitudes towards certain topics. There is dissent within the game research community as to whether opportunity for social interaction qualifies as a game element. While obviously not present in single-player games, the vast majority of multi-player games do offer some kind of opportunity to communicate with fellow players.

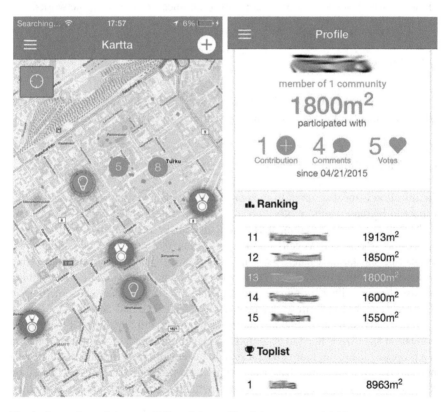

Fig. 1 Screenshots of the map (*left*) and the profile of the prototype (*right*)

3.3 Study Participants

For the Game trial, which took place in late summer, we had nine participants (3f, 6 m; mean age: 34.9) and for the Non-Game trial, taking place in late winter, eleven people (5f, 6 m; mean age: 42.8) agreed to take part. Different people participated in the two trials; only one participant took part in both trials. While the majority of participants used their own device, we also gave out some. In both studies we had quite homogeneous demographics including both digital natives (20–29 years old; Prensky 2001) and people over the age of 50 as well as a mix of students, full-time workers and unemployed. All participants stemmed from the middle-class and had no apparent migration background.

Neither participants from the Non-Game trial nor from the Game trial had used a similar or for that matter any civic or location-based public participation platform before. Participants of both trials use their mobile devices multiple times a day to constantly. While participants from the Non-Game trial often use apps for way-finding and orientation purposes, participants from the Game trial mostly use their smartphones for browsing the Internet, sending text-based messages and taking photos. Both groups only rarely to sometimes browse social networks.

3.4 Measures

For the evaluation of the two field trials we were interested in what factors motivated participants to actively use the application and engage in discussions.

3.4.1 Motivations to Participate

Gamification is most commonly applied in order to change user behavior and in this respect to increase users' motivation to use the system. Based on the reasons and goals that give rise to action, it is distinguished between intrinsic and extrinsic motivation. Being intrinsically motivated means that someone does something because he/she finds it inherently enjoyable and truly believes in the underlying values. On the other hand, there is extrinsic motivation, where people do something because they are either trying to avoid something or because they get something in return (e.g. reward in form of payment, fame) (Tiam-Lee and See 2014).

Active involvement in public affairs is predominantly triggered by extrinsic motivation as citizens hope to accomplish a change in policy or the implementation of a proposed idea (e.g. more bike racks) and hence anticipate an outcome that fuels their motivation. We argue that location-based public participation is also based on intrinsic motivation. Some citizens might participate because they see it as their democratic duty and want to be a part of it as they believe in the values of this form of government. Others again might be interested in the topics being currently discussed in their city in order to stay informed as well as learn new facts. Yet others participate with the intention to get to know their neighborhood and the people living in it. Hence, location-based public participation is constituted of both intrinsic as well as extrinsic motivation, the composition of those two types might however differ among individuals. With this research we want to explore a) which type of motivation plays a greater role in encouraging engagement and b) whether game aspects in the participation method can provide (additional) motivation for those citizens who do not feel any motivation to become engaged (or not enough to become active).

As a means to collect data on study participants' motivation for actively engaging with the application, participants of both trials were asked to answer a survey after the study. One question block in the survey inquired participants to rank various reasons of using the *b-Part* application according to their relevance. Items provided for this question regarding motivation to participate can be mapped to either intrinsic or extrinsic motivation. Those two categories were again divided into three sub-categories (cf. Table 2). These sub-categories are loosely based on Barbuto and Scholl's (1998) typology of motivation sources. The category *pleasure* represents their "intrinsic process" source and reflects on motivational factors where people do things for the fun they experience while doing it. We extended Barbuto and Scholl's typology by splitting their "internal self-concept" into *social* and *learning*. Under *learning* we understand processes connected to learning new things and overcoming challenges, which are also connected to achieving new (knowledge) levels in a particular domain. For the extrinsic motivational sources we used the descriptors *reputation*, *institutional* and *personal*. *Reputation* is based on "goal internalization" and represents reasons where people aim to fulfill internalized values or moral goals. *Institutional* stands for "instrumental" motivation sources where people wish to fulfill (legal) expectations or comply with external requirements. The descriptor *personal* characterizes "external self-concept" representing concepts such as interpersonal relations, relatedness and other social satisfiers.

Table 2 provides an overview of the wording of the question items and how they were mapped to the sub-categories of intrinsic and extrinsic motivation respectively. For instance, the concept pleasure for intrinsic motivation was assessed with

Table 2 Framework for assessing motivational factors; survey items in third column

Intrinsic	Social	"Socializing"
	Pleasure	"It was fun"
	Learning	"Discussing relevant issues"
Extrinsic	Reputation	"Making others aware of a problem"
	Institutional	"I believe my post has an impact"
	Personal	"Voicing my opinion"

the survey item "Using the app was fun" and "meeting new people" as social motivation. For measuring the impact on extrinsic motivation we asked participants to state whether they believed their input had an impact for "institutional" and "voicing my opinion" as personal factor. For each aspect the personal relevance was assessed on a 5-point Likert-scale ranging from $1 =$ "no relevance at all" to $5 =$ "very relevant". We did not include a section asking about the convenience of the method. We did however give participants the opportunity to state other motivational aspects by adding an "Other" option. If ticked, participants were asked to specify what else had motivated them.

The game elements included in the mobile participation platform do not necessarily match those motivational factors completely. While *social* obviously links to the opportunity to communicate with other users (social interaction) and *reputation* might be connected to a user's ranking on the highscore list based on the number of achieved points, *pleasure* could sum up all of elements referring to the intended gameful experience. At this stage of this research we were not such much interested in the influence of specific elements on participant's motivation, but more in whether the general motivations of engaging with a location-based public participation platform differ and in what way when game aspects are introduced. Linking motivational factors to specific game elements requires further research and arguably a slightly different study design.

3.4.2 Levels of Participation

On a very abstract level, we distinguish between participation and engagement. By participation we refer to a user's quantitative use of a location-based public participation system. In the case of the b-Part app, participation could be measured by counting the number of contributions and posts a user posted. In contrast to this quantitative aspect, under engagement we understand the quality of a user's participation. Here we define quality as a combination of a variety of factors such as the clarity of the posts (e.g. amount of details given) and the fitness to the underlying participation process. Regarding the latter, a post would be of low quality if addressing the presence of gnomes in someone's garden but the process being about making a city friendlier for bicyclists.

For this research, we focused on the participation aspect and investigated whether certain motivations influences the quantity of participation in a mobile location-based participation application. Quantity was measured by drawing on data from the backend logs and simply counting and adding up the number of contributions and comments a user had posted. Taking findings from literature into account, we postulate that especially social aspects and factors related to learning would support participation. In case there were a high trust of among participants, what we summarized as *institutional* motivation should also at least somehow encourage participation.

4 Results

4.1 Motivations to Participate

In order to investigate whether the types of motivation differed among the two trials and therefore prototype versions, we ran an independent sample t-test using intrinsic and extrinsic motivation (average of all item belonging to that type for each user) as our dependent variable. The results show that participants in the Game trial had statistically significantly lower intrinsic motivation (2.19 ± 0.43) compared to in the Non-Game trial (3.58 ± 0.62), $t(18) = -5.68$, $p = 0.000$. While extrinsic motivation was slightly higher among Game trial participants ($M = 4.04$) than in the Non-Game trial ($M = 3.94$), this difference was not significant ($p = 0.770$). Running the same test with individual motivational factors (see Table 3), we found that *pleasure* was significantly less relevant in the Game trial ($p = 0.002$).

Among the most relevant motivational aspects participants from the Non-Game trial stated to be related to *pleasure* (mean: 4.45), *personal* (mean: 4.00) and *reputation* extrinsic motivation (mean: 4.00). The results are also summarized in Table 3. Ten out of eleven participants rated the opportunity to voice their opinion as at least relevant. Only one stated that this aspect was not relevant at all for his/her motivation to engage with the app. Making others aware of an issue was further rated as relevant among the vast majority. The third most important factor encouraging the use of the location-based public participation platform for participants of the Non-Game trial was the belief that their contribution would make an impact (mean: 3.82). Only for one participant this was slightly relevant. The least relevant item was socializing (mean: 2.64), whereas four participants put that this aspect was somewhat relevant to them. Four participants further added own items: curiosity, ability to learn about events, proposing own ideas and organizing initiatives. Whereas curiosity was only rated as somewhat relevant, the other items were stated to be at least relevant for those participants.

The most important motivational aspects for participants of the Game trial were related to *reputation* (mean: 4.33) and *personal* extrinsic motivation (mean: 4.22). No participant rated these items as less than somewhat relevant. Another aspect that was also rated relatively relevant in the Game trial, is the opportunity to discuss relevant issues (mean: 3.78) with others, which we associated with the intrinsic motivation of *learning* from discussions. Again not relevant for their motivation to

Table 3 Importance of individual factors on motivation to participate (mean values, standard deviation in brackets)

		Game trial	Non-game trial
Intrinsic	Social	2.11 (0.78)	2.64 (1.21)
	Pleasure	2.89 (1.05)	4.45 (0.69)
	Learning	3.78 (0.67)	3.64 (0.92)
Extrinsic	Reputation	4.33 (0.71)	4.00 (0.89)
	Institutional	3.56 (0.88)	3.82 (0.98)
	Personal	4.22 (0.83)	4.00 (1.1)

become engaged was the opportunity to socialize with other users (mean: 2.11). Participants of the Game trial rated this aspect as even less relevant as participants from the Non-Game trial. The aspect *pleasure* was rated in the mid-range of relevant motivational factors. However with three participants stating that the fun aspect was relevant to them, also three saying it was somewhat relevant and the rest (3) indicating it was slightly or not relevant at all, *pleasure* was differently perceived in the Game trial. In the Non-Game trial no one rated the fun aspect less than somewhat relevant with six stating it was very relevant for them. From the seven items listed, the belief that one's post has an impact was rated the fourth important factor in motivating them to become engaged. Five of nine participants of the Game trial indicated their intention to make a difference as at least relevant.

Three participants further noted down own aspects: improving things, discovering new parts of the city and "letting my ideas out into the community". All of them can be linked to a willingness to actively participate in urban topics. With the prototype version of the Game trial containing game elements, we added another survey item to the motivational aspects: "I wanted to succeed in the game." The desire or motivation to be successful in the game (i.e. collect more points than others; keep as many contributions alive as possible), was the least relevant. Six out of nine participants indicated thriving in the game was not relevant at all for them. Only one indicated it to be somewhat relevant.

Those participants from the Game trial who believed their posts to have impact, also rated making others aware of problems as a strong motivation to engage with the prototype. Those participants who did not rate fun as being an important motivation, were also not very keen on succeeding in the game. On the other side, for those fun was a motivation, also found discussing relevant issues as relevant. This suggest that fun did not relate to "playing" around with the app or spending one's time walking around outside, but was in fact more related to the fact that one was taking part in discussions relating urban and public topics. In the Non-Game trial those who were strongly motivated by experiencing fun, also rated voicing owns opinion and making others aware of issues as relevant. While not completely irrelevant, discussing topics with others was less relevant for participants of the Non-Game trial.

4.2 Levels of Participation

In a previous analysis we investigated whether gamification leads to more participation using descriptive statistics by comparing activity counts (= sum of *all* in-app activities such as voting, posting, commenting) of all users in both studies (Thiel and Lehner 2015). Several scholars argue that handing out "likes" is a "watered down form of participation" (Foth and Brynskov 2016) (also referred to as "slacktivism") but does not contribute to an effective and sustainable public engagement. Following this thought, we excluded the number of times a user voted on either a comment or contribution for this analysis. Instead we concentrated on the sum of contributions and comments posted (referred to as *activity count* hereafter). In the Game trial the *activity count* of participants totaled to 138, in the Non-Game trial this number was 97. On average participants in the Game trial had an activity count of 17.3, in the Non-Game trial 8.8 respectively.

Looking at the scores of individual users, we ran an independent samples t-test with activity count as the dependent variable and the corresponding trial (Game trial or Non-Game trial) as independent variable. The result shows that participation (based on activity counts) in the Game trial was significantly higher ($M = 17.3$, $SD = 7.3$) than in the Non-Game trial, ($M = 8.8$, $SD = 3.5$), $t = 3.02$, $p = 0.014$. This suggests that gamification does significantly increase participation. Due to the small n in both studies (n < 10 in the Game trial), one should be careful when interpreting this result. However, calculating the effect size according to Cohen's (1988) method, attests a rather high effect of gamification on participation ($d = 1.47$).

We further looked at whether certain motivations impacted the level of constructive activity (i.e. posting contributions and comments) in the application. For each motivational factor we investigated whether they correlate with the activity count of the respective user. In order to check for applicability of non-parametric tests, we created scatter plots of each motivational factor (grouped by trial). Except for the factor reputation ("Making others aware of a problem") we did not find any linear (or any other common) relationship between the two variables. The same results occur when using relative activity (i.e. dividing individual's activity by the maximum activity for the trial). Running a Spearman's rank-order correlation to determine the relationship between the motivational aspect relating to *reputation* benefits and level of participation did not reveal a correlation ($rs = 0.129$, $p = 0.598$). The scatter plots further suggested a possible relationship between the factor learning ("Discussing relevant issues") and level of participation for participants in the Non-Game trial. Again the Spearman's correlation did not confirm such a relationship ($rs = 0.204$, $p = 0.471$). Regarding the additional motivational item for the Game trial relating to playing the game ("I wanted to succeed in the game"), we did not find any (significant) relationships between neither level of participation (activity count) nor any other motivational factors. There was however a strong, negative correlation between succeeding in the game and learning, which was not significant ($rs = -0.546$, $p = 0.129$).

Using the average of all factors in the categories intrinsic and extrinsic motivation for each user, we re-ran the same analysis process. For neither of the trials did the scatter plot of extrinsic motivation and level of participation suggest any common (particularly not linear) relationship. Assuming a relationship between intrinsic motivation and level of participation in the Game trial, a Spearman's rank-order correlation was run. There was a strong, positive correlation between intrinsic motivation and activity level, which was significant ($rs = 0.875$, $p = 0.022$).

When computing the non-parametric test we also created correlation matrices including all motivational factors and level of participation. For the Non-Game trial this matrix revealed a strong, positive and significant correlation between the factors *reputation* and personal benefits ($rs = 0.676, p = 0.022$) as well as a strong, positive and significant correlation between the factors *institutional* and *personal* benefits ($rs = 0.606, p = 0.048$). In contrast, analysis of the responses of participants in the Game trial additionally revealed a very strong, positive correlation between *social* and *personal* factors, which was statistically significant ($rs = 0.900, p = 0.001$).

5 Discussion

Comparing the activity levels of participants in both trials, our results show that participation was higher when using an application with game elements. While we did not find any significant relationships between level of participation and specific motivational factors, results imply that intrinsic motivation in total (which was actually lower than in the Non-Game trial, *RQ1*) positively affected activity in a gamified location-based public participation application. This suggests that gamification positively influenced people's willingness to engage with a location-based public participation platform.

5.1 Comparing Motivations

The belief that participation will have an influence on the current status of public and urban topics (*institutional*) came third when rating factors that motivate them to engage in these topics. This could mirror the reported general mistrust people seem to have in politics (Harding et al. 2015). On the other hand, our finding could also be due to the trial not involving any authorities and participants reflecting on the limited impact content they posted during the trial can cause. However, if there really were a deep-rooted mistrust in society, most participants would have rated this aspect as not relevant at all. Considering that the most important motivation for participants across both trial was making others aware of a problem and voicing their opinion, citizens are not that disenchanted with politics that they would turn away from participation altogether, not even bothering to bring their opinions to the attention of others. This finding refutes arguments from previous studies which claimed that citizens have become used to the in-active state of "consuming" rather than participating in urban life and politics (Vigoda 2002). It also undermines predictions that technological interventions can shift citizens' role transforming them from "users and choosers" to "makers and shapers" of policies and decisions (Lukensmeyer and Torres 2008).

Comparing the importance of bringing about change (*institutional*) as a motivation for both trials, this factor was almost equally ranked in the Game and the Non-Game trial. While having been rated slightly less important in the Game trial, this minor difference cannot be taken as an indication that game elements caused a perceived devaluation of the content. Yet, one participant noted that tagging contributions with emoticons let the content appear less serious. With also others not seeing the value of being able to state one's mood, this element made them skeptical of whether the process of using this participation tool will have an effect. Hence, such a game element (= emoticons) should not be used to tag content in location-based public participation tools. Overall, in order to avoid a perceived devaluation, Werbach and Hunter (2012) advise to only apply the key concepts of a game.

The finding that *social* interaction did not seem to be quite relevant for any of the two study groups, suggests that the game elements integrated in this prototype did not have any impact on this particular motivational aspect. In fact, the vast majority of participants stated that socializing was at most somewhat relevant for their motivation to engage. In the Non-Game study the influence of the opportunity to interact with other users varied greatly, indicating that for some people this can indeed be a motivation to engage. Future studies over a longer time period should investigate, whether this finding only applies for these short-term settings (in which participants simply did not have time to communicate with others) or whether social interaction is not a motivating factor in the domain of public participation at all. Comparing it to other aspects assessed in the two studies (listed in Table 2), the main motivation to become active seems to be to genuine interest in urban topics. This interest is fostered by both intrinsic (i.e. learning about urban topics) and extrinsic motivation (i.e. gaining recognition as attentive citizen).

Some scholars consider social interaction as another game-related element (cf. Bowser et al. 2014, Chou 2015, Hunicke et al. 2004). According to the constrained definition of gamification, social interaction does not classify as game element, as it is not unique to games. Referring back to the definition by Deterding et al. (2011), classifying social interaction as game element is debatable, as it is not strictly characteristic to games as several none-gamified application include this feature. However, also other game elements (e.g. those listed in Table 1) would not automatically identify an application to be gamified. For the purpose of studying effects on motivation we included the aspect in our evaluation.

Bringing the aspect of social interaction in relation to the motivation to experience fun, we can draw different conclusions for the two trials. While for participants of the Non-Game trial fun played the most important role in their motivation to engage, this aspect played only a minor role for participants of the Game trial. This is foremost astonishing, as games have always been associated with fun where game elements should make non-game products more enjoyable (cf. Zichermann and Cunningham 2011). As the prototype version in the Non-Game trial did not include any game elements, participants must have derived their fun from other aspects. As argued before, one of those aspect could be that participants enjoyed voicing their concerns and suggestions. Regarding the acceptance and thus also effectiveness of gamification, it should be added that when asking about the aspects participants did not like about the application, three participants of the Game trial referred to game elements. While one user did not like the gamification in general, the others commented on specific elements. One participant rated the ranking as negative and the other was adverse to the idea of adding emotions (e.g. funny, sad) to contributions.

Pleasure or fun was at least somewhat relevant in both studies. With us not specifically having asked what exactly caused participants to enjoy the trial or the participation, we can only speculate about the factors. When asked what the participants liked most about the application, half of them listed the usability of the app. Other participants further listed specific features, where the opportunity to post contributions was mentioned the most often. Together with the most relevant motivations, we conclude that participants experienced fun in openly raising their voice and engaging in urban topics.

Based on our classification of motivational aspects into intrinsic and extrinsic motivation, the findings show that for both trials overall extrinsic factors seemed to have been more relevant in motivating participants to engage than intrinsic factors. An exception is only the motivational factor pleasure in the Non-Game trial. It has been argued that the danger of extrinsic motivation is that it can harm intrinsic factors or even replace it altogether. This is even more critical, when using incentives. Removing those might dramatically and irrevocably decrease the overall motivation. In the case of these case studies however, participants stated that the prospective of winning or succeeding in the game did only play a minor role in motivating them to become active. Their main motivation lay in the chance to participate in urban topics and the opportunity to bring about change. One participant even noted that "It felt like we are changing the world a bit". Furthermore, the factors linked to intrinsic motivation were not completely irrelevant for participants. Apart from wanting to succeed in the game, all intrinsic factors were seen as at least slightly relevant. Learning was even ranked more than somewhat relevant in both trials (Game trial: 3.78; Non-Game trial: 3.64).

Overall, our results show that intrinsic motivation to use the application was significantly lower in the Game trial than in the Non-Game trial *(RQ1)*, echoing previous findings (Nicholson 2012). It should be noted however, that considering the similar levels of extrinsic motivation in both trials, intrinsic motivation was not suppressed or replaced by extrinsic motivation as suggested by (Zichermann and Cunningham 2011). Participants of both trials are indeed pursuing a reward or outcome: causing impact with their input in urban politics. Seeing that this is the sole purpose of location-based public participation and therefore of e-participation platforms, this "incentive" will arguably never diminish, unless authorities loose interest and stop providing feedback—a scenario that would arguably also stop the usage of such a platform altogether.

5.2 Influence of Motivation on Participation

Our analysis showed that participation was higher when using a location-based public participation tools with game elements compared to one without. For this evaluation we considered active or constructive participation, meaning that users posted content instead of just viewing or rating posts. Our findings however do not show whether the quality of this posted content is the same in both studies. Considering critics that point to the likelihood of users "gaming the system" and not adhering to its underlying purpose (here posting content that is relevant for the community and/or administration), investigating the potential adverse effect on data quality is highly relevant but will be covered to future research (Bowser et al. 2013b).

This research further investigated the existence of any relationships between certain motivations and users' participation rates. Our results suggest that none of the motivational factors analyzed in this study directly influenced active participation (*RQ2*). This provides further evidence that participation underlies a more complex motivational framework that assumed linear relationships between individual factors and sum of activity. In their proposed "motivational arc" Crowston and Fagnot (2008) for instance identify separate motives for different contribution level (= engagement stages), where curiosity together with an evaluation of costs and benefits of contributing makes up motivation in the initial phase of participation.

With regards to the effectiveness of gamification, we found no indication that seeking success in the game (e.g. gaining more points) influenced participation; nor did striving for success correlate with any other motivational factors. Yet, we found a weak, negative non-significant relationship between wanting to discuss urban topics and playing the game. This could be explained by citizens being motivated by the game aspects ("players") not being interested in the underlying issues/purposes, which could be an indication for players producing content of lower quality than from citizens that do want to be part of discussions. Together with our finding that participants were in general highly interested in urban topics, might suggest that there is no need for additional (game related) motivators if people are already motivated by other aspects. This would go in line with a similar result found when evaluated motivations of participating in a gamified citizen science application (Bowser et al. 2013a). On the other hand our results further indicate that intrinsic motivation with game aspects present was lower, what seems to confirm concerns that achievement-based gamification strategies reduces intrinsic motivation (Kohn 1999).

These results leads us to conclude that introducing game elements to a location-based public participation platform can decrease existing (intrinsic) motivation to participate while at the same time increasing participation rates. In the Game trial extrinsic motivation was slightly but not significantly higher than in the trial without game aspects. This fact suggests that the motivation boost (potentially) caused by game elements cannot make up for the decrease in intrinsic motivation. This is likely due to the way gamification was implemented. A different combination or implementation of the game elements might result in an overall increased motivation. On the other hand, reflecting on our finding that intrinsic motivation seemed to significantly increase the level of participation in the Game trial, a lower intrinsic motivation might counterbalances the fact that gamification can support participation. This findings confirms our assumption that location-based public participation can also be triggered by intrinsic motivation.

6 Limitations

When reflecting on these findings, it is important to note that they are based on an evaluation of a specific mobile participation platform that integrates a list of specific game elements. Therefore, the influences on motivational aspects might be characteristic for this set of game elements and their implementation, but they might not be generalizable for all instances of location-based public participation tools. The choice of game elements certainly influences user perception, which finally alters conclusions to be drawn from such an analysis. For instance, in other studies, competition has been shown to not be favored by everyone (Eveleigh et al. 2013). This paper should therefore be regarded as a piece in a bigger puzzle of our research undertaking, which is essentially to understand the influence of specific game elements and their combinations. Following this thought, future studies will focus more on understanding which game elements can evoke or increase what types of motivations. This will make the link between motivational aspects and specific game elements more apparent.

Although both trials were conducted "in the wild", they do not entirely reflect real-world conditions of location-based public participation platforms. One aspect biasing the outcomes is certainly the duration of the study, which did not allow participants to explore the opportunities of this application in a broader range of situations and contexts. Another, related aspect is the influence of the so-called novelty effect. Foremost caused by factors such as curiosity and lack of experience with the system, users would interact differently with the system than with one that they had known and used for a long time. Studies in HCI are arguably almost in all cases subjected to this novelty effect, but its impact can be reduced by prolonging the duration of the study.

Due to these consequential limitations, findings from this user study should be seen as trends only that need to be validated in a longitudinal field trial that can better simulate real-world settings. After all, the main problem with location-based public participation platforms is to sustain engagement in the long term. On the other hand, some participatory processes might only be designed to take place in a short time frame (i.e. one afternoon), for instance when a city department seeks early feedback or input for an upcoming planning session. For these condensed and focused methods of public involvement, evaluations conducted in exactly these short term settings are relevant as well.

7 Conclusions

With the objective to increase location-based public participation, numerous digital platforms allowing the public to raise their voice have been created. Until now, an increase in the level of participation has not been reported yet. Assuming that these novel platforms are not (engaging) enough, new strategies are being investigated.

One idea is to foster motivation to participate by adding game-related elements to those platforms. While the success of gamification in this respect has been confirmed in several studies for other domains, it is yet to investigate what effects game elements have in the domain of e-participation.

We conducted two trials in order to gain insights regarding the motivations people have more engaging in public affairs. Both studies explored the usage of game-inspired elements in a mobile application designed for fostering location-based public participation. This paper focused specifically on motivation to engage and whether game elements would have an impact on motivational factors. Our findings indicate that gamification has the potential to increase participation within mobile location-based public participation apps, but also decreases the overall intrinsic motivation.

Judging by the stated motivations to actively engage, we only found significant differences for the perceived *pleasure* (fun) between the two trials. This suggests that the tested gamification (= selection of game elements) did not impact the motivation for location-based public participation. While almost all motivational aspects to actively engage were not affected by the inclusion of game elements, participants who used the non-gamified version seemed to enjoy participating more than the other group with game elements. We further found that the main motivation to become active was genuine interest in urban topics and the desire to raise one's voice. Hence in contrast to the commonly believed political apathy, people are interested in urban topics and are willing to actively engage by proposing their own ideas. In case of some participants, they even enjoy raising their voice. Overall, our findings show that while intrinsic motivation certainly plays a role, extrinsic motivation is more relevant for encouraging citizens to become active in urban decision-making and public topics.

Future work will entail a similar study focusing on the effect of game elements in location-based public participation, but with a bigger sample, over a longer time period and in closer cooperation with a local authority.

References

Åström, J and Grönlund, Å (2012) Online consultations in local government: What works, when, and why. Connecting democracy: Online consultation and the flow of political communication, page 75.

Barbuto Jr, J. E. and Scholl, R. W. (1998) Motivation sources inventory: Development and validation of new scales to measure an integrative taxonomy of motivation. Psychological Reports, 82(3):1011–1022.

Bowser, A., Hansen, D., He, Y., Boston, C., Reid, M., Gunnell, L., and Preece, J. (2013a). Using gamification to inspire new citizen science volunteers. In Proceedings of the First Int. Conference on Gameful Design, Research, and Applications, pages 18–25. ACM.

Bowser, A., Hansen, D., and Preece, J. (2013b). Gamifying citizen science: Lessons and future directions. In Workshop on Designing Gamification: Creating Gameful and Playful Experiences.

Bowser, A., Hansen, D., Preece, J., He, Y., Boston, C., and Hammock, J. (2014). Gamifying citizen science: a study of two user groups. In Proceedings of the companion publication of the 17th ACM conference on Computer supported cooperative work & social computing, pages 137–140. ACM.

Brady, H. E., Verba, S., and Schlozman, K. L. (1995) Beyond SES: A resource model of political participation. American Political Science Review, 89(02):271–294.

Chou, Y. (2015). Actionable gamification: Beyond points. Badges, and Leaderboards, Kindle Edition, Octalysis Media (Eds.).

Cohen, J. (1988). Statistical power analysis for the behavioral sciences. N.J: L. Erlbaum Associates, Hilsdale, 2 edition.

Coronado Escobar, J. E. and Vasquez Urriago, A. R. (2014). Gamification: an effective mechanism to promote civic engagement and generate trust? In Proceedings of the 8th International Conference on Theory and Practice of Electronic Governance, pages 514–515. ACM.

Crowston, K and Fagnot, I (2008) The motivational arc of massive virtual collaboration. In Proceedings of the IFIP WG, volume 9. Citeseer.

Deterding, S., Dixon, D., Khaled, R., and Nacke, L. (2011) From game design elements to gamefulness: defining gamification. In Proceedings of the 15th international academic MindTrek conference: Envisioning future media environments, pages 9–15. ACM.

Devisch, O., Poplin, A., & Sofronie, S. (2016). The Gamification of Civic Participation: Two Experiments in Improving the Skills of Citizens to Reflect Collectively on Spatial Issues. Journal of Urban Technology, 1–22.

Ertiö, T.-P. (2015). Participatory apps for urban planning - space for improvement. Planning Practice & Research, 30(3):303–321.

Eveleigh, A., Jennett, C., Lynn, S., and Cox, A. L. (2013). "I want to be a Captain! I want to be a Captain!": Gamification in the Old Weather Citizen Science Project. In Proceedings of the First International Conference on Gameful Design, Research, and Applications, pages 79–82. ACM.

Flatla, D. R., Gutwin, C., Nacke, L. E., Bateman, S., and Mandryk, R. L. (2011). Calibration games: making calibration tasks enjoyable by adding motivating game elements. In Proceedings of the 24th annual ACM symposium on User interface software and technology, pages 403–412. ACM.

Foth, M. and Brynskov, M. (2016). Participatory action research for civic engagement. In Civic Media: Technology, Design, Practice, pages 563–580, Cambridge, MA. MIT Press.

Gordon, E., & Baldwin-Philippi, J. (2014). Playful civic learning: Enabling lateral trust and reflection in game-based public participation. International Journal of Communication, 8, 28.

Harding, M., Knowles, B., Davies, N., & Rouncefield, M. (2015). HCI, Civic Engagement & Trust. In Proceedings of the 33rd Annual ACM Conference on Human Factors in Computing Systems, pages. 2833–2842. ACM.

Hamari, J., Koivisto, J., and Sarsa, H. (2014). Does gamification work?–a literature review of empirical studies on gamification. In System Sciences (HICSS), 2014 47th Hawaii International Conference on, pages 3025–3034. IEEE.

Hunicke, R., LeBlanc, M., and Zubek, R. (2004). Mda: A formal approach to game design and game research. In Proceedings of the AAAI Workshop on Challenges in Game AI, volume 4, page 1.

Islam, M. S. (2008). Towards a sustainable e-participation implementation model. European Journal of ePractice, 5(10).

Itoko, T., Arita, S., Kobayashi, M., & Takagi, H. (2014). Involving senior workers in crowdsourced proofreading. In International Conference on Universal Access in Human-Computer Interaction, pages 106–117. Springer International Publishing.

Kohn, A. (1999). Punished by rewards: The trouble with gold stars, incentive plans, A's, praise, and other bribes. Houghton Mifflin Harcourt.

Krek, A. (2005). Rational ignorance of the citizens in public participatory planning. In 10th symposium on Information-and communication technologies (ICT) in urban planning and spatial development and impacts of ICT on physical space, CORP, volume 5.

Lepper, M. R., Greene, D., and Nisbett, R. E. (1973). Undermining children's intrinsic interest with extrinsic reward: A test of the" overjustification" hypothesis. Journal of Personality and social Psychology, 28(1):129.

Lukensmeyer, C. J. and Torres, L. H. (2008). Citizensourcing: Citizen participation in a networked nation. Civic engagement in a network society, pages 207–233.

Macintosh, A. (2004). Characterizing e-participation in policy-making. In System Sciences, 2004. Proceedings of the 37th Annual Hawaii International Conference on, pages 10–pp. IEEE.

Nam, T. (2011). Suggesting frameworks of citizen-sourcing via Government 2.0. Government Information Quarterly (29), 12–20.

Nicholson, S. (2012). A user-centered theoretical framework for meaningful gamification. Games + Learning + Society 8.0, 8.

Poplin, A. (2012). Playful location-based public participation in urban planning: A case study for online serious games. Computers, Environment and Urban Systems, 36(3):195–206.

Poplin, A. (2014). Digital serious game for urban planning:"b3 -design your marketplace!". Environment and Planning B: Planning and Design, 41(3):493–511.

Preist, C., Massung, E., and Coyle, D. (2014). Competing or aiming to be average?: normification as a means of engaging digital volunteers. In Proceedings of the 17th ACM conference on Computer supported cooperative work & social computing, pages 1222–1233. ACM.

Prensky, M. (2001). Digital natives, digital immigrants part 1. On the horizon, 9(5):1–6.

Rehm, S. (2015). DoGood: A gamified mobile app to promote civic engagement. Master thesis, Ludwig-Maximilians-Universität München.

Shkabatur, J. (2011). Cities@ crossroads: Digital technology and local democracy in America. Brooklyn Law Review, 76(4): 11-11.

Stembert, N. and Mulder, I. J. (2013). Love your city! An interactive platform empowering citizens to turn the public domain into a participatory domain. In International Conference Using ICT, Social Media and Mobile Technologies to Foster Self-Organisation in Urban and Neighbourhood Governance, Delft, The Netherlands, 16–17 May 2013.

Thiel, S.-K. (2016). A Review of introducing Game Elements to e-Participation. In Proceedings of CeDEM16: Conference for E-Democracy and Open Government, pages 3–9. IEEE.

Thiel, S.-K. and Lehner, U. (2015). Exploring the effects of game elements in m-participation. In Proceedings of the 2015 British HCI Conference, British HCI'15, pages 65–73, New York, NY, USA. ACM.

Thiel, S.-K., Lehner, U., Sturmer, T., and Gospodarek, J. (2015). Insights from an m-participation prototype in the wild. In IEEE International Conference on Pervasive Computing and Communication Workshops, pages 166–171. IEEE.

Thom, J., Millen, D. & DiMicco, J. (2012). Removing gamification from an enterprise SNS. In Proceedings of CSCW 2012. New York: ACM Press.

Thomas, J. C. and Streib, G. (2003). The new face of government: citizen-initiated contacts in the era of e-government. Journal of public administration research and theory, 13(1):83–102.

Tiam-Lee, T. J. and See, S. (2014). Building a sentiment corpus using a gamified framework. In Int. Conference on Humanoid, Nanotechnology, Information Technology, Communication and Control, Environment and Management (HNICEM), 2014, pages 1–8. IEEE.

Torres, L. H. (2007). Citizen sourcing in the public interest. Knowledge Management for Development Journal, 3(1):134–145.

Ueyama, Y., Tamai, M., Arakawa, Y., & Yasumoto, K. (2014). Gamification-based incentive mechanism for participatory sensing. In IEEE International Conference on Pervasive Computing and Communications Workshops, pages 98–103. IEEE.

Vigoda, E. (2002). From responsiveness to collaboration: Governance, citizens, and the next generation of public administration. Public administration review, 62(5):527–540.

Werbach, K. and Hunter, D. (2012). For the win: How game thinking can revolutionize your business. Wharton Digital Press.

Zichermann, G. and Cunningham, C. (2011). Gamification by design: Implementing game mechanics in web and mobile apps. O'Reilly Media, Inc.

Printed by Books on Demand, Germany